Withdrawn
University of Waterloo

# Electron Microdiffraction

# Electron Microdiffraction

J. C. H. *Spence* and J. M. *Zuo*

*Arizona State University*
*Tempe, Arizona*

Plenum Press • New York and London

Library of Congress Cataloging-in-Publication Data

Spence, John C. H.
  Electron microdiffraction / J.C.H. Spence and J.M. Zuo.
     p.   cm.
  Includes bibliographical references and index.
  ISBN 0-306-44262-0
  1. Thin films--Optical properties.  2. Crystal optics.
  3. Electrons--Diffraction.  4. Solid state physics.  5. Soild state
  chemistry.    I. Zuo, J. M.   II. Title.
  QC176.84.O7S65  1992
  530.4'275--dc20                                              92-33963
                                                                   CIP

ISBN 0-306-44262-0

© 1992 Plenum Press, New York
A Division of Plenum Publishing Corporation
233 Spring Street, New York, N.Y. 10013

All rights reserved

No part of this book may be reproduced, stored in a retrieval system, or transmitted in any form or by any means, electronic, mechanical, photocopying, microfilming, recording, or otherwise, without written permission from the Publisher

Printed in the United States of America

Large-angle convergent beam pattern recorded at 100 kV from silicon in the [111] orientation. The Omega filter of the Zeiss 912 has been used to remove most inelastic scattering, together with a very small selected area aperture, resulting in high contrast. (Courtesy of Dr. J. Mayer, M.P.I., Stuttgart, Germany)

# Preface

Much of this book was written during a sabbatical visit by J.C.H.S. to the Max Planck Institute in Stuttgart during 1991. We are therefore grateful to Professors M. Ruhle and A. Seeger for acting as hosts during this time, and to the Alexander von Humbolt Foundation for the Senior Scientist Award which made this visit possible. The Ph.D. work of one of us (J.M.Z.) has also provided much of the background for the book, together with our recent papers with various collaborators.

Of these, perhaps the most important stimulus to our work on convergent-beam electron diffraction resulted from a visit to the National Science Foundation's Electron Microscopy Facility at Arizona State University by Professor R. Høier in 1988, and from a return visit to Trondheim by J.C.H.S. in 1990. We are therefore particularly grateful to Professor Høier and his students and co-workers for their encouragement and collaboration. At ASU, we owe a particular debt of gratitude to Professor M. O'Keeffe for his encouragement. The depth of his understanding of crystal structures and his role as passionate skeptic have frequently been invaluable. Professor John Cowley has also been an invaluable sounding board for ideas, and was responsible for much of the experimental and theoretical work on coherent nanodiffraction. The sections on this topic derive mainly from collaborations by J.C.H.S. with him in the seventies. Apart from that, we have tried to review the literature as impartially as possibly and, at the same time, bring out the underlying concepts in a clear and unified manner, so that the book will be useful for graduate students. We are particularly grateful to Dr. J. A. Eades for his critical review of Chapter 7 and to Dr. J. Mayer. We apologize to those authors whose work may have been overlooked among the many hundreds of papers. In order to make the book more practically useful, we have included some FORTRAN source listings, together with POSTSCRIPT code, which allows the direct printing of Kikuchi and HOLZ line patterns on modern laser printers from the programs. Support from NSF award DMR-9015867 ("Electron Crystallography") and the facilities of the NSF-ASU National Center for High Resolution Electron Microscopy is gratefully acknowledged.

# Contents

1. **A Brief History of Electron Microdiffraction** ....... 1

2. **The Geometry of CBED Patterns** ....... 7
   - 2.1. One-Dimensional (Systematics) CBED ....... 9
   - 2.2. Two-Dimensional CBED ....... 15
   - 2.3. The Geometry of HOLZ and Kikuchi Lines (Kinematic) ....... 20

3. **Theory** ....... 29
   - 3.1. Electron and X-Ray Structure Factors. Temperature Factors ....... 30
   - 3.2. Many-Beam Theory ....... 35
   - 3.3. Two-Beam Theory and the Bethe Potentials. How Bragg's Law May Fail If More Than Two Beams Are Excited ....... 40
   - 3.4. The Concept of the Dispersion Surface ....... 44
   - 3.5. Degenerate and Nondegenerate Perturbation Theory ....... 48
   - 3.6. Three-Beam Theory and Particular Solutions for Centrosymmetric Crystals ....... 52
   - 3.7. Three-Beam Theory—Noncentrosymmetric Crystals and the Phase Problem ....... 56
   - 3.8. Dynamic HOLZ Intensities and Positions. Dispersion Surfaces for HOLZ Lines. How the Bragg Law Depends on Local Composition ....... 63
   - 3.9. Absorption and Its Effects ....... 75
   - 3.10. Channeling, Bound States, and Atomic Strings ....... 82

4. **The Measurement of Low-Order Structure Factors and Thickness** ....... 85

|  |  |  |
|---|---|---|
| 4.1. | Two-Beam Methods for Thickness Determination ......... | 86 |
| 4.2. | Automated Dynamical Least-Squares Refinement Methods ................................................................ | 88 |
| 4.3. | Error Propagation in Refinement by CBED .................. | 98 |
| 4.4. | Measurement of Debye–Waller Factors by CBED ........ | 100 |
| 4.5. | What Is Measured. Charge Density and the Total Crystal Energy ................................................................ | 101 |
| 4.6. | The Significance of the Mean Inner Potential $V_0$ ........... | 104 |

## 5. Applications of Three- and Many-Beam Theory .................... 107

| | | |
|---|---|---|
| 5.1. | Finding Atom Positions. Structure-Factor Phase Measurement. Enantiamorphs .................................... | 107 |
| 5.2. | Critical Voltages ...................................................... | 120 |
| 5.3. | The Intersecting Kikuchi-Line and HOLZ-Line Methods | 123 |
| 5.4. | Local Measurement of Strains, Composition, and Accelerating Voltage Using HOLZ Lines ..................... | 125 |
| 5.5. | Polarity Determination in Acentric Crystals ................. | 131 |
| 5.6. | The Inversion Problem .............................................. | 134 |

## 6. Large-Angle Methods ...................................................... 137

| | | |
|---|---|---|
| 6.1. | Large-Angle Methods for the ZOLZ .......................... | 137 |
| 6.2. | Large-Angle Methods for the HOLZ .......................... | 143 |

## 7. Symmetry Determination ................................................. 145

| | | |
|---|---|---|
| 7.1. | Point Groups .......................................................... | 148 |
| 7.2. | Bravais Lattice Determination ................................... | 158 |
| 7.3. | Space Groups ......................................................... | 160 |
| 7.4. | Phase Identification ................................................. | 166 |

## 8. Coherent Nanoprobes. STEM. Defects and Amorphous Materials ..................................................................... 169

| | | |
|---|---|---|
| 8.1. | Subnanometer Probes, Coherence, and the Best Focus for CBED ............................................................... | 170 |
| 8.2. | Lattice Imaging in STEM Using Overlapping CBED Orders ................................................................... | 177 |
| 8.3. | CBED from Defects—Coherent Nanoprobes. Site Symmetry, Atom Positions ....................................... | 181 |

|           |                                                                                              |      |
|-----------|----------------------------------------------------------------------------------------------|------|
| 8.4.      | Ronchigrams, Point Projection Shadow Lattice Images, and Holograms                           | 191  |
| 8.5.      | CBED from Defects—Incoherent. Dislocations and Planar Faults. Multilayers. CBIM              | 201  |
| 8.6.      | Microdiffraction from Amorphous Materials                                                    | 210  |

## 9. Instrumentation and Experimental Technique ... 213

| 9.1. | Instrumentation | 214 |
| 9.2. | Experimental Technique | 217 |
| 9.3. | Detectors and Energy Filters | 219 |
| 9.4. | Electron Sources | 233 |

**References** ... 245

**Appendix 1. Useful Relationships in Dynamical Theory** ... 255

**Appendix 2. Electron Wavelengths, Physical Constants, etc.** ... 257

**Appendix 3. Crystallographic Data** ... 259

| A3.1. | The Reciprocal Lattice | 259 |
| A3.2. | The Seven Crystal Systems | 260 |
| A3.3. | Interplanar Spacings | 261 |
| A3.4. | Extinction Conditions Resulting from Screw and Glide Symmetry | 261 |
| A3.5. | Symmetries in Zone-Axis CBED Patterns | 263 |
| A3.6. | Height $H$ of HOLZ Planes for Zone $(u, v, w)$ | 265 |
| A3.7. | The Use of a Metric Matrix for Crystallographic Calculations | 265 |

**Appendix 4. Indexed Diffraction Patterns with HOLZ** ... 267

**Appendix 5. Computer Programs** ... 277

| A5.1. | Plotting HOLZ Lines | 277 |

|        |                                                         |     |
| ------ | ------------------------------------------------------- | --- |
| A5.2.  | Bloch-wave Dynamical Programs                           | 296 |
| A5.3.  | Multislice Programs                                     | 308 |

**Appendix 6. Crystal Structure Data** ........................................... 323

**Appendix 7. A Bibliography of CBED Applications Indexed by Material** ........................................... 329

**References for the Appendixes** ........................................... 353

**Index** ........................................... 355

# Symbols, Units, Sign Conventions, and Terminology Used in This Book

## TERMINOLOGY

The terms *convergent-beam electron diffraction* (CBED) and *microdiffraction* are used interchangeably throughout the book, according to current usage. Both refer to the electron diffraction patterns which are formed when an electron source is focused onto a thin sample. "Coherent CBED" and "incoherent CBED" refer specifically to the degree of coherence in the illumination aperture. (In the first case the probe is necessarily coherent. In the second it is partially coherent. An incoherent probe is impossible in practice.) Roughly speaking, the coherent case is usually obtained only on field-emission electron microscopes.

A recommendation for future usage might be that the coherent case be known as nanodiffraction and the incoherent case as CBED. Both are examples of electron microdiffraction. CBED patterns are also known as Kossel–Möllenstedt patterns. A Kossel pattern is an incoherent CBED pattern with overlapping orders. A Kikuchi pattern is generated by inelastic scattering (often covering a range greater than the Bragg angle) from within a crystal. A ronchigram is a coherent CBED pattern with overlapping orders (often formed with the illumination aperture removed entirely), and is usually obtained from rather thin crystals. LACBED means large-angle convergent-beam electron diffraction, and refers to incoherent CBED patterns in which special techniques have been used to prevent overlap of the orders, so that an angular view of an order may be obtained which is larger than the Bragg angle. The terms "sample," "crystal," "specimen," and "foil" are used synonymously throughout the book. The uncrystallographic terms "horizontal" and "vertical" have sometimes been used for brevity, and are defined by the direction of the electron beam.

We adopt the crystallographic conventions for planes and directions, thus $(hkl)$ denotes a set of parallel crystal planes, $\{hkl\}$ a family of planes equivalent by symmetry, $[hkl]$ a direction, and $\langle hkl \rangle$ a set of equivalent directions.

## SIGN CONVENTIONS

A plane wave is taken to have the form $\exp(+2\pi i \mathbf{K} \cdot \mathbf{r})$ throughout this book, in accordance with the most widespread usage in quantum mechanics texts and in the dynamical electron diffraction literature. (Since inelastic scattering is not considered, the time dependence of the wavefunction is suppressed.) This convention is the opposite to that used throughout the *International Tables for Crystallography*, in X-ray crystallography generally, in Cowley (1981b), Spence (1988a), and in many of the multislice multiple electron scattering programs. The choice of signs thus presents an unavoidable dilemma in electron crystallography, as discussed in Saxton *et al.* (1983) and Spence (1988a). Our choice means that our formulas for structure factors [such as Eq. (3.3)] are unconventional. [The minus signs in Eqs. (3.3) and (3.8) would be replaced by plus signs in X-ray crystallography.] Broadly speaking, reversing the sign of the phase of the incident plane wave requires a change of sign in the exponential kernel of the Fourier transform between real and reciprocal space, which has the effect of conjugating all structure factors. For historical reasons, the Bloch-wave program listed in Appendix 5.2 uses the crystallographic convention, $\exp(-2\pi i \mathbf{K} \cdot \mathbf{r})$.

For the determination of structure-factor phase invariants, either convention, if used consistently, leads to the same conclusions. Thus, for example, the use of an incident plane wave $\exp(-2\pi i \mathbf{K} \cdot \mathbf{r})$ would reverse the sign of every term in the three-phase invariant given in Eq. (3.59). As shown at the end of Section 5.1, a compensating sign change elsewhere then results in the same predicted intensity for all crystals. (Two crystals which are enantiomorphous pairs, however, differ only in the sign of the phase invariant, and can therefore be distinguished using CBED intensity measurements.)

Interferometry and holography experiments (of which three-beam dynamical scattering is an example) always measure magnitudes of phase differences, which are again independent of the sign of the phase of the incident wave.

Section 5.1 describes the "absolute" determination of individual structure-factor phases in BeO, for an origin taken on the Be site. Reversing the sign of the phase of the incident wave would reverse the sign of these phases, and this must be considered (together with the choice of origin) when comparing this work with results from other researchers.

## UNITS

SI units are used throughout, except where stated. Angstrom (Å), centimeter (cm), and esu units are also given occasionally for convenience. Electron scattering factors are specified in both the old (Born approximation) system in units of angstroms and the new SI system in units of volt meter$^3$ (V m$^3$). See Section 3.1 for more details.

## SYMBOLS

In the old system, based on the Born approximation and given in the *International Tables for Crystallography* before 1989, electron structure factors $F_g^B$ were defined in terms of atomic scattering factors $f_i^B$ by

$$F_g^B = \sum_j f_j^B(s) \exp(-2\pi i \mathbf{g} \cdot \mathbf{r}_j) \quad \text{in angstroms}$$

In the new system (*International Tables* since 1989) SI units have been used, and the structure factors made a property of the material alone, independent of accelerating voltage or scattering theory, so that

$$F_g = \sum_j f_j(s) \exp(-2\pi i \mathbf{g} \cdot \mathbf{r}_j) \quad \text{in volt meter}^3$$

where

$$f_i(s) = \int V_i(\mathbf{r}) \exp(-4\pi i \mathbf{s} \cdot \mathbf{r}) \, d\tau$$

Here $V(\mathbf{r})$ is the atomic potential (in volts), and $2\mathbf{s} = \mathbf{u} = \mathbf{K}' - \mathbf{K}$ with $|\mathbf{u}| = 2\sin\theta/\lambda$. In addition $\mathbf{K}$ and $\mathbf{K}'$ are the incident and elastically scattered electron wavevectors, and $\theta$ is half the total scattering angle $\Theta$. The symbol $f_i^e(s)$ has also been used for the new electron atomic scattering factor.

Fourier transforms are defined with signs such that, for example, the (positive) crystal potential is

$$V(\mathbf{r}) = \sum_\mathbf{g} V_\mathbf{g} \exp(-2\pi i \mathbf{g} \cdot \mathbf{r})$$

in volts, therefore

$$V_g = \frac{F_g}{\Omega} \quad \text{in volts}$$

where $\Omega$ is the volume of the unit cell in meter$^3$.

The old system is related to the new by

$$f_j^B(\mathbf{s}) = \left(\frac{2\pi m_0 |e|}{h^2}\right) f_j(\mathbf{s})$$

Here $|\mathbf{s}| = \sin\theta/\lambda \approx \Theta/2\lambda$ is half the scattering vector, as used in the *International Tables for Crystallography*. Scattering factors are expressed in terms of $f(s)$ rather than $f(u)$.

| | |
|---|---|
| $B$ | $= 8\pi^2 \langle u^2 \rangle$ The temperature factor is $\exp(-Bs^2)$ for amplitudes |
| $\lambda$ | Relativistic electron wavelength $= hc/W\beta$ |
| $\mathbf{K}_0$ | Incident electron wavevector in vacuum, $|\mathbf{K}_0| = 1/\lambda$ |
| $\mathbf{K}_0'$ | Scattered wavevector |
| $\mathbf{K}$ | Wavevector corrected for mean inner potential, $\mathbf{K}^2 = \mathbf{K}_0^2 + U_0$ |
| $\mathbf{k}^{(j)}$ | Labeling wavevector for Bloch wave $j$ inside the crystal |
| $S_\mathbf{g}$ | Excitation error, positive if $\mathbf{g}$ is inside the Ewald sphere |
| $\theta_c$ | Incident beam divergence, the semiangle subtended by the effective source at the sample |
| $\theta_B$ | The Bragg angle |
| $\Theta$ | Total scattering angle, the angle between $\mathbf{K}'$ and $\mathbf{K}$, twice $\theta$ and $\theta_{\text{Bragg}}$ |
| $\mathbf{g}$ | Reciprocal lattice vector, $|\mathbf{g}| = 1/d_{hkl}$ |
| $W$ | Total relativistic energy of an electron, $W = m c^2 = \gamma m_0 c^2$ |
| $m$ | Relativistic mass of an electron $= \gamma m_0 = \gamma m_e$ |
| $\sigma$ | Interaction constant $= 2\pi m |e| \lambda/h^2$ |
| $U_g$ | Dynamical structure factor $= \sigma V_g/\lambda\pi = \gamma F_g^B/\pi\Omega$ in units of length$^{-2}$. In noncentrosymmetric crystals, $U_g = U_g^c + iU_g'$, where $U_g^c$ and $U_g'$ are each the complex Fourier coefficients of real potentials |
| $V_g$ | $= h^2 U_g/(2m|e|)$, Fourier coefficient of the crystal potential, in volts |
| $V(\mathbf{r})$ | Crystal potential, in volts |
| $\xi_g$ | Two-beam extinction distance, $\xi_g = 1/(\lambda |U_g|) = \pi/(|V_g| \sigma) = \pi\Omega/(\gamma\lambda |F_g^B|)$ |
| $t$ | Sample thickness |

$E_0$     Accelerating voltage
$E_r$     Relativistic accelerating voltage $[= E_0 + (e/2m_0c^2)E_0^2]$
$E$     An electric field
$\gamma$     Relativistic factor $(1 - v^2/c^2)^{-1/2}$
$\gamma^i$     Eigenvalue of the structure matrix
$\alpha_c$     Coherence angle. $\alpha_c = \lambda/d_s$

In expressions such as Eq. (8.1) we adopt the convention that all quantities following the solidus are to be divided into those preceeding it.

# Introduction: A Note to Students in Materials Science

Electron microdiffraction is a technique for extracting crystal-structure information from submicron regions of material, using a transmission electron microscope (TEM). Information on the atomic structure of defects, on local crystal symmetry, and on strain fields may be extracted from regions as small as a nanometer in diameter or less by this method. This book is about the interpretation of these transmission electron diffraction patterns. In particular, we have attempted to summarize and develop all the useful knowledge which has been gained over the years from the study of the multiple electron scattering problem, and to apply it to practical problems in the interpretation of microdiffraction patterns. The book covers theory, applications, and experimental methods. It has been said that it is now impossible to write a book on electron microscopy—the subject is simply too large. Several volumes would be required to cover in detail all the relevant knowledge, from ultrahigh vacuum systems to X-ray detectors, dynamical scattering theory in transmission and reflection, electron optics, and electron crystallography for organic materials. We have concentrated in this book on the signal which is available from most modern TEM instruments that we feel is most accurately quantifiable—the transmission electron diffraction pattern.

From the beginning it was realized that the convergent-beam electron microdiffraction method in particular is an extraordinarily efficient technique. The wide range of illumination angles used provides a large amount of information in the form of rocking curves for many diffracted orders, presented simultaneously from the smallest possible region of crystal. In addition, different regions of the pattern are sensitive to different crystal parameters—the outer reflections and HOLZ line intensities to atomic position parameters and temperature factors, the inner reflections for bonding effects, and HOLZ line positions for lattice parameters and strain. This book shows how this information may be extracted by a process of automated dynamical refinement, weighted according to the type of information required and the portion of the CBED pattern used.

Students embarking on a course in materials science at the graduate level will want to know where this book fits into the overall scheme of modern materials science. It is often said that materials science is the study of the structure, properties, and synthesis of materials. The relationship between structure and properties is heavily emphasized in recent work. Characterization forms an essential part of all the pertinent analysis. This book concerns one mode (microdiffraction) of operation for an electron microscope. It therefore treats a rather specialized topic, which is part of the overall characterization effort in materials science. A considerable effort is now under way in many countries aimed at understanding and predicting the properties of new materials in terms of atomic mechanisms. Microdiffraction contributes to this effort by giving us crystallographic information on a near-atomic scale. The following brief list of recent applications gives some impression of the contribution microdiffraction has made to our understanding of atomic processes in solids.

1. In the semiconductor industry, considerable effort has gone into the production of multilayer devices using molecular beam epitaxy in order to grow devices at low temperatures, where atomically abrupt interfaces can be generated and diffusion avoided. For silicon–germanium multilayers, for example, it has proven possible to grow metastable films in which there are no misfit dislocations to interfere with device performance, provided the layers are sufficiently thin. Then the electronic structure of the layers depends critically on the strains stored in them. Only by nanodiffraction has it been possible to measure this strain within a single layer (Pike *et al.*, 1991) and at these interfaces (Eagelsham *et al.*, 1989). In another example of the application of the same method [Zuo, 1991a; see also Boe and Gjønnes (1991) for related work], the change in lattice constant at two points a few hundred nanometers apart in the oxide superconductor $YBa_2Cu_3O_{6-x}$ was measured, and the result related to changes in the local oxygen concentration $x$, using data from neutron diffraction. The onset of superconductivity is known to be strongly correlated with oxygen concentration and ordering.
2. For many applications it is required to form a strong bond

between a ceramic and a metal. The strength of such a bond depends crucially on the interface energies involved. If a thin layer of a third phase is present at the interface, the relevant interface energies will differ from those at a bilayer. Thus the resistance to crack propagation at the interface may be controlled by a layer of material just a few unit cells thick. Only by nanodiffraction has it been possible to identify the crystal structure of such an intergranular phase (Lodge and Cowley, 1984).

3. The performance of the new high-temperature superalloys used in turbine blade applications frequently depends on the presence of many fine intermetallic precipitates and a complex microstructure. These microcrystals may be as small as a few nanometers in diameter. There now exist several examples of phase identification in these and similar alloys using convergent-beam diffraction to identify the space group of the microcrystal; see, for example, Fraser (1983).

4. In the field of catalysis, information is required on heavy metal particles, perhaps 5 nm in diameter or smaller, lying on a low-atomic-number substrate. The method of high-angle annular detector (HAAD) scanning transmission electron microscopy (STEM) combined with microdiffraction has proven extremely powerful for the study of these particles, since it is possible to suppress diffraction contrast effects from the substrate (Treacy *et al.*, 1978) and to combine the method with secondary electron imaging (Liu and Cowley, 1990). The HAAD signal depends on the atomic number of the sample, its thickness, and the vibration amplitude. The secondary electron image provides topographic information on a subnanometer scale. The microdiffraction technique is indispensable for the identification of individual particles and their crystallographic relationship with the substrate.

More generally, the need for the information which microdiffraction provides arises in many fields, from fundamental experiments in condensed matter physics, through mineralogy and solid state chemistry, to physical metallurgy and materials science generally. In all these fields there has been a growing recognition of the importance of the effect of crystalline microstructure on the behavior and properties of materials. In condensed matter physics, for example, the method has been used to study changes in symmetry during phase transitions on a local scale, and to map out the distribution of valence electrons in semiconductors. Other uses of the technique include the very accurate measurement of electron microscope accelerating voltage, of sample thickness, of crystal lattice spacings, and of crystal structure-factor amplitudes and phases. Recent developments, using an automated refinement cycle similar to the Rietvelt method in neutron diffraction, suggest that atomic positions and lattice strains may be measured with high accuracy, allowing deductions to be made concerning ordering and impurity segregation in crystals. The method also provides a powerful technique for microphase identification, through its unique ability to determine the space group of very small microcrystals. When the smallest (subnanometer) coherent electron probes are used, the theory of CBED becomes the theory of scanning transmission electron microscopy (STEM) and, for very thin crystals, of in-line electron holography. This technique, while yet to be fully developed, possesses in principle the same power to determine the atomic structure of crystalline defects as high-resolution electron microscopy, while allowing simultaneous microanalysis.

Microdiffraction patterns are obtainable on all modern analytical electron microscopes. They differ from conventional selected area ("spot" or "point") diffraction patterns in that a small electron probe is focused onto the sample, causing the diffraction spots to broaden out into disks. We shall use the terms convergent-beam electron diffraction (CBED) and microdiffraction synonymously throughout this book to refer to all electron diffraction patterns in which a submicron electron probe is focused onto a sample. The term "nanodiffraction" will be used to refer to similar patterns, in which the probe size is smaller than about a nanometer in diameter. These patterns are also known as Kossel–Möllenstedt patterns, after their discoverers. (A "Kossel pattern" is a CBED pattern whose orders overlap.) The book covers four main topics: Space-group determination by CBED, structure-factor measurement by CBED, coherent CBED (holograms, ronchigrams) from defects using subnanometer probes, and strain measurement using HOLZ-line shifts. A chapter on instrumentation and experimental technique is included. We have also tried to provide a unified and coherent account of all the relevant theoretical background. We have emphasized fundamentals in the theory chapters, because our experience has been that this is the way in which progress is made. We hope that our unified treatment of the theory will suggest new experiments and new signals which can be used for materials characterization. For example, only quite recently has it been widely appreciated in the materials community that Bragg's law can be in error by as much as 5% for electron diffraction (due to multiple scattering), and that this has important consequences for the accurate measurement of strains near interfaces.

Working in an electron microscopy laboratory for many years has convinced us that most of the enormous amount of information contained in electron diffraction patterns is simply thrown away. This book is an attempt to reverse that trend, and to promote the development of a truly quantitative electron crystallography method. The book is intended for graduate students and professional research workers in materials science and condensed matter physics. We have tried to cover the theory, the practice, and some applications of the method, together with a very brief account of the history of the subject. An undergraduate knowledge of crystallography and some familiarity with electron microscopy are assumed.

The first half of the book is concerned mainly with theory, the second with applications, techniques, computer algorithms, and instrumentation. Students completing graduate courses in materials science may wish to read only the following: Chapter 2, Sections 3.3 and 3.4, Section 4.1, Section 5.5, and Chapters 7 and 9. For those wanting only an introduction to the subject, Chapters 2 and 9 are the most important.

While a large number of papers, a few review chapters, an atlas of CBED patterns, and two outstanding volumes of annotated examples of CBED patterns have appeared, there seems to be no other unified description of the theory and practice of convergent-beam electron diffraction (CBED) which aims at pedagogic soundness in an integrated treatment, using a consistent formalism throughout. Some useful companion review papers include: The two numbers of *J. Microsc. Tech.*, (1989) Vol. 13, Nos. 1 and 2 devoted to CBED, and Cowley (1978), Eades (1988), Muddle (1985), Spence and Carpenter (1986), Steeds (1979), and Williams (1987). The most important books on the subject are Cowley (1992) and Tanaka *et al.* (1988).

A recent literature survey (Sung and Williams, 1991) lists over 200 publications on CBED in 1991, as against 10 in 1975. This growth in the popularity of the method is due to advances both in computing power and in the performance of modern analytical electron microscopes. These include the appearance of convenient and reliable field-emission guns for TEM, of Omega energy-filtering devices, of cooled CCD detectors, and stable heating and cooling double-tilt specimen holders, of the new RISC work stations, together with the improved vacuum systems on modern instruments which reduce contamination and so make CBED possible. We shall see, however, that the basic theoretical work on which these modern developments are based occurred mainly in the 1960s and earlier.

We have written this book with the strong conviction that the intensity distribution in elastic (energy-filtered) electron microdiffraction patterns now provides the most accurately quantifiable information available on a modern analytical transmission electron microscope (TEM). For example, using appropriate dynamical computations, it is possible to measure strains as small as $10^{-4}$, structure-factor phases with an accuracy of much less than one degree, and sample thickness to an accuracy of a few angstroms. This information is obtainable from regions as small as a few nanometers in diameter. Our second motivation has been the fascinating and rich history of the subject, briefly outlined in Chapter 1.

On reading the literature, one is struck by the enormous variety of applications of CBED. These include studies of charge-density waves in layered structures, analyses of the symmetry changes which accompany low-temperature phase transitions, accurate mapping of the distribution of valence electrons in semiconductors, and phase identification and strain measurement around precipitates in alloys or semiconductors. To review all this work, published in over one-thousand papers, and to extract its implications for materials physics would be a herculean task. (We have, however, included a bibliographic index of applications indexed according to material in Appendix 7.) Thus, to experts in the field, the examples in this book may seem somewhat oversimplified. But our limited aim has been only to explain the principles of CBED, to provide all the useful theory in a consistent format, and to convey enough understanding to students and researchers to let them get started with microdiffraction. In order to make the book more practically useful, we have included some FORTRAN source code listings, together with the POSTSCRIPT code which will allow the direct printing of Kikuchi- and HOLZ-line patterns on modern laser printers.

# 1

# A Brief History of Electron Microdiffraction

The history of convergent-beam electron microdiffraction is a fascinating blend of elegantly derived results from the basic physics of electron scattering theory, together with their application to subtle problems in understanding real crystalline solids. The following is a very brief review of some of the main historical developments in convergent-beam electron diffraction, with apologies to those whose work may have been overlooked. Literature references are given elsewhere in the relevant section of this book.

In 1928, the results of Hans Bethe's "Ph.D." thesis (under the direction of A. Sommerfeld) were published, describing the use of "Bloch waves" to solve the high-energy electron diffraction problem for the reflection case, in order to account for the observations of Davisson and Germer in the U.S.A. and Thomson and Reid in the U.K. Bethe's work was based on Ewald's earlier treatment of the multiple scattering problem in X-ray diffraction. A perturbation method (the "Bethe potentials," on which three-beam theory is based) was also given. This work was contemporaneous with, and apparently independent of, Bloch's introduction of "Bloch waves" into band theory. Bethe's theory was extended to the transmission case by Blackman in 1939.

The origins of the convergent-beam concept lie in the discovery of Kossel patterns in X-ray diffraction in 1937 and of Kikuchi patterns in 1928. Both may be thought of as due to internal sources of radiation from atomic sites in the crystal. In order to understand this process in more detail, the young G. Möllenstedt was asked in 1937 by W. Kossel in Danzig to build a 45-kV convergent-beam electron diffraction camera [see *Phys. Status Solidi A*, **116**, 13 (1989)]. This would supply an external source of diverging radiation. Möllenstedt's design [for his diploma thesis (M.Sc.) project] is shown in Fig. 1.1. It used a plasma discharge in a wine

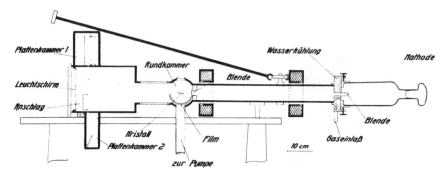

*Figure 1.1.* Convergent-beam diffraction camera designed by G. Möllenstedt in 1937 for operation at 45 kV. A wine bottle was used to contain a gas-discharge electron source. The vacuum was about $10^{-3}$ torr and the probe size about 40 $\mu$, a combination of conditions which apparently minimized contamination. A later instrument was built to run at 750 kV in 1944.

bottle as the electron source. The vacuum at the specimen was $10^{-3}$ torr and the probe size was about 40 $\mu$. With such a large probe, contamination was not a problem, despite the poor vacuum. Using flakes of mica as a sample, CBED patterns such as that shown in Fig. 1.2 were obtained, exactly like the patterns obtainable on the most modern machines. Deviations from the kinematic theory were immediately noted. Improved 65-kV and 750-kV machines were also built, before the war brought developments to an end. In 1940, MacGillavry used two-beam theory to fit experimental CBED patterns in the first attempt to measure structure factors using dynamical electron diffraction theory.

Theoretical work on the dynamical theory with emphasis on the symmetry properties of the scattering was continued throughout the 1950s by researchers such as Niehrs, Fukahara, Fues, Howie and Whelan, Fujimoto, Miyake, Tournarie, Sturkey, and Cowley and Moodie. In 1957, K. Kambe, in his study of three-beam theory showed that the intensity depends on a certain sum of three structure-factor phases (the three-phase invariant), which is independent of the choice of origin and so might be measured. Throughout the sixties, the CBED method was developed almost solely by Lehmpfuhl in Berlin, and by Goodman and Moodie in Melbourne using the unsatisfactory (but modifiable) instruments available to them. In 1965, Gjønnes and Moodie (building on earlier work) explained the occurrence of forbidden reflections in the presence of strong multiple scattering (as previously observed by Goodman and Lehmpfuhl). These could then be used to identify translational symmetry elements. The implications of combining results

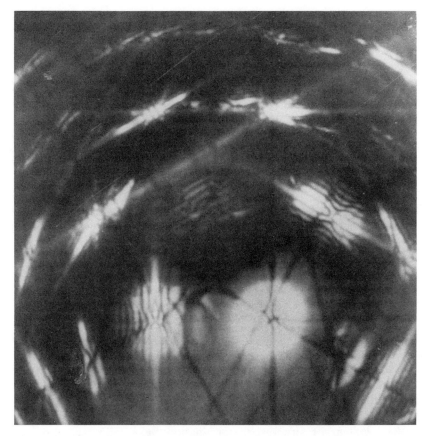

*Figure 1.2.* CBED pattern obtained from mica in 1937 at 45 kV using the apparatus shown in Fig. 1.1. This pattern was analyzed using two-beam theory by MacGillavry and others (Kossel and Möllenstedt, 1942; MacGillavry, 1940).

from the reciprocity theorem with crystal symmetry elements were first appreciated in 1968 by Pogeny and Turner, working in John Cowley's group. Work by Uyeda and Høier during this period showed how the position of Kikuchi lines may be used to determine accelerating voltages and lattice constants, and the importance of dynamical corrections to the line positions understood. [This would later be studied in great detail for the closely related problem of the higher-order Laue zone (HOLZ) lines used for strain measurement.] The critical voltage effect on Kikuchi lines was discovered at about this time by Uyeda, Watanabe, and co-workers in Japan. In their 1971 three-beam analysis of this effect, Gjønnes and Høier showed that an eigenvalue degeneracy existed in three-beam theory, so that the absence of intensity at certain points in these patterns may be used to determine the three-phase invariant for centrosymmetric crystals. Thus, with the dynamical problem "solved," the implications of crystal symmetry and reciprocity understood, dynamical corrections to the Bragg law understood, and the three-phase invariants defined (with their effects elucidated for centric crystals), we might say that the heroic age of CBED theory came to an end.

By the early seventies, then, systematic procedures for point-group and space-group determination by CBED had begun to emerge from the groups in Melbourne (Goodman, Moodie, *et al.*) and Bristol (Steeds, Buxton, and co-workers), and a lively debate ensued on the possible symmetry-breaking effects of boundary conditions. The theoretical foundations for point-group determination were firmly established by Buxton and co-workers in the context of group theory; they went on to develop the perturbation theory for HOLZ interactions (with later elegant contributions from Portier and Gratias, Tinnable, Kogiso, Kastner, and others). The Bristol group then embarked on a systematic application of the CBED method to a wide variety of problems in materials science and condensed matter physics. This focused effort in Bristol over many years (resulting among other things in the publication of an atlas of CBED patterns for alloy phases in 1984) was perhaps responsible more than anything else for establishing the success of the method.

As one example among many, the work of this group on phase transformations in layer compounds supporting charge-density waves brought the CBED technique to the attention of a much larger audience of solid state physicists for the first time. The most comprehensive attack on the problem of structure determination by CBED has also been described by this group in their successful determination of the structure of AuGeAs. In the United States, the subsequent popularity of the method owed most to Eades' work at Illinois. Throughout the late 1970s and early eighties, Cowley (together with one of the authors) in Arizona were developing the coherent CBED method using subnanometer probes. The application of the superlattice method within the multislice algorithm was developed for coherent CBED patterns from defects in 1977, and the theory of STEM lattice imaging developed as a result. In 1981, Cowley produced some remarkable CBED patterns using the Vacuum Generators HB5 instrument from regions of crystal smaller than the unit cell, which were seen to repeat with the period of the lattice as the probe was moved across the crystal. Similar work on nanodiffraction, including the use of imaging energy filters and novel detectors, was later developed by Brown and co-workers in Cambridge, U.K. Both groups subsequently developed techniques for the study of defects in crystals using subnanometer probes. A nonscanning alternative to the Eades/Tanaka scanning method for avoiding overlap of adjacent orders was developed by Tanaka, whose group also produced (starting in 1985) two invaluable volumes of beautiful CBED patterns covering a wide range of applications and "case studies." This group has since produced much of the highest quality work in the field. The CBED method had otherwise been slow to develop in Japan prior to Tanaka's efforts.

In Berlin, quantitative work on structure factor measurement continued under Lehmpfuhl in the early eighties. The use of CBED patterns to study line and planar defects also first began to be studied at about this time. Large-angle CBED methods (LACBED) for HOLZ and ZOLZ reflections were then developed in 1986 by Taftø, Vincent, and co-workers. HOLZ effects from artificial superlattices appeared first in the work of Cherns in 1987. The persistent rediscovery of the value of shadow imaging in CBED (producing various hybrid modes such as CBIM and LACBED) and of the value of HOLZ lines (because their intensities are frequently kinematic, allowing simple rules to be derived) is a feature of work during this period. Research on structure-factor phase measurement in noncentrosymmetric crystals was begun in earnest in the mid-1980s by Marthinsen, Høier, and later Bird and others, resulting finally in experimental structure-factor phase measurements with an accuracy of better than one degree by Zuo and co-workers in 1989. During the same period, many measurements of local strains began to appear, based on measurements of HOLZ-line positions with various approximate dynamical correction schemes (recently summarized by Lin and Bird) in development of the earlier work by Gjønnes, Høier, and Olsen on dynamical shifts on Kikuchi lines.

Only very recently have the computing times for whole CBED patterns been reduced to less than an hour or so for simple crystals, and this progress in computer hardware explains much of the recent renewed interest in CBED. The recent use of elastic energy filtering has greatly increased accuracy (see, for example, the work of Mayer *et al.* cited in Chapter 9 for an evaluation of the Zeiss Omega filter for CBED work). This, together with the use of cooled CCD cameras, on-line work stations, and field-emission guns, brings our subject to the threshold of its most exciting era, in which the techniques of quantitative electron microcrystallography will be applied to a wide range of problems in materials science, solid state chemistry, mineralogy, and condensed matter physics.

# 2

# The Geometry of CBED Patterns

This chapter describes the geometry of CBED patterns, and discusses the practical techniques used to assign a beam direction to each point in the central (000) disk. This "calibration" information is needed for all the techniques described in the remaining sections of this book. A thorough grasp of the figures in this chapter is absolutely essential for any understanding of the CBED technique. Experience shows that few electron microscopists understand the geometry of CBED patterns. This chapter and Chapter 9 are the most important chapters in this book.

We must first define two important wavevectors and the wavelength of the electron in the electron microscope. The electron wavelength $\lambda$ in vacuum may be derived formally from Eq. (3.14) [with $V(\mathbf{r}) = 0$], but for the present we simply quote the relativistically correct result in numerical form. Thus

$$\lambda = 12.2643/(E_0 + 0.97845 \times 10^{-6} E_0^2)^{1/2} \qquad (2.1)$$

where $E_0$ is the electron microscope accelerating voltage in volts (V) and $\lambda$ is obtained in angstroms (Å). Appendix 2 tabulates values of $\lambda$ for many common microscope voltages. The magnitude of the wavevector of the beam electron in vacuum is thus $K_0 = 1/\lambda = \beta \gamma m_0 c^2/hc$. The direction of $\mathbf{K}_0$ is that of the collimated incident beam.

Although we will see in later chapters that the effects of inclined boundary conditions and multiple scattering can "shift" the Bragg condition, in this chapter we will ignore these small effects. But the mean "refractive index" effect of the average crystal potential must now be considered for consistency with later results derived in Chapter 3. The electron speeds up as it enters the crystal, being attracted to the positve atomic nuclei, thereby gaining kinetic energy and a longer wavevector. We define $V_g$ as the complex Fourier coefficients of the (positive) crystal potential, in volts. Then the magnitude $K$ of the mean wavevector inside the crystal is given by

$$K^2 = K_0^2 + (2m\,|e|\,V_0/h^2) = K_0^2 + U_0 \qquad (2.2)$$

so that

$$K \approx K_0 + U_0/2K_0 \qquad (2.3)$$

Here $e$ is the charge on the electron, $m$ the relativistic electron mass, and $h$ is Planck's constant, while $U_g$, the "structure factor" of electron diffraction, has the dimension of length$^{-2}$. Values of $V_0$ are tabulated in Spence (1992c) for various crystals, and vary from 10 to about 50 V with increasing atomic number and density (see Section 4.5). In the reflection geometry, the Bragg condition depends strongly on $U_0$, but for most transmission microscopy on untilted slab samples we will see that it is the components of the wavevectors $\mathbf{K}$ and $\mathbf{K}_0$ in the zero-order Laue zone (ZOLZ) which determine the Bragg condition, and these are equal.

Ordinary selected-area or "point" transmission electron diffraction patterns consist of bright spots of electron intensity which occur in directions satisfying the small-angle Bragg law

$$\Theta = 2\theta_B \approx \lambda/d_{hkl} \qquad (2.4)$$

where $\Theta$ is the total scattering angle between the incident and diffracted beam, $\theta_B$ is the Bragg angle, and $d_{hkl}$ is a crystal interplanar spacing. Appendix 3 describes the relationship between $d_{hkl}$ and the unit cell dimensions for the various crystal systems. We will see in this book that Eq. (2.4) is an approximation, since it is based on the wavevector $|K_0| = 1/\lambda$ for the beam electron *outside* the crystal. It is usually sufficiently accurate for diffraction by planes containing the beam direction in a horizontal slab-shaped sample. In the reflection geometry the corrections due to $U_0$ cannot be ignored. In the transmission case, also, for planes not containing the beam direction the errors resulting from the use of Eq. (2.4) can be appreciable. (For example, measurements of the accelerating voltage obtained by applying this equation to the positions of HOLZ lines in the central disk of a CBED pattern may be in error by as much as several kilovolts. For accurate work, dynamical calculations are required.)

Convergent-beam diffraction patterns are formed with the electron probe focused onto the sample, causing the diffraction spots to broaden out into disks. When confronted with a new CBED pattern, the first problem is to index it and to determine the range of incident beam directions covered. In order to illustrate the principle of the CBED method, we will assume in this chapter that the crystal structure is known and that the pattern has been indexed. (A list of common crystals and their structures is given in Appendix 6.) Methods used to index "point" electron diffraction patterns are given in Section 7.2, in Loretto (1984), and in Appendix 4.

By the zero-order Laue zone (ZOLZ), we specifically mean in this book the plane of reciprocal lattice points which passes through the origin. A higher-order Laue zone (HOLZ) is any other reciprocal lattice plane parallel to this, not passing through the origin. (Other definitions are used in the crystallographic literature, however the preceding conforms to common usage in electron crystallography.) To simplify the mathematics, we consider here only thin crystals in the form of parallel-sided slabs, whose surface normal is approximately antiparallel to the beam direction. Samples of similar shape, but inclined to the beam, are considered in later chapters.

## 2.1. ONE-DIMENSIONAL (SYSTEMATICS) CBED

Figure 2.1 shows a simplified ray diagram for a CBED pattern in the systematics or "one-dimensional" case. Here, by choice of orientation, the electron beam is permitted to "see" only a single family of parallel crystal planes, so that the point diffraction pattern would consist of a single line of bright spots. The electron source is focused to a small probe on the surface of a thin crystalline sample. Throughout this book (except in Chapter 8, where we deal with subnanometer probes) we will assume that the final illumination aperture C2 is perfectly incoherently filled with electrons, that is, it acts as an ideally incoherent effective source. (Chapter 8 contains a discussion of coherence.) Then each point P within the final condenser aperture acts as an independent point electron emitter, which defines the direction of a planewave at the sample. If we further assume that the crystal is a parallel-sided slab of perfectly crystalline material in which no inelastic scattering or defects occur, then each such incident plane wave can only be scattered by multiples of twice the Bragg angle. Thus a source point such as P gives rise to a set of scattered waves, which reach the detector at a family of points such as P'. A different source point Q similarly results in a different family of diffracted beams Q'. Since the angles are small, the distances $X$ between points P' on the film are proportional to reciprocal lattice vectors $g$ according to

$$X = L2\Theta_B \approx Lg\lambda \qquad (2.5)$$

where $L$ is the electron microscope camera length.

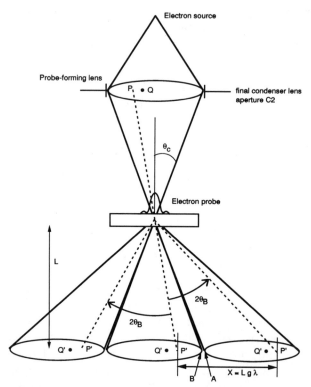

*Figure 2.1.* Simplified ray diagram for convergent-beam electron diffraction. If only elastic Bragg scattering is allowed, source point P gives rise to conjugate points P', one in each disk. Source point Q defines a different incident beam direction and set of diffracted beams Q'. The camera length is $L$.

Since the distances between points P' are fixed by the crystal structure for given experimental conditions, we see from the figure that *the identification of a point in the central CBED disk defines a complete point diffraction pattern, with one point taken from each CBED disk.* In two dimensions, we will see that these conjugate points lie on the two-dimensional reciprocal lattice. A CBED pattern may thus be thought of as a set of point diffraction patterns, laid side by side. The set of points P' defines one such point pattern, while Q' defines another. The set Q' refers to a different incident beam direction Q from that (P) which gives rise to the points P'. To test your understanding of Fig. 2.1, find the diffraction conditions at point A. Where is the corresponding incident beam direction? What might the intensities be at these two points? What might the intensity be at point B? (Assume a zone axis pattern.)

## GEOMETRY OF CBED PATTERNS

Now the variation of intensity for a particular diffracted beam with the direction of the incident beam is known as a *rocking curve*. Thus, we may say that *the CBED method displays a rocking curve simultaneously in every diffracted order*. It will be evident in later chapters (see, e.g., Fig. 5.2a) that it is possible to satisfy the Bragg condition for at most two reflections in a single one-dimensional (systematics) CBED pattern. Since the incident beam divergence $\theta_c$ may be as large as the Bragg angle $\theta_B$ before the orders overlap, an angular range as large as twice the Bragg angle may be obtained in each diffracted order. (Chapter 6 describes techniques for increasing this range.) *The great power of the CBED method derives from the fact that so much crystal structure information (i.e., many rocking curves) can be recorded simultaneously, and that this information is obtainable from very small (nanometer-sized) regions of crystal.*

Thus each point in the central CBED disk corresponds to a different incident beam direction, and defines a family of conjugate points differing by reciprocal lattice vectors, one in each CBED disk. We now consider how these directions can be related to the Ewald sphere construction, in order to determine which portions of the pattern are at the Bragg condition and will therefore appear bright. We also need a simple way of specifying the incident beam direction for each point in the central disk, for later use in computer programs. Following the convention established by Ewald, wavevectors are drawn toward the origin in order to define a sphere, which defines the locus of all possible elastic scattering events such that $|\mathbf{K}| = |\mathbf{K}'|$, where $\mathbf{K}'$ is the scattered wavevector. Since $\mathbf{K}$ varies in direction over the incident cone of illumination, there is a different Ewald sphere orientation for each point in the central disk. This is illustrated in Fig. 2.2a, which should be studied carefully.

Two related quantities are commonly used to specify the beam direction in CBED—the component $\mathbf{K}_t$ of the incident wavevector $\mathbf{K}$ in the zero-order Laue zone, and the excitation errors $S_g$. Our aim will be to assign one of these to each point in the central CBED disk for an experimental CBED pattern. The deviation angle $\alpha$ shown and the deviation parameter $w = S_g \xi_g$ are also used ($\xi_g$, the extinction distance, is introduced in Chapter 3). These quantities indicate deviation from the exact Bragg condition, at which $\alpha = 0$. We first consider the relationship between these quantities. By introducing a small-angle approximation and using the fact that, in Fig. 2.2a, $\sin(\theta_B - \alpha) = K_t/K$, we find that

$$K_t = -g/2 - KS_g/g \qquad (2.6)$$

so that

$$2KS_g = -2K_t g - g^2 \qquad (2.7)$$

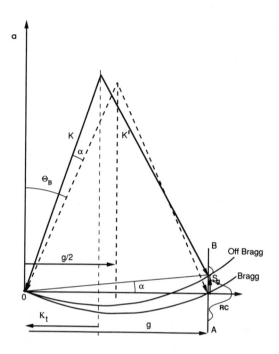

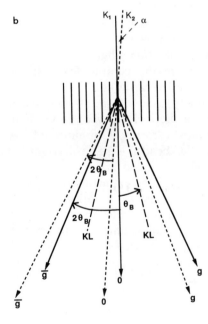

*Figure 2.2.* (a) Two Ewald sphere orientations differing by $\alpha$, just off (continuous lines) are on (dashed lines) the Bragg condition. Note the direction of $K_t$. The excitation error is $S_g$. (b) Beams of electrons diffracted by a thin crystal, shown for two incident beam directions (continuous and short-dashed lines). Longer dashes indicate Kikuchi line directions, which are "attached" to the crystal. Intensity of diffracted beams varies with incident beam direction.

for a reflection **g** in the zero-order Laue zone (ZOLZ) parallel to $\mathbf{K}_t$. In addition, Fig. 2.2a shows that

$$S_g \approx 2\theta_B \alpha / \lambda \approx g\alpha \tag{2.8}$$

At the Bragg condition for reflection $g$, we have $K_t = -g/2$ and $\alpha = 0$, so that $S_g = 0$. "Inside" the Bragg condition (as drawn, with $\theta < \theta_B$), $|K_t| < g/2$, so that $S_g$ is negative. If the beam runs parallel to a crystal zone axis, $K_t = 0$. Figure 2.2b shows a simplified ray diagram in real space, showing how CBED beam directions $g$ vary when the incident beam $K_1$ is moved to orientation $K_2$. (The intensity of the diffracted beams also varies as this is done.) We note that the directions of the Kikuchi lines KL due to inelastic scattering do not change. Unlike the Ewald sphere diagram, on which all wavevectors meet at a common origin, experimental CBED patterns focus each wavevector to a different point on the film, displaced by an amount proportional to $\alpha$.

Figure 2.3 shows an experimental systematics CBED pattern obtained from the (111) planes of MgO at 120 kV. The broad vertical dark

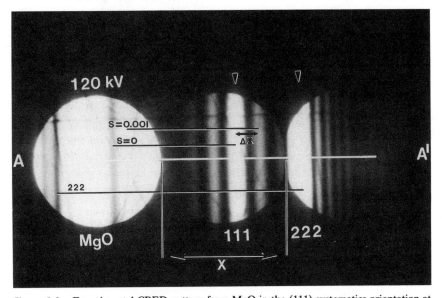

*Figure 2.3.* Experimental CBED pattern from MgO in the (111) systematics orientation at 120 kV. The distance corresponding to the (222) reciprocal lattice vector is shown. The line labeled $S = 0$ connects an incident beam direction [in the (000) disk] with the corresponding point in the (111) disk which is at the Bragg condition. The second line connects similar points for which the excitation error is 0.001, as marked. Distances $\Delta X$ and $X$ needed for assigning these excitation errors to the film are indicated. The arrowheads above indicate the (111) and (222) Bragg directions.

bands of intensity result from diffraction by all planes ($hhh$) type (written $\{hhh\}$). The oblique fine lines ("HOLZ lines") show the trajectories of points along which the Bragg condition is satisfied for a higher-order Laue zone (HOLZ) reflection with different (nonsystematics) indices. These reflections are discussed in more detail in later chapters. To calibrate the pattern, we must first measure the distance $X$ in Eq. (2.5) corresponding to the first-order (111) reciprocal lattice vector. This may be done by measuring the distance between the edges of the disks as shown on the figure. We find (on the original print) $X = 5.6$ cm. This is the transverse distance on the print which corresponds to scattering through twice the Bragg angle, and it can be used to scale other measurements since it fixes $L$, the camera length, in Eq. (2.5).

The center of the vertical band of maximum intensity in the (111) disk corresponds to the (111) Bragg condition. (We will see in Chapter 3 that this may also be a band of minimum intensity at certain thicknesses, as shown in Fig. 4.1.) The line (of length $X$) labeled $S = 0$ on Fig. 2.3 can thus be drawn. The end of this line in the (111) disk shows the point where the diffracted beam intensity is at the Bragg condition, while the start of the line in the (000) disk indicates the position of the corresponding (000) (plane-wave) beam. We note a darkening at this point, since most energy is diffracted into the first-order reflection at this orientation. Using $\lambda = 0.033491$ Å at 120 kV and $d_{111} = 2.42487$ (from Appendices 3 and 6, where the cell constant of 0.42 nm is given for MgO) Eq. (2.4) then gives $\Theta = 2\theta_B = 13.81$ milliradians (mrad) as the total scattering angle for this Bragg condition, where $S_g = \alpha = 0$.

We now consider another pair of points (marked $S = 0.001$), corresponding to a different incident-beam direction. Equation (2.5) and Fig. 2.2a show that, in general, distances $\Delta X$ measured on the enlarged print of the CBED pattern are proportional to the corresponding change $\alpha$ in scattering angle. Thus, by proportion,

$$\Delta X / X \approx \alpha / 2\Theta_B \qquad (2.9)$$

Using Eq. (2.8), we obtain

$$S_g \approx (\Delta X / X) g^2 / \lambda \qquad (2.10)$$

Measurement from the print as shown on Fig. 2.3 gives $\Delta X = 1$ cm. With $g = (d_{111})^{-1}$ for MgO we find $S_g = S = 0.001$ Å$^{-1}$, as indicated. Thus the line marked $S = 0.001$ corresponds to an orientation "outside" the Bragg condition, where the total scattering angle is greater than the Bragg condition and the excitation error is positive. It would be necessary to reduce the exposure for the central beam in order to observe the corresponding reduction in intensity in the central disk at this first subsidiary maximum.

In a similar way, an excitation error may be assigned to every point across the first-order disk. (This will be needed for the specimen thickness procedure described in Section 4.1 and for comparisons with computed patterns.) Excitation errors may also be assigned to the higher-order disks. For example, it is important to understand that the bright band in the second-order (222) disk marked with an arrowhead above it corresponds to a point at which the (222) Bragg condition is satisfied. The corresponding direct (000) beam is indicated at the left end of the line labeled $g$ (222). At the right-hand end of this line, $S_{222} = 0$. It is important to appreciate that in this pattern, the Bragg condition has therefore been satisfied at *two* points, corresponding to the first- and second-order reflections in the systematics row.

Using Eq. (2.6), values of $K_t$, the component of the wavevector in the plane of the ZOLZ, could also be assigned to every point across Fig. 2.3 along the line AA'. In this way, a zero-loss energy-filtered scan along this line could be compared with the results of numerical computations. These require values of $\mathbf{K}_t$ as input.

## 3.2. TWO-DIMENSIONAL CBED

In this section we will learn to trace out the locus of the Bragg condition in a two-dimensional pattern, and to establish a coordinate system that will be employed in the remainder of the book. In the pattern shown in Fig. 2.3, the Bragg condition was satisfied along two lines for two different reflections. If we now allow the incident beam to excite many reflections in the ZOLZ, it becomes possible to satisfy many Bragg conditions simultaneously. Figure 2.4 shows this situation. The incident beam is not parallel to a major crystal zone axis. The Ewald sphere ES, defining the Bragg condition, now intersects reciprocal lattice points in the ZOLZ on a circle LS, known as the Laue circle. We now consider the coordinate system to be used in more detail.

Consider the family of points in the CBED disks which are at the exact Bragg condition. Treat these as a scaled replica of the reciprocal lattice construction for the ZOLZ (see Appendix 3). Then the two-dimensional vector $\mathbf{K}_t$ is the component of the incident wavevector in this plane. We take the **z** axis in the direction of the ZOLZ zone axis and the **x** axis in the direction of one of the reciprocal lattice vectors. The **y** axis runs in direction $\mathbf{z} \times \mathbf{x}$. The origin is taken at the center of the Laue circle [not at the center of the (000) disk]. This origin point can sometimes be identified in experimental CBED patterns from the symmetrical pattern of Kikuchi lines which are seen to cross at this point, as shown in Fig. 2.5. (These lines may be thought of as being attached to the crystal.) Since they do not move as the beam direction varies across the central disk, they provide a fixed origin (see Fig. 2.2b). We will also show that Laue circles are concentric on experimental CBED patterns. It is convenient to normalize the length of these vectors to unity when conducting measurements from a print, namely, we call the distance $X$ in Fig. 2.3 in the **x** direction one unit.

With this fixed choice of origin, the variation of incident-beam directions across the (000) disk can now be specified by variations in the length and direction of $\mathbf{K}_t$, as shown in Fig. 2.5. A conjugate point for this beam direction is indicated by the reciprocal lattice vector **g** shown. (The length of this must again be determined from measurements at the edge of the disk.) Physically, as we explore different points in the central disk, the incident (plane-wave) beam direction varies, the directions of the corresponding diffracted beams vary, but the direction of the crystal

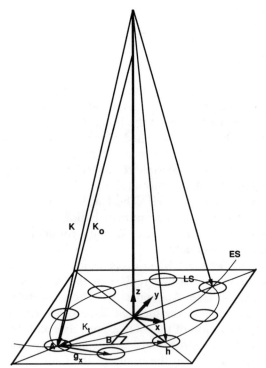

*Figure 2.4.* The Laue circle in the zero-order Laue zone. The center of this circle is taken as the origin of coordinates throughout this book. The two-dimensional Bragg condition requires that distance Ah be twice AB, as shown; **K** is the electron wavevector after correction for the mean inner potential of the crystal.

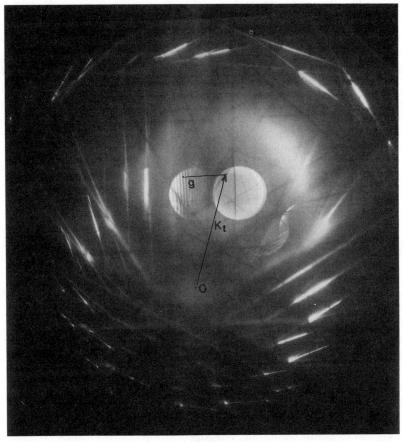

*Figure 2.5.* Experimental CBED pattern recorded near the [111] zone axis in silicon at 120 kV. The center of the zone is marked at O. One incident-beam orientation is marked $\mathbf{K}_t$ together with a conjugate point in the neighboring disk (courtesy of C. Deininger and J. Meyer, unpublished work).

zone axis (indicated by inelastic scattering) remains fixed and so can be used to define an origin.

Equation (2.7) for the excitation errors must now be modified for the two-dimensional case. The Bragg condition may be understood by requiring constructive interference from a stack of half-silvered mirrors, resembling crystal planes. But we can understand much more about Bragg scattering in three dimensions, and in the process define rigorously the excitation erros $S_g$, by applying the laws of conservation of energy and crystal momentum. For "elastic" Bragg scattering, energy conservation requires that $|\mathbf{K}| = |\mathbf{K}'|$, where $\mathbf{K}'$ is the scattered wavevector, if the very small energy $h^2\mathbf{g}^2/2M$ carried off by the recoiling crystal (of mass $M$) is neglected. This equation defines the Ewald sphere shown in Fig. 2.2a. Conservation of crystal momentum requires in addition that, at the Bragg condition, the change in wavevector of the beam electron should be equal to *any* reciprocal lattice vector, so that $\mathbf{K} = \mathbf{K}' - \mathbf{g}$. Note the signs here. The total momentum before scattering is $h\mathbf{K}$, while after scattering it is $h(\mathbf{K}' - \mathbf{g})$, because of Ewald's convention for drawing $\mathbf{g}$ away from the origin. Thus $h\mathbf{g}$ represents physically the negative of the momentum carried off by the crystal. On squaring both sides we obtain $2\mathbf{K}_t \cdot \mathbf{g} + g^2 = 0$ if $\mathbf{g}$ is in the ZOLZ. If a small deviation from the Bragg condition is allowed, so that $\mathbf{K}' = \mathbf{K} + \mathbf{g} + \mathbf{S}$ (with $\mathbf{S}$, by convention, in the $z$ direction along the beam), the same process gives, for small $\mathbf{g}$ and $\mathbf{s}$,

$$2KS_g = K^2 - (\mathbf{K} + \mathbf{g})^2 \qquad (2.11)$$

$$= -2\mathbf{K}_t \cdot \mathbf{g} - g^2 \qquad (2.12)$$

if $\mathbf{g}$ has no component out of the ZOLZ. Equation (2.11) therefore holds for HOLZ and ZOLZ reflections, while Eq. (2.12) applies only to ZOLZ reflections. Equation (2.11) is used to define $S_g$. In the dynamical theory, a similar term (with the second $K$ replaced by a Bloch-wave wavevector) will occur, which we will express in terms of this same excitation error.

The excitation error $S_g$ has a simple interpretation in terms of the uncertainty principle. For components in the $z$ direction, this becomes $\Delta K_z \Delta Z = 1$ (since $|K| = 1/\lambda$). The elastic scattering event is known to occur within a distance $\Delta Z = t$ (the thickness of the sample), so the spread in $z$ components of the distribution of elastically scattered wavevectors must be $\Delta K_z = 1/t = \Delta S_g$, which is just equal to the width of the kinematic rocking curve. In two-beam theory, the thickness is replaced by the extinction distance.

Equation (2.12), with $S_g = 0$, was first given by P. P. Ewald (Kittel, 1976) as an alternative form of Bragg's law. For a ZOLZ reciprocal lattice vector $\mathbf{h}$ to be at the Bragg condition, we require that the component $\mathbf{K}_t$ of either $\mathbf{K}$ or $\mathbf{K}_0$ in the ZOLZ when projected in direction $\mathbf{h}$ be equal to $|\mathbf{h}/2|$. This distance is indicated as AB in Fig. 2.4. We note that $\mathbf{h}$ need not now lie on a diameter of the Ewald sphere, as for the systematics case.

Figure 2.6 shows a real-space diagram, indicating how two sets of nonparallel crystal planes $\mathbf{g}$ and $\mathbf{h}$, each at the Bragg condition, can give rise to three strong spots if $\mathbf{g}$ and $\mathbf{h}$ lie on the Laue circle. The form of the intensity distributions in the CBED disks will be discussed in Chapter 3.

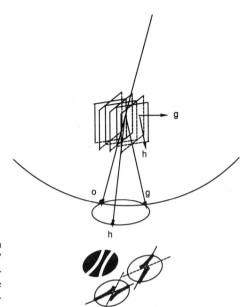

*Figure 2.6.* Three-beam electron diffraction. The electron beam "sees" two distinct sets of planes in the crystal, each at the Bragg condition. The loci of the Bragg condition are indicated below in CBED disks.

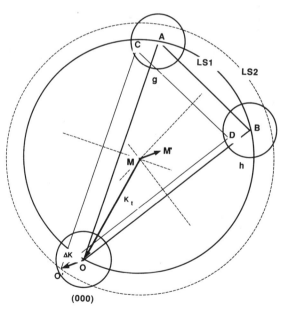

*Figure 2.7.* Two different "point" electron diffraction patterns, OAB and O'CD, are shown within the CBED disks 0, g, and h. Their Laue circles are marked LS1 and LS2, respectively. The zone axis occurs at M, and the Bragg condition is satisfied at A and B. Excitation errors are positive for points C and D, which lie outside the Ewald sphere.

In Fig. 2.2a, the distance $K_t$ is proportional to $\alpha$, and the diameter of the Laue circle is always $2K_t$, which varies with $\alpha$. In two dimensions, a different Laue circle (of different diameter) passes through each point in the central disk. A little thought should convince you that these circles are concentric about the zone center, as shown in Fig. 2.7. If we imagine this figure to be illuminated from above, the choice of a new source point in C2 slightly to the right of O at $-\Delta\mathbf{K}_t$ actually produces a point of intensity to the left at O', where OO' = $\Delta\mathbf{K}_t$, as shown. This is due to the inversion which occurs as rays pass through the sample. The point O' may now be taken as the new origin of a point diffraction pattern, whose conjugate points C and D in the **g** and **h** CBED disks differ by reciprocal lattice vectors. If, however, the change in beam direction $\Delta\mathbf{K}_t$ were referred to the original origin at O, this would move the center of the Laue circle from M to M'. This displacement is exactly cancelled by the origin shift from O to O', due to the inversion mentioned above. Note that the intensities at C and D due to orientation LS2 do not lie on this Laue circle—such a situation only occurs for the exact Bragg condition.

## 2.3. THE GEOMETRY OF HOLZ AND KIKUCHI LINES (KINEMATIC)

In this section we consider the geometry of HOLZ and Kikuchi lines in the kinematic approximation. We will learn how to index and plot these lines, and how to identify incident beam directions from two intersecting Kikuchi or HOLZ lines. The equations for the lines will be derived in the kinematic approximation. We assume for the moment that the crystal structure is known.

A HOLZ line is the locus of the Bragg condition for a HOLZ reflection **g**. The lines therefore occur in pairs, a maximum of intensity in the outer HOLZ ring (excess line) and a corresponding minimum of intensity in the incident-beam disk (deficiency line). In the kinematic approximation, the line positions are given by the simple Bragg law, Eq. (2.4). The geometry of HOLZ lines is thus the geometry of the Bragg condition projected onto the plane of observation. This construction also applies to high-index ZOLZ reflections. Kikuchi lines arise from the elastic scattering of inelastically scattered electrons. A discussion of the origin of Kikuchi lines can be found in Hirsch *et al.* (1977). The inelastic scattering has an angular distribution which is peaked around the electron propagation direction. Thus we may think of the inelastically scattered electrons as originating from an electron source with a large convergence angle. An X-ray pattern with a very large convergence angle is known as

# GEOMETRY OF CBED PATTERNS

a Kossel pattern. If we are only concerned with the geometry of the lines, we may treat Kikuchi patterns as a kind of "thickness averaged" electron Kossel pattern, since the inelastic electrons are generated continuously throughout the sample. It is also customary to neglect the small change in electron energy due to the energy loss which results from inelastic scattering. Hence both Kikuchi lines and HOLZ lines arise from the same elastic Bragg-scattering mechanism. The difference between them lies in the source of electrons in each case—wide cones of inelastically scattered electrons inside the crystal for Kikuchi lines, and a smaller cone, generated by an external source, for HOLZ lines. In the following we will discuss only the generation of HOLZ lines, but the results apply equally well to Kikuchi lines.

Figure 2.3 shows, in addition to coarse modulations of intensity, a set of fine lines crossing the central (000) disk. These define orientations where the Ewald sphere crosses reciprocal lattice points in higher-order Laue zones, as shown in Fig. 2.8, which has been drawn for silicon. A HOLZ layer $n$ may be defined by the equation $n = \mathbf{B} \cdot \mathbf{h}$, where $\mathbf{h}$ is any point in the HOLZ reciprocal lattice plane, and $\mathbf{B}$ is the zone-axis direction, written without common factors. If the beam is aligned with a zone axis, then $\mathbf{B}$ is antiparallel to the beam direction. Thus we may confirm that $n = 1$ for the first-order laue zone (FOLZ) shown in Fig. 2.8,

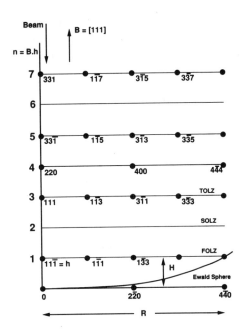

*Figure 2.8.* Side view of reciprocal lattice for silicon with Ewald sphere. The quantities $H$ and $R$ in Eq. (2.13) are indicated for the axial-beam direction $\mathbf{B}$, where $K_t = 0$. The order $n$ of the HOLZ layers are indicated, and absent layers are seen for $n = 2$ and $n = 6$.

and $n = 3$ for the third-order Laue zone (TOLZ). In addition, for certain structures, all points in the $n$th-order HOLZ layer may be forbidden due to the presence of screw or glide symmetry elements (see Chapter 7). Thus, because of the diamond glide element, the $n = 2$ and $n = 6$ layers are absent in Fig. 2.8. Here, all structure factors are zero since, on these planes, $h + k + l = 2p$, where $p$ is odd, which is a condition for extinction in the diamond lattice. The indexing of HOLZ lines is further discussed below, while indexed patterns are given in Appendix 3.

The higher-order reflections themselves may be visible as a bright outer ring of reflections if a small camera length is used, as shown in Fig. 2.10a. These reflections, and their complements in the zero-order disk, are extremely useful in CBED, as we shall see later. For example, they may be used to confirm an orientation determination in non-centrosymmetric crystals, to check for stacking disorder in the beam direction, to observe tetragonal distortions in otherwise cubic crystals, to measure strains in crystals, and for the measurement of structure factors.

The radius $R$ of this HOLZ ring is given from the Ewald sphere construction (see Fig. 2.8) as

$$R \approx (K_t^2 + 2H/\lambda)^{1/2} \qquad (2.13)$$

in reciprocal length units, where $H$ is the height of the HOLZ layer considered and $K_t$ is the component of the incident-beam wavevector in the ZOLZ plane. (In zone-axis orientations, $K_t = 0$.) The approximation in (2.13) assumes that the wavevector $1/\lambda$ is much larger than both $K_t$ and $H$. The height $H$ may thus be obtained from a measurement of $R$ if $\lambda$ is known. In the absence of extinctions due to screw and glide symmetry elements, $H$ may be taken as a measure of the spacing between reciprocal lattice planes in the beam direction (see Section 7.2). For the $[h, k, l]$ zone axis of a cubic crystal,

$$H_{h,k,l} = \frac{n}{a\sqrt{h^2 + k^2 + l^2}}$$

For fcc crystals, as a result of the centering of the lattice, $n = 1$ if $(h + k + l)$ is odd, and $n = 2$ if $(h + k + l)$ is even. For a bcc crystal, $n = 2$ if $h, k, l$ are all odd integers, otherwise $n = 1$. Appendix 3 contains expressions for the other crystal classes, however the centering of these lattices must also be considered.

Figure 2.9 shows the geometric construction of a HOLZ line. All incident-beam directions **K** satisfying the Bragg law for planes **g** form a cone. The length of the side of the cone is $1/\lambda$, and Bragg scattered

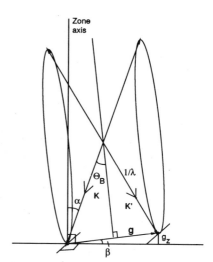

Figure 2.9. The two cones shown define the locus of all incident (**K**) and diffracted (**K'**) rays which can satisfy the Bragg condition for a given HOLZ reflection. A portion of these cones intersects the film plane to give an approximately straight HOLZ line.

beams **K'** form a second cone. A HOLZ-line pair is given by the projection of these cones onto the observation plane. In an experiment, the observation plane is approximately parallel to the ZOLZ for the nearest zone axis. We therefore assume this ZOLZ as the observation plane throughout this book. In high-energy electron diffraction, the length of the side of the cone is very large compared to the length of a typical **g** vector, so that only a very small portion of the cone is observed. Thus the projection of the cone may be approximated by a straight line, and this straight line is perpendicular to the **g** vector, as shown in Fig. 2.9. The position of the HOLZ line is defined by the angle $\alpha$, the angle between the incident beam and the zone axis, and is given by

$$\alpha = 90° - (90° - \Theta_B) - \beta = \Theta_B - \beta = \sin^{-1}(g\lambda/2) - \sin^{-1}(g_z/g) \quad (2.14)$$

If this line is translated by the vector **g**, we obtain the corresponding line in the outer HOLZ ring, or "dark-field disk."

We now return to Eq. (2.11) in order to obtain a description of a HOLZ line (in the central disk) in our standard coordinate system (see Fig. 2.4). On expanding this equation in terms of components, we find that the geometric (kinematic) trajectory of the HOLZ line **g** is given by

$$K_y = -\frac{g_x}{g_y} K_x + \frac{g_z}{g_y} K_z - \frac{g^2}{2g_y} \quad (2.15)$$

with respect to an origin at the zone center. Here, the $z$ component of $K$ is given to a good approximation by

$$K_z \approx \sqrt{K^2 - K_{xc}^2 - K_{yc}^2} \qquad (2.16)$$

where $K_{xc}$ and $K_{yc}$ are the $x$ and $y$ components of a vector drawn from the center of the zone to the center of the zero-order CBED disk. This approximation is equivalent to approximating a section of a cone by a straight line. It will be needed in Section 3.8 for our analysis of strain measurements [where a dynamical correction to Eq. (2.15) is given] and is used in the HOLZ simulation program listed in Appendix 5.

The indexing of HOLZ lines is performed in two stages. The first step is to index the diffraction pattern and identify possible zone axes. The second step is to simulate and index the HOLZ lines. The procedure for indexing a diffraction pattern is described in several texts (Hirsch et al., 1977; Loretto, 1984; Williams, 1987). Appendix 4 of this book shows indexed reciprocal lattices for many of the commonly encountered Bravais lattices and orientations. Appendix 6 gives the structure and cell constants for many commonly encountered crystals. On the figures in Appendix 4, sufficient information is given to allow indexing of the entire three-dimensional lattice, and hence the HOLZ lines. Fournier et al. (1989) also describe a systematic technique for indexing diffraction patterns. A worked example of indexing is given in Steeds and Evans (1980). Figure 2.10a and 2.10b show experimental and computer-simulated patterns for the [−211] zone axis of silicon at 100 kV.

Assuming the crystal structure to be known, the simplest indexing procedure is first to determine the indices of two reciprocal lattice basis vectors in the ZOLZ. The vector cross product of these then gives the indices of the zone axis. If one of these vectors is known, then only the length ratio and the angle between the two vectors will be needed to determine the indices of the second vector. Otherwise, the camera length must be known. In practice, the indexing of a high-index zone axis can often be simplified by finding the nearest identifiable low-index zone axis, and measuring the angle between these two zone axes. This can be done by tilting along a low-index Kikuchi band, and reading off the angular change on the goniometer. Once the zone axis is determined using the vector cross product of the two basis vectors, the geometric diffraction pattern can be simulated, including HOLZ reflections. Appendix 5 lists a suitable program for plotting these patterns on a postscript laser printer. A photocopy of the FOLZ lattice produced on transparent plastic sheet may then be laid over a print of the experimental ZOLZ CBED pattern. This will confirm the indexing, and reveal the misregistry between the

# GEOMETRY OF CBED PATTERNS

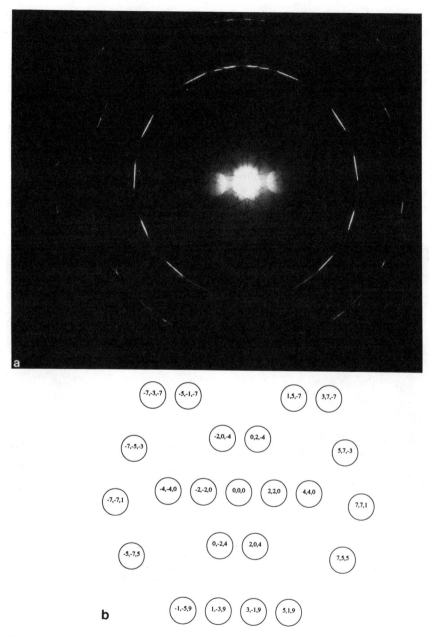

*Figure 2.10.* (a) Experimental Si (−221) CBED pattern recorded at 100 kV, showing three HOLZ rings. (b) Computer-simulated pattern corresponding to Fig. 2.10a. The ZOLZ reflections have been indexed.

FOLZ and ZOLZ reciprocal lattice layers. (The geometry of reciprocal lattices is discussed more fully in Section 7.2.) For example, Fig. 2.10a is found to be a $[-2, 2, 1]$ zone, for which the simulated and indexed pattern is shown in Fig. 2.10b. The initial determination for Fig. 2.10a was completed by finding two neighboring zone axes, since all the (204)-type nonsystematic reflections in the zone were very weak.

In the central disk the "deficit" HOLZ line may now be indexed by identifying its pair **g** at the Bragg condition in the outer ring. This is easily done, since both lines run approximately normal to **g**. Figure 2.11a shows an enlarged view of the central disk of the pattern shown in Fig. 20a. Figure 2.11b shows a simulated HOLZ-line pattern for this case obtained using the program given in Appendix 5 at 100 kV. The simulation is based on Eqs. (2.15) and (2.16) and is seen to agree well with the experimental pattern. (Minor discrepancies will be discussed in

*Figure 2.11.* (a) Zero-order CBED disk enlarged from the experimental $[-221]$ HOLZ pattern shown in Fig. 2.10a. Many fine HOLZ lines can be seen crossing the disk. We find the incident-beam direction at the point C, using point R as a reference. (b) Computer-simulated HOLZ lines corresponding to Fig. 2.11a. Kinematic calculation. The lines have been indexed.

## GEOMETRY OF CBED PATTERNS

later chapters. For this sparse zone, dynamical effects are small.) We note that the entire two-dimensional pattern of lines should agree, and that discrepancies are most readily observed at intersections.

The position of HOLZ lines depends on the microscope accelerating voltage and on the lattice constants through the Bragg law. The distance between the line intersections therefore provides a sensitive parameter for strain measurement, as discussed in Section 5.4. Unlike X-ray or neutron diffraction, it is difficult to measure the Bragg angle very accurately by electron diffraction using spot patterns, because of the electron-optical distortions in the projector lenses. The relative movement of three or more HOLZ lines in the zero disk avoids this difficulty, and is therefore the best method for the measurement of high-voltage, lattice constants and local strains.

HOLZ lines are narrow for the following reasons. As indicated on Fig. 2.2a, Bragg scattering occurs strongly only when the Ewald sphere

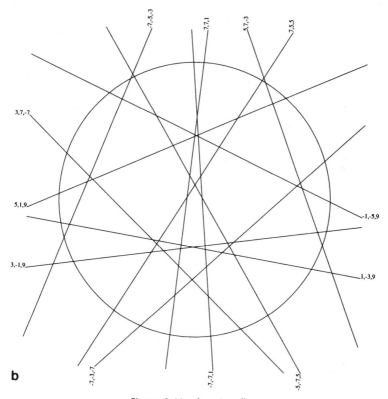

*Figure 2.11.* (*continued*)

passes near the central maximum of the rocking curve RC. In two-beam theory, the *envelope* of this has width approximately $1/\xi_g$, and $\xi_g$ is large for high-order reflections. For HOLZ reflections, the sphere cuts this distribution at a steeper angle than in the ZOLZ, and intensity is therefore observed over a smaller angular range.

The indexed HOLZ and/or Kikuchi-line pattern may now be used for the determination of the incident-beam direction. This is defined by Eq. (2.15) along a line. The intersection of two HOLZ or Kikuchi lines may be used as a reference point, and the beam direction at this point can easily be found by solving Eq. (2.15). For example, we may wish to determine the incident-beam direction at the center of the zero-order disk in Fig. 2.11a at the point marked C. The intersection of lines (7, 5, 5) and (−5, −7, 5), marked R, will be used as a reference point (see also Fig. 2.11b). The equations for these two lines are

$$K_y = 1.6364 K_x + 0.04025$$

and

$$K_y = -1.6364 K_x + 0.04025$$

respectively. The two basis vectors which have been used are $\mathbf{x} = g(2, 2, 0)$ and $\mathbf{y} = (-0.6667, 0.6667, -2.6667)$. The incident-beam direction at the intersection of these two lines is

$$\mathbf{K}_t^R = 0.0\mathbf{x} + 0.04025\mathbf{y}$$

The vector from the center of the zero disk to the reference point R is measured to be

$$\mathbf{D} = -0.0247\mathbf{x} + 0.00617\mathbf{y}$$

Thus the incident-beam direction at the center of the zero disk of Fig. 2.11a is

$$\mathbf{K}_t^C = \mathbf{K}_t^R + \mathbf{D} = -0.0247\mathbf{x} + 0.04642\mathbf{y} = (-0.08023, -0.01857, -0.1238)$$

As a result of multiple electron scattering, we will find in Section 3.5 that the simple Bragg law requires a correction, and that these HOLZ lines are actually shifted slightly from the positions given above by an amount which depends on the ZOLZ structure factors.

# 3

# Theory

In this chapter we outline the Bloch-wave dynamical theory of high-energy transmission electron diffraction (HEED), after defining some useful quantities and the relationships between them. We then specialize to the two-beam case, which provides the quickest and most useful guide to CBED intensity distributions. However, the study of non-centrosymmetric crystals requires the three-beam theory, so this is also treated. Both nondegenerate and degenerate perturbation theory are introduced and serve as the foundation for our treatment of three-beam theory, dynamic HOLZ effects, and absorption in both centric and acentric crystals. The purpose of this chapter is to collect together useful results to be used in the later chapters on applications, and to derive these otherwise scattered results in a consistent formalism.

Historically, there have been two main (related) approaches to the subject. The first is based on the study of few-beam solutions (and cases reducible to them) in arbitrary orientations, following Bethe's work. This powerful approach was developed first in Europe and Japan for the study of Kikuchi lines and, more recently, for HOLZ lines and phase measurement. It has been most successful in small-unit cell crystals. The second approach, developed by the Bristol group and more useful for large-unit cell materials in orientations of high symmetry, treats layers of atoms, with many ZOLZ reflections excited simultaneously, as in the multislice formulation. HOLZ lines are then treated as a perturbation. The multislice approach, developed in Australia, has also proven useful for this case, but is less well suited to the incorporation of HOLZ reflections (see Appendix 5 for a comparison of computational algorithms). Thus, one can either start out with the idea of a strongly excited ZOLZ perturbed by HOLZ lines (which works best in orientations of high symmetry for a dense ZOLZ), or start with a few strong ZOLZ reflections, again perturbed by the HOLZ. (This works best in off-axis orientations, and may be solved in greater detail.)

Students new to the field looking for pedagogically sound reviews are particularly referred to Hirsch *et al.* (1977), Humphreys (1979), Metherall (1975), and Reimer (1984)

## 3.1. ELECTRON AND X-RAY STRUCTURE FACTORS. TEMPERATURE FACTORS

The purpose of this section is to relate the quantities measured in electron crystallography to the corresponding quantities in X-ray crystallography. The results will be needed in Chapter 4, and for the remainder of this chapter. The sign convention used in the following is the quantum mechanical standard, which takes an incident plane wave to have the form $\exp(2\pi i k \cdot \mathbf{r})$. For a discussion of the various sign conventions used in electron diffraction, see Saxton *et al.* (1983), Spence (1988a), and the list of symbols at the beginning of this book. We have used the values for the fundamental physical constants, which are listed in Appendix 2. In dynamical theory, changes in these quantities will redistribute intensity among the multiply-scattered beams, so that the values of these constants are important if, for example, small changes in crystal charge density are to be measured.

The diffracted beams in an electron diffraction experiment are solutions of the Schrödinger equation (see Section 3.2). Their intensities are therefore related to the total crystal electrostatic potential $V(\mathbf{r})$ as seen by the beam electron (not by the crystal electrons). This quantity $V(\mathbf{r})$ includes the nuclear potential contribution. Maxwell's equations, however, give the corresponding intensities for the X-ray case, and these therefore depend on the electronic charge density $-|e|\rho(\mathbf{r})$, excluding the unpolarized nuclear contribution $|e|\rho_n(\mathbf{r})$. Both $\rho(\mathbf{r})$ and $V(\mathbf{r})$ may be derived from the crystal electron wavefunction. Exchange and virtual inelastic scattering effects between the beam and crystal electrons are negligible (Rez, 1978), as discussed in Section 4.5.

Since $\rho_{\text{Tot}}(\mathbf{r}) = \rho_n(\mathbf{r}) - \rho(\mathbf{r})$ and $V(\mathbf{r})$ are then related by Poisson's equation, a relationship may be obtained between the Fourier coefficients $V_{\mathbf{g}}$ of $V(\mathbf{r})$ and the Fourier coefficients $F_{\mathbf{g}}^X/\Omega$ of $\rho(\mathbf{r})$. These quantities $F_{\mathbf{g}}^X$ are the dimensionless X-ray structure factors, measured in number of electrons for a given cell.

First, we must consider individual atoms rather than the whole crystal. Let the total ground-state charge density (including the nuclear contribution $\rho_n$) for one atom be $\rho_{\text{Tot}}(r) = \rho_n(r) - \rho(r)$ in units of "number of electrons per unit volume." The X-ray atomic scattering factor $f_{(s)}^X$ is the Fourier transform of atomic charge density at $\mathbf{u} = 2\mathbf{s}$ in reciprocal space, where **s** is the scattering vector, defined as half the difference vector between scattered wavevector $\mathbf{K}_0'$ and incident wavevector $\mathbf{K}_0$, i.e., $2\mathbf{s} = \mathbf{K}_0' - \mathbf{K}_0$. For the Bragg condition $\mathbf{K}_0' = \mathbf{K}_0 + \mathbf{g}$, we have $|s| = g/2 = 1/(2d_g)$, and $s = \sin\theta_B/\lambda$ according to the Bragg law. The dimensionless scattering factor $f_{(s)}^X$ has the units of number of electrons and is given by

$$f^X(s) = \int \rho(r)\exp[-2\pi i(\mathbf{K}'-\mathbf{K}_0)\cdot\mathbf{r}] = \int \rho(r)\exp(-4\pi i\mathbf{s}\cdot\mathbf{r})$$

The atomic potential whose Fourier transform $f_i(s)$ is measurable by electron diffraction is $V(\mathbf{r})$, where (in SI units)

$$\nabla^2 V(\mathbf{r}) = -\frac{|e|}{\varepsilon_0}(\rho_n - \rho) \tag{3.1}$$

Hence, by analogy with the definition of the X-ray atomic scattering factor, we have in SI units

$$\begin{aligned}
f^e(s) &= \int V(\mathbf{r})\exp(-4\pi i\mathbf{s}\cdot\mathbf{r})\,d\tau \\
&= -\frac{1}{16\pi^2 s^2}\int V(\mathbf{r})\nabla^2\exp(-4\pi i\mathbf{s}\cdot\mathbf{r})\,d\tau \\
&= -\frac{1}{16\pi^2 s^2}\int \exp(-4\pi i\mathbf{s}\cdot\mathbf{r})\nabla^2 V(\mathbf{r})\,d\tau \\
&= \frac{|e|}{16\pi^2\varepsilon_0 s^2}\int[Z\delta(\mathbf{r})-\rho(\mathbf{r})]\exp(-4\pi i\mathbf{s}\cdot\mathbf{r})\,d\tau \\
&= \frac{|e|}{16\pi^2\varepsilon_0 s^2}[Z - f^X(s)] \tag{3.2a}
\end{aligned}$$

where we have used integration by parts and taken $V(\mathbf{r}) = 0$ for large $r$. The definition of electron atomic scattering factor here adopts the convention in the new version of the *International Tables for Crystallography*. The traditional definition is based on the Born approximation and takes

$$f^B = \frac{2\pi m_e |e|}{h^2}f^e = \frac{m_e e^2}{8\pi\varepsilon_0 h^2}\frac{Z-f^X(s)}{s^2} = 0.02393367\frac{Z-f^X(s)}{s^2} \tag{3.2b}$$

with $s$ in units of Å$^{-1}$, giving $f^B$ in units of Å.

For a crystal with several atoms in the unit cell, the electron atomic scattering factor $f(s)$ can be summed over the contents of the unit cell (with a suitable phase factor) to give values of the Fourier coefficients of crystal potential

$$V_\mathbf{g} = \frac{1}{\Omega} \sum_i f^e(s) \exp(-2\pi i \mathbf{g} \cdot \mathbf{r}_i) \quad (3.3)$$

where $s = \sin\theta_B/\lambda = |\mathbf{g}|/2$. Using this result in Eq (3.2) gives the familiar Mott–Bethe relationship (Mott, 1930), which may be written in SI units as

$$V_\mathbf{g} = \frac{|e|}{16\pi^3\varepsilon_0\Omega} \sum_i \left(\frac{[Z_i - f_i^X(s)]}{s^2}\right) \exp(-B_i s^2) \exp(-2\pi i \mathbf{g} \cdot \mathbf{r}_i)$$

$$= \frac{h^2}{8\pi\varepsilon_0 m_e |e| \Omega} F_\mathbf{g}^B \quad (3.4)$$

for atoms at $\mathbf{r}_i$ of atomic number $Z_i$. Here the sum is over the unit cell of volume $\Omega$ (m³), $\theta_B$ is the Bragg angle, and $F_\mathbf{g}^B$ is the electron structure factor according to the old (Born approximation) system (*International Tables* before 1990); $V_\mathbf{g}$ is given in volts while $m_e$ is the rest mass of the electron.

We have also introduced into Eq. (3.4) the Debye–Waller factor $B_i = 8\pi^2 \langle u^2 \rangle$ for species $i$, with $\langle u^2 \rangle$ the mean-square vibrational amplitude of the atom. In the Debye model for lattice vibrations, a linear dispersion relationship is assumed and the excitation spectrum parameterized using a Debye temperature $\Theta_D$. Then the mean-square vibrational amplitude is given by

$$\bar{u}^2 = \frac{3h^2}{4\pi^2 M k \Theta_D} \left(\frac{1}{4} + \frac{T^2}{\Theta_D^2} \int_0^{\Theta_D/T} \frac{x\,dx}{\exp(x) - 1}\right)$$

where $M$ is the atomic mass. Values of $B_i$ and the integral above are tabulated in the *International Tables for Crystallography*, and a recent tabulation for elemental crystals and comparison with measurements can be found in Sears and Shelley (1991). The first temperature-independent term in the parentheses results from zero-point motion, and this produces appreciable effects in electron diffraction which cannot be neglected or removed by cooling (Humphreys and Hirsch, 1968). The role of the Debye–Waller factor in the "absorption" of electrons that results from inelastic scattering is discussed in Section 3.9.

# THEORY

If $B$, $s$, and $\Omega$ are instead given in angstrom units, then $V_g$ is given in volts as

$$V_g = \frac{1.145896}{\Omega} \sum_i \frac{[Z_i - f_i^X(s)]}{s^2} \exp(-B_i s^2) \exp(-2\pi i \mathbf{g} \cdot \mathbf{r}_i)$$

$$= \frac{47.878009}{\Omega} F_g^B \qquad (3.5)$$

Equation (3.5) allows electron "structure factors" $V_g$ and $F_g$ to be evaluated from tabulations of X-ray atomic scattering factors $f_i^X(s)$ if $B_i$ is known. Two other quantities commonly used in the literature are $U_g$ and $\xi_g$, given by

$$U_g = \frac{\gamma F_g^B}{\pi \Omega} = \frac{2m|e|V_g}{h^2} \qquad (3.6a)$$

and

$$\xi_g = 1/(\lambda |U_g|) = \pi/(|V_g|\sigma) = \pi\Omega/(\gamma \lambda |F_g^B|) \qquad (3.6b)$$

where $\gamma$ is the relativistic constant. Many of these useful relationships are collected together in Appendix 1. The definition of extinction distance $\xi_g$ is based on the two-beam intensity expression at the Bragg condition; it depends on the amplitude of the structure factors.

Since these quantities depend on the details of the scattering experiment (accelerating voltage) they are not true "structure factors," unlike $V_g$ and $F_g$, which are properties of the crystal alone, as defined above. For $V_g$ in volts and $U_g$ in angstroms$^{-2}$, we then have

$$U_g = 0.006648352(1 + 1.956934 \times 10^{-6} E_0) V_g \qquad (3.7)$$

If the X-ray structure factor is defined by

$$F_g^X = \sum_i f_i^X(s) \exp(-B_i s^2) \exp(-2\pi i \mathbf{g} \cdot \mathbf{r}) \qquad (3.8)$$

with the electronic charge density (in electrons per cell) as

$$\rho(\mathbf{r}) = \frac{1}{\Omega} \sum_g F_g^X \exp(2\pi i \mathbf{g} \cdot \mathbf{r}) \qquad (3.9)$$

then Eqs. (3.8), (3.5), and (3.7) give the following expression for the retrieval of an X-ray structure factor $F_g^X$ from electron diffraction data:

$$F_g^X = \sum_i Z_i \exp(-B_i s^2) \exp(-2\pi i \mathbf{g} \cdot \mathbf{r}) - \left(\frac{8\pi^2 \varepsilon_0 h^2 \Omega s^2}{\gamma m_e e^2}\right) U_g \quad (3.10)$$

$$= \sum_i Z_i \exp(-B_i s^2) \exp(-2\pi i \mathbf{g} \cdot \mathbf{r}) - \left(\frac{C \Omega s^2}{\gamma}\right) U_g \quad (3.11)$$

Here the numerical constant $C = 131.2625$ if $s$, $\Omega$, and $U_g$ are given in angstrom units.

The role of the temperature factor in conversions between X-ray and electron structure factors is important. The atomic vibrational amplitude $u$ is appreciable even at 0 K, where in many materials it falls to only about half its room-temperature value. Thus the observable crystal potential is a temperature-dependent quantity—the static potential computed from band-structure calculations is not an experimental observable. From Eqs. (3.5) and (3.10) we may draw the following conclusions:

1. A knowledge of the Debye–Waller factor $B_i$ is essential in order to convert a structure factor $U_g$ measured by electron diffraction at temperature $T$ into the corresponding X-ray structure factor $F_g^X$, at temperature $T$. A knowledge of $B_i$ is similarly required to convert measured X-ray scattering factors into electron scattering factors for the same temperature.
2. Debye–Waller factors must be known for conversions involving calculated scattering factors.
3. In electron diffraction computer programs used for the refinement of a few low-order reflections, the Debye–Waller factor should be included in all reflections *except* those $U_g$ values being refined. Then a measurement of these $U_g$ values (including the effects of temperature) is obtained for one particular temperature.
4. Debye–Waller factors must be known in order to compare the results of measurements taken at different temperatures. Since the neutron diffraction studies or lattice dynamical calculations required to determine $\Delta_i$ have been completed for relatively few crystals, this has created considerable difficulty in the past for comparisons of structure-factor measurements by different groups, who may work at different temperatures. Before starting a project on structure-factor measurement for a particular

material, a survey of the existing literature on Debye–Waller factors is strongly recommended. Many are listed in the *International Tables*. See also Sears and Shelley (1991).

5. From Eq. (3.2) we see that, at small scattering angles, small changes in $f^X$ result in large changes in $f_i$. Thus if $f_i$ is known to a particular percent error, $f^X$ may be deduced to a greater accuracy.

The asymptotic behavior of the scattering factors for large and small values of $s$ must be considered (Peng and Cowley, 1988). Expansions of both $f^X$ and $f_i$ in terms of Gaussian functions have appeared (Doyle and Turner, 1968; also *International Tables for Crystallography*, 1989) of the form

$$f_j^X(s) = \sum_{i=1}^{4} a_i \exp(-b_i s^2) + c \qquad (3.12)$$

To obtain the desired asymptotic behavior in which the electron scattering factor converges to the Rutherford scattering form at high angles, while ensuring that $f_i^X(0) = Z$, it is convenient to use Eq. (3.3) [rather than a Gaussian expansion of $f_i(s)$] written in the form

$$V_g = \frac{1.145896}{\Omega} \sum_i \frac{\left[ f_i^X(0) - \sum_{j=1}^{4} a_j \exp(-b_j s^2) - c \right]}{s^2} \exp(-B_i s^2) \exp(-2\pi i \mathbf{g} \cdot \mathbf{r}_i)$$

$$= \frac{1.145896}{\Omega} \sum_i \frac{\left\{ \sum_{j=1}^{4} a_j [1 - \exp(-b_j s^2)] \right\}}{s^2} \exp(-B_i s^2) \exp(-2\pi i \mathbf{g} \cdot \mathbf{r}_i)$$

$$= \frac{1.145896}{\Omega} \sum_{j=1}^{4} a_j b_j \qquad \text{for } s = 0 \qquad (3.13)$$

Section 4.6 discusses the significance of $V_0$ in terms of the diamagnetic susceptibility (Miyake, 1940).

## 3.2. MANY-BEAM THEORY

Reviews of the Bloch-wave method for solving the problem of high-energy transmission electron diffraction in the ZOLZ approximation can be found in Humphreys (1979) and Metherall (1975). We now extend these treatments by the renormalized eigenvector method of Lewis *et al.* (1978) to include HOLZ effects, acentric crystals, absorption, and inclined boundary conditions. Our aim is to provide the theoretical basis for computer algorithms, from which specialized two- and three-beam cases may be extracted in subsequent sections. We consider a collimated incident electron beam of the form $\exp(2\pi i \mathbf{K}_0 \cdot \mathbf{r})$. Then the "relativistically corrected" Schrödinger equation describing high-energy electron diffraction is

$$\frac{-h^2}{8\pi^2 m} \nabla^2 \Psi(\mathbf{r}) - |e| V(\mathbf{r})\Psi(\mathbf{r}) = \frac{h^2 K_0^2}{2m} \Psi(\mathbf{r}) \qquad (3.14)$$

Justification for the use of Eq. (3.14) and implied lack of spin effects can be found in the reviews mentioned above and in Section 4.5. We define three-dimensional reciprocal lattice vectors $\mathbf{g}$, and expand the wavefunction inside the crystal as a sum of Bloch waves

$$\Psi(\mathbf{r}) = \sum_i c_i \exp(2\pi i \mathbf{k}^{(i)} \cdot \mathbf{r}) \sum_\mathbf{g} C_\mathbf{g}^i \exp(2\pi i \mathbf{g} \cdot \mathbf{r}) \qquad (3.15)$$

and expand the scaled crystal potential [see Eq. (2.2)] as a Fourier series

$$U(\mathbf{r}) = \sum_\mathbf{g} U_\mathbf{g}^c \exp(2\pi i \mathbf{g} \cdot \mathbf{r})$$

The result of inserting these two equations into Eq. (3.14) and equating coefficients yields the standard dispersion equation of high-energy electron diffraction (HEED)

$$[\mathbf{K}^2 - (\mathbf{k}^{(j)} + \mathbf{g})^2] C_\mathbf{g}^{(j)} + \sum_\mathbf{h} U_{\mathbf{g}-\mathbf{h}}^c C_\mathbf{h}^{(j)} = 0 \qquad (3.16)$$

The complex electron "structure factors" are, as in Section 2.1,

$$U_\mathbf{g}^c = 2m |e| V_\mathbf{g}/h^2$$

with $V_\mathbf{g}$ a Fourier coefficient of total crystal potential in volts. The "two-beam" extinction distance is $\xi_\mathbf{g} = |K|/|U_\mathbf{g}^c| = 1/(\lambda |U_\mathbf{g}^c|)$. The general approach is to force Eq. (3.16) into the form of an eigenvalue equation, and then solve to yield eigenvalues and eigenvectors. The constants $c_i$ must be obtained from boundary conditions, and so depend on the shape of the crystal. The tangential components of the incident and Bloch wavevectors inside the crystal must be matched at the crystal entrance

# THEORY

surface. Thus, for a crystal slab, for all $j$,

$$\mathbf{k}_t^{(j)} = \mathbf{K}_t = \mathbf{K}_{0,t}$$

An additional imaginary potential may be added to describe the depletion of the elastic wavefield by inelastic scattering (absorption). The total potential is then known as an optical potential, given by

$$U(\mathbf{r}) = U^c(\mathbf{r}) + iU'(\mathbf{r})$$

An additional correction to the real potential representing virtual inelastic scattering has been neglected, since this is a very small effect (Rez, 1978). The Fourier components of the total potential are

$$U_\mathbf{g} = U_\mathbf{g}^c + iU_\mathbf{g}' \qquad (3.17)$$

In noncentrosymmetric (acentric) crystals, both $U_\mathbf{g}^c$ and $U_\mathbf{g}'$ are the complex Fourier coefficients of real potentials with the period of the lattice. The ratio $U_\mathbf{g}'/U_\mathbf{g}^c$ is known as an absorption coefficient, which is real for centric crystals and complex for acentric crystals. In the presence of absorption, the structure matrix is then no longer Hermitian.

We now let

$$\mathbf{k}^{(j)} = \mathbf{K} + \gamma^{(j)}\mathbf{n} \qquad (3.18a)$$

where $\mathbf{n}$ is a unit vector out of, and normal to, the slab (against the beam), and both $\mathbf{k}^{(j)}$ and $\gamma^{(j)}$ may be complex, to allow for absorption. If we introduce the expansion

$$\begin{aligned} K^2 - (\mathbf{k} + \mathbf{g})^2 &= K^2 - (\mathbf{K} + \mathbf{g})^2 - 2(\mathbf{K} + \mathbf{g}) \cdot \mathbf{n}\gamma - \gamma^2 \\ &= 2KS_g - 2(\mathbf{K} + \mathbf{g}) \cdot \mathbf{n}\gamma - \gamma^2 \end{aligned} \qquad (3.18b)$$

then

$$[2KS_g - 2(\mathbf{K} + \mathbf{g}) \cdot \mathbf{n}\gamma^{(j)} - \gamma^{(j)^2}]C_\mathbf{g} + \sum_\mathbf{h} U_{\mathbf{g}-\mathbf{h}} C_\mathbf{h} = 0 \qquad (3.19)$$

is obtained with excitation errors $S_g$ defined in Eq. (2.11). Thus

$$2KS_g = K^2 - (\mathbf{K} + \mathbf{g})^2 \qquad (2.11)$$

in general, and

$$2KS_g = -2\mathbf{K}_t \cdot \mathbf{g} - \mathbf{g}^2 \qquad (2.12)$$

for ZOLZ reflections.

As discussed in Chapter 2, we may think of the CBED pattern as a function of $\mathbf{K}_t$, a vector which originates in the center of the zone axis and extends to a point of interest in the central disk.

Four diffraction conditions in particular are commonly encountered in CBED: the axial (systematics) orientation, the zone-axis orientation [for which $\mathbf{K}_t = 0$ at the center of the (000) disk], the two-beam Bragg condition ($\mathbf{K}_t = -\mathbf{g}/2$), and the three-beam condition (described in Section 3.4) which is important for phase determination.

When using a program, the range (area) of values of $\mathbf{K}_t$ to be covered in the ZOLZ must be defined in fractional reciprocal space coordinates—the program then prints the intensities of a requested subset of the beams in the calculation over the corresponding specified angular range. The full set of beams at all conjugate points (one in each CBED disk) is computed for every value of $\mathbf{K}_t$.

Equation (3.19) includes both the forward scattered waves of interest for CBED, and the back scattered waves important for RHEED. We define $g_n = \mathbf{g} \cdot \mathbf{n}$ and $K_n = \mathbf{K} \cdot \mathbf{n}$. In transmission diffraction, $K_n$ is large, $\gamma$ is small ($\gamma \ll K_n$) for the forward-scattered waves and large ($\gamma \approx -2K_n$) for the back-scattered waves. The excitation coefficients of the back-scattered waves are very small and usually neglected in high-energy transmission electron diffraction. Neglect of the back-scattered waves ($\gamma \ll K_n$) and neglect of $\gamma^2$ terms leads to

$$2KS_g C_\mathbf{g}^{(j)} + \sum_\mathbf{h} U_{\mathbf{g}-\mathbf{h}} C_\mathbf{h}^{(j)} = 2K_n\left(1 + \frac{g_n}{K_n}\right)\gamma^{(j)} C_\mathbf{g} \qquad (3.20)$$

This with Eq. (2.11) includes all HOLZ effects, boundary inclination effects, and absorption terms, and may be applied to acentric crystals. The most important approximation has been the neglect of back scattering. To transform this equation into a linear eigenvalue equation, we define new eigenvector elements

$$B_\mathbf{g}^{(j)} = \left(1 + \frac{g_n}{K_n}\right)^{1/2} C_\mathbf{g}^{(j)} \qquad (3.21)$$

so that, using Eq. (3.20),

$$\frac{2KS_g B_\mathbf{g}^{(j)}}{(1 + g_n/K_n)} + \sum_\mathbf{h} \frac{B_\mathbf{h}^{(j)} U_{\mathbf{g}-\mathbf{h}}}{\sqrt{1 + g_n/K_n}\sqrt{1 + h_n/K_n}} = 2K_n \gamma^{(j)} B_\mathbf{g}^{(j)} \qquad (3.22)$$

This is the fundamental eigenvalue equation to be solved. If the surface normal is approximately antiparallel to the beam, so that

# THEORY

$K_n \gg g_n$, then $g_n/K_n$ is negligible and we have

$$2KS_g C_{\mathbf{g}}^{(j)} + \sum_{\mathbf{h}} U_{\mathbf{g-h}} C_{\mathbf{h}}^{(j)} = 2K_n \gamma^{(j)} C_{\mathbf{g}}$$

The latter equation may be written in matrix form

$$\mathbf{AC}^i = 2K_n \gamma^i \mathbf{C}^i \tag{3.23}$$

where the off-diagonal entries of the "structure matrix" $\mathbf{A}$ are $U_{\mathbf{gh}}$, while the diagonal entries are the excitation error terms $2KS_g$. Here $\mathbf{C}^j$ is a column vector. Equation (3.23) includes HOLZ effects in an approximate way, and the crystal tilt through the term in $K_n$. The distances $\gamma^j$ have a geometric interpretation as the displacement, in the direction of the surface normal, of the true dynamical dispersion surface from spheres drawn about every reciprocal lattice point. These spheres have radii $K$. We will use Eq. (3.23) mostly throughout this book because of its simplicity.

For centrosymmetric crystals without absorption $\mathbf{A}$ is real, symmetric, and Hermitian. For centrosymmetric crystals with absorption $\mathbf{A}$ is symmetric, complex, and not Hermitian.

If $n$-beams are included, structure matrix $\mathbf{A}$ is $n \times n$, and Eq. (3.23) gives $n$ eigenvalues and $n$ eigenvectors. These define the wavefield inside the crystal according to Eq. (3.15). Alternatively, Eq. (3.15) may be regrouped according to the "Darwin representation" (Hirsch et al., 1977) of $n$ plane waves propagating in the crystal, each in the direction of $\mathbf{K} + \mathbf{g}$, namely,

$$\Psi(\mathbf{r}) = \sum_{\mathbf{g}} \phi_{\mathbf{g}} \exp[2\pi i (\mathbf{K} + \mathbf{g}) \cdot \mathbf{r}]$$

Then the wave amplitude at crystal thickness $t$ becomes

$$\phi_{\mathbf{g}}(t) = \sum_{i=1}^{n} c_i C_{\mathbf{g}}^i \exp(2\pi i \gamma^i t) \tag{3.24}$$

The excitation coefficients $c_i$ are to be determined by matching the incident waves with waves inside the crystal at the entrance surface. That is, we set $t = 0$ in Eq. (3.24) and then solve the resulting linear equation for $c_i$. This can be most elegantly expressed by writing Eq. (3.24) in matrix form

$$\begin{pmatrix} \phi_0(t) \\ \phi_g(t) \\ \vdots \end{pmatrix} = \begin{pmatrix} C_0^1 & \cdots & C_0^n \\ C_g^1 & \cdots & C_g^n \\ \vdots & \ddots & \vdots \end{pmatrix} \begin{pmatrix} \exp(2\pi i \gamma^1 t) & \cdots & 0 \\ \vdots & \ddots & \vdots \\ 0 & \cdots & \exp(2\pi i \gamma^n t) \end{pmatrix} \begin{pmatrix} c_1 \\ \vdots \\ c_n \end{pmatrix}$$

At the sample entrance surface, $t = 0$, in which case we have

$$\begin{pmatrix} \phi_0(0) \\ \phi_g(0) \\ \vdots \end{pmatrix} = \mathbf{C} \begin{pmatrix} c_1 \\ \vdots \\ c_n \end{pmatrix} \quad (3.25)$$

The excitation coefficients $c_i$ are found by premultiplying by $\mathbf{C}^{-1}$ on both sides of Eq. (3.25). Upon substituting $c_i$ into Eq. (3.25), the wavefield inside the crystal is found to be

$$\begin{pmatrix} \phi_0(t) \\ \phi_g(t) \\ \vdots \end{pmatrix} = \mathbf{C} \begin{pmatrix} \exp(2\pi i \gamma^1 t) & \cdots & 0 \\ \vdots & \ddots & \vdots \\ 0 & \cdots & \exp(2\pi i \gamma^n t) \end{pmatrix} \mathbf{C}^{-1} \begin{pmatrix} \phi_0(0) \\ \phi_g(0) \\ \vdots \end{pmatrix} \quad (3.26)$$

The matrix $\mathbf{S} = \mathbf{C}\{\exp(2\pi i \gamma t)\}\mathbf{C}^{-1}$ is called the "scattering matrix" and relates the incident waves to the scattered waves at crystal thickness $t$.

If absorption is not included, the inverse of the eigenvector matrix is the transposed conjugate of the eigenvector matrix, i.e., $\mathbf{C}^{-1} = \mathbf{C}\dagger$. If there is only one incident plane wave [$\phi_0(0) = 1$ and $\phi_g(0) = 0$], we find $c_i = C_0^{i*}$ without absorption, and $c_i = C_0^{i-1}$ with absorption. Quantities $C_0^{i-1}$ are the elements of the first column of the inverse of the matrix whose elements are $C_g^i$ (column $i$, row $g$). The intensity of a particular Bragg beam (for a given incident plane wave $\mathbf{K}$) is then found from

$$I_g(K_x, K_y) = |\phi_g(t)|^2 = \left| \sum_{i=1}^{n} C_0^{i-1} C_g^i \exp(2\pi i \gamma^i t) \right|^2 \quad (3.27)$$

where $t$ is again the crystal thickness.

## 3.3. TWO-BEAM THEORY AND THE BETHE POTENTIALS. HOW BRAGG'S LAW MAY FAIL IF MORE THAN TWO BEAMS ARE EXCITED

In this section we give the results of the two-beam theory of electron diffraction. We then introduce the Bethe potentials, which together with the two-beam approximation provide the simplest account of perturbing many-beam effects. These results are used extensively in later sections. The Bethe potentials allow weak beam effects to be included and offer the simplest account of three-beam effects. Apart from the understanding it provides, the main use of two-beam theory in CBED comes from its

# THEORY

ability to provide a quick estimate of sample thickness. The two-beam theory may be obtained from Eq. (3.23) if only one Fourier coefficient $V_g$ is retained. (Terms in $V_0$ affect the phase of all the diffracted beams equally, and so may only be observed in interference experiments.) If absorption is also neglected, Eq. (3.23) becomes

$$\begin{pmatrix} -2K_n\gamma & U_{-g} \\ U_g & 2KS_g - 2K_n\gamma \end{pmatrix} \begin{pmatrix} C_0 \\ C_g \end{pmatrix} = 0 \qquad (3.28)$$

The values of $\gamma$ are found by setting the determinant of the structure matrix equal to zero, i.e., $|A| = 0$. This leads to the following quadratic equation:

$$(2K_n\gamma)^2 - 2KS_g(2K_n\gamma) - |U_g|^2 = 0$$

which gives

$$2K_n\gamma^{1,2} = KS_g \pm \sqrt{(KS_g)^2 + |U_g|^2} \qquad (3.29)$$

where superscript 1 refers to $+$ and 2 to $-$. If we set $\omega = KS_g/|U_g| = S_g\xi_g$ and $\cot\beta = \omega$, then Eq. (3.28) yields the ratio of $C_0$ and $C_g$ in the form

$$C_0/C_g = \omega \pm \sqrt{1+\omega^2}\exp(-i\varphi_g)$$

$$= \begin{cases} \dfrac{\cos(\beta/2)}{\sin(\beta/2)}\exp(-i\varphi_g) & \text{for the } + \text{ sign} \\ -\dfrac{\sin(\beta/2)}{\cos(\beta/2)}\exp(-i\varphi_g) & \text{for the } - \text{ sign} \end{cases}$$

Since $\cos^2\beta/2 + \sin^2\beta/2 = 1$ and is normalized, we may take the eigenvector matrix to be

$$C_0^1 = \cos(\beta/2)\exp(-i\varphi_g), \qquad C_g^1 = \sin(\beta/2)$$
$$C_0^2 = -\sin(\beta/2), \qquad C_g^2 = \cos(\beta/2)\exp(i\varphi_g) \qquad (3.30)$$

When the results of Eqs. (3.29) and (3.30) are substituted in Eq. (3.24), we obtain the intensity $I_g$ as a function of sample thickness $t$, structure factor $U_g$, accelerating voltage, and excitation error $S_g$:

$$I_g = \frac{|U_g|^2 \sin^2\left(\frac{\pi t}{K_n}\sqrt{K^2 S_g^2 + |U_g|^2}\right)}{K^2 S_g^2 + |U_g|^2} \quad (3.31)$$

$$= \frac{|U_g|^2 \sin^2(\pi t \Delta \gamma)}{(K_n \Delta \gamma)^2} \quad (3.32)$$

with

$$I_0 = 1 - I_g$$

Here $\Delta \gamma$ is the gap between the dispersion surfaces, discussed in the next section. Figure 3.1 shows this function plotted as a function of both thickness and orientation. The dimensionless parameter $\omega = S_g \xi_g$ has been used, with $\xi_g = 1/(\lambda U_g)$. The variation of the intensity with thickness is known as "Pendellosung," or thickness fringe oscillations. At a particular thickness, the intensity variation with excitation error shown gives a good impression of the intensity variations seen in an experimental systematics CBED disk around the Bragg condition, as shown in Fig. 2.3. The fine HOLZ lines seen crossing the (111) disk in Fig. 2.3, however, cannot be reproduced in a two-beam theory based solely on $V_{111}$. The calibration precedure for $S_g$ given in Chapter 2 could enable data taken from Fig. 2.3 along the line AA' to be compared directly with

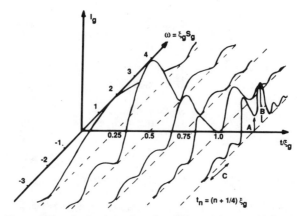

*Figure 3.1.* The variation in the intensity of a diffracted beam with both thickness and incident-beam direction (represented by $\omega$) in the two-beam approximation. Structure factors are best determined using the ratio $A/B$ of intensities shown, at which thickness the sensitivity is greatest.

Eq. (3.31), and values of $U_g$ and $t$ found to produce the best fit. This is the basis of the techniques described in Chapter 4.

The intensity minima of Eq. (3.31) occur at $S_g$ values given by

$$S_g^2 = \frac{n^2}{t^2(K/K_n)^2} - \frac{|U_g|^2}{K^2} \qquad (3.33)$$

which may be related to the incident beam angle $\alpha$ through Eq. (2.8). The factor $K/K_n$ in Eq. (3.28) is often neglected in the literature. If so, then the thickness measured is the effective crystal thickness, $t^{\text{eff}} = t(K/K_n) = t/\cos\theta$, where $\theta$ is the angle between the beam direction and the surface normal. The intensity at the intensity maxima is given by

$$I_g = \left[\frac{\pi t |U_g|}{K_n}\right]^2 \Big/ (1+x^2) \qquad (3.34)$$

where $x$ is given by the solutions of $x = \tan(x)$.

At the Bragg condition $S_g = 0$, the intensity becomes

$$I_g = \sin^2\left(\frac{\pi t |U_g|}{K_n}\right) \qquad (3.35)$$

The periodicity of the intensity with thickness when $S_g \neq 0$ is $L = \xi_g(1+\omega^2)^{-1/2}$.

The above discussion assumes an ideal two-beam case, in which all other beams have zero intensity. In reality, there are always weak beams present. One way to include these weak-beam effects on two-beam diffraction is to use the perturbation method of Bethe. The criteria for classifying a beam $h$ as a weak beam is that $K|S_h| \gg |U_g|$. Thus $|S_g|$ must be comparable with $|U_g|/K$ to justify treating $g$ as a strong beam. Applying these conditions to Eq. (3.23), we have

$$C_h = \frac{\sum_{\mathbf{h}'} U_{\mathbf{h}-\mathbf{h}'} C_{\mathbf{h}'}}{2KS_h - 2K_n\gamma} \approx \frac{U_h C_0}{2KS_h} + \frac{U_{h-g} C_g}{2KS_h}$$

with second-order terms neglected. Substitution of this equation into (3.23) yields a modified two-beam equation

$$\begin{pmatrix} 2KS_0^{\text{eff}} - 2K_n\gamma & U_{-g}^{\text{eff}} \\ U_g^{\text{eff}} & 2KS_g^{\text{eff}} - 2K_n\gamma \end{pmatrix} \begin{pmatrix} C_0 \\ C_g \end{pmatrix} = 0$$

where

$$2KS_0^{\text{eff}} = -\sum_{\mathbf{h}} \frac{|U_{\mathbf{h}}|^2}{2KS_h}, \quad 2KS_g^{\text{eff}} = 2KS_g - \sum_{\mathbf{h}} \frac{|U_{\mathbf{g-h}}|^2}{2KS_h} \quad (3.36)$$

and

$$U_g^{\text{eff}} = U_g - \sum_{\mathbf{h}} \frac{U_{\mathbf{h}} U_{\mathbf{g-h}}}{2KS_h} \quad (3.37)$$

This effective structure factor, first introduced by Bethe, incorporates weak-beam effects within the two-beam approximation. It is therefore known as the Bethe potential. The two-beam intensity, now using this Bethe potential, is given by

$$I_g = \frac{|U_g^{\text{eff}}|^2 \sin^2\left[\frac{\pi t}{K_n} \sqrt{K^2(S_g^{\text{eff}} - S_0^{\text{eff}})^2 + |U_g^{\text{eff}}|^2}\right]}{K^2(S_g^{\text{eff}} - S_0^{\text{eff}})^2 + |U_g^{\text{eff}}|^2} \quad (3.38)$$

Two new effects are seen to result from the inclusion of additional weak beams. First, the Bragg law no longer strictly applies, as we see from the appearance of the effective excitation error in Eq. (3.38). This is the origin of the displaced Kikuchi lines and HOLZ lines, which will be discussed in Sections 3.6 and 5.4. The precise position of the intensity maximum near the Bragg angle thus depends on the values of the other structure factors. Second, the extinction distance now also depends on orientation. Other effects, including those involving structure-factor phases, will be discussed later. The effective potential is sometimes called the second Bethe approximation in the literature. It is the simplest approximation for many-beam effects and has been widely used to explain, for example, the critical voltage effect (see Section 5.2).

The two-beam approximation is most accurate in small unit cell crystals where it is possible, because of the sparse reciprocal lattice, to minimize the excitation of other beams. However, a glance at an experimental pattern may give a misleading impression of the validity of this approximation, since two-beam conditions should exist for all thicknesses.

## 3.4. THE CONCEPT OF THE DISPERSION SURFACE

In this section, we introduce the dispersion surface construction. The dispersion surfaces are formed by the locus of allowed wavevectors for a

# THEORY

given total beam electron energy, as determined by the accelerating voltage. All wavevectors inside the crystal are restricted to lie on the dispersion surfaces. Mathematically, the dispersion surfaces are described by Eq. (3.16). The two-beam results of the last section will be used as an example to illustrate the dispersion surface construction. The concept of dispersion surfaces in electron diffraction is similar to the concept of energy bands in solid state physics. They differ, however, in the physical quantities represented. The energy bands plot the allowed energy $E$ of a crystal electron for a given crystal momentum $h\mathbf{k}^j$. (Here $\mathbf{k}^j$ labels the Bloch-wave state $j$.) The dispersion surfaces plot the opposite, that is, they plot the locus of allowed momentum $h\mathbf{k}^j$ for the incident-beam electron of given energy $E_0$. For elastic electron diffraction, the beam electron has a constant total energy $E_0$. As a result of the interaction of the incident electron with the crystal potential, the allowed momentum (and corresponding kinetic energy) of an incident electron varies between Bloch-wave states $\mathbf{k}^j$. The difference between the kinetic and total energy of a Bloch-wave state is taken up by potential energy.

As an example, the two-beam dispersion surfaces shown in Fig. 3.2 were constructed according to the following procedure:

1. Approximate dispersion surfaces are plotted, using the "empty lattice" approximation. Here all the interaction parameters are set to zero: $U_{gh} = 0$. From Eq. (3.15), this gives

$$(\mathbf{k}^j + \mathbf{g})^2 = K^2$$

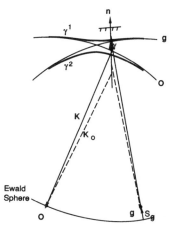

Figure 3.2. Dispersion surfaces in the two-beam approximation. Here $\gamma$ and $S_g$ are measured in the direction of the surface normal $n$. Surfaces are numbered from the top down. 0 and g are spheres of radius $K$ drawn (not to scale) about 0 and g.

Hence, in the empty-lattice approximation, the dispersion surfaces are a set of spheres of radius $K$ centered on each reciprocal lattice point $g$. These are shown as thin lines in Fig. 3.2. (The aspect ratio of the figure has been exaggerated for clarity—in high-energy electron diffraction the ratio $g/K$ in particular is much smaller than shown in the figure)

2. A vector is drawn in the direction normal to the entrance surface of the crystal, and intersecting the vector **K**, as shown by the arrow in Fig. 3.2.
3. The dynamical dispersion constants $\gamma^i$ are calculated from Eq. (3.16), now using nonzero $U_{gh}$ values. In the two-beam case, $\gamma$ may be obtained from Eq. (3.29). According to Eq. (3.18a), the values of $\gamma^j$ are measured along the surface-normal direction, starting from a point on that $K$ sphere which is centered on the origin of the reciprocal lattice. Thus the incident-beam direction **K** must be drawn first. Then points on the dispersion surfaces are drawn at distances $\gamma^j$ measured from the end point of **K** along the surface-normal direction. The complete dispersion surfaces are obtained by repeating this for each possible beam direction, and these are shown as bold lines in Fig. 3.2.

Two methods have been used in the literature to present calculated or measured dispersion surfaces. The first is similar to Fig. 3.2—here the momenta **k** are plotted against the $x$ coordinate of the incident-beam direction $\mathbf{K}_t$ (as defined in Chapter 2). The plotting can be made easier by assuming that the surface normal is along the $z$ axis, with the $x$ axis as $\mathbf{K}_t$. Fig. 3.3a shows the two-beam dispersion surfaces plotted in this fashion. The alternative method, mostly used by the Norwegian groups, uses $\gamma$ as the $y$ axis, assuming that the surface normal is antiparallel to the beam direction and the incident beam's $K$ sphere is taken as the $x$ axis. Only a plane section of the dispersion surfaces are plotted, thus the $K$ spheres are shown as circles. In such a diagram the $K$ spheres appear flattened. The departure from flatness is actually very small in high-energy electron diffraction within the angular range typically shown (about 0.1 rad). Figure 3.3b shows the two-beam dispersion surfaces plotted in this way. The bold lines are the dispersion surfaces, the thin line is the $K$ sphere drawn about reflection **g**. The $x$ axis is the incident-beam $K$ sphere. Each point on the incident-beam $K$ sphere corresponds to a different incident-beam direction.

For the two-beam dispersion surfaces, a gap opens near the Bragg condition for reflection **g**, where the $K$ spheres about **0** and **g** intersect. The gap has width (measured vertically) $|U_g|/K$ at the Bragg condition.

# THEORY

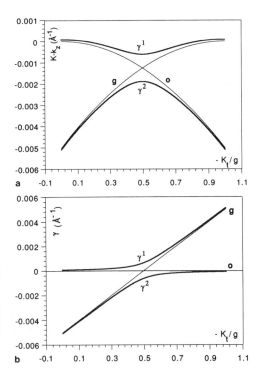

Figure 3.3. (a) Calculated two-beam dispersion surfaces. The abscissa is dimensionless and the ordinate is given as $(K - k_z)$ in reciprocal angstroms. (b) Similar to Fig. 3.3a, but plotted in such a way that the asymptotic sphere about the origin is taken as a straight line (an excellent approximation at 100 kV or above). Two-beam theory.

The gap between the dispersion surfaces and the $K$ spheres decreases as one moves away from the Bragg condition.

The momenta of the incident-beam electron states inside the crystal, as represented by the dispersion surfaces, are also physical observables, as are energies in crystal band theory. One way to observe these momenta is to use a wedge-shaped crystal, since this has a different boundary condition at its entrance and exit surfaces (Lehmpfuhl and Reissland, 1968). The principle is demonstrated in Fig. 3.4, where $n_i$ and $n_e$ are the entrance and exit surface-normal directions, respectively. The Bloch-wave vectors $\mathbf{k}^1$ and $\mathbf{k}^2$ are determined by the entrance surface boundary condition. However, the transmitted wave vectors $\mathbf{K}_0^1$ and $\mathbf{K}_0^2$ are determined by the boundary condition at the exit surface. The difference between the entrance and exit surface boundary condition and the difference in the Bloch-wave vectors causes the angular splitting in the transmitted beam. This splitting is proportional to the gap between the dispersion surfaces. The same splitting is expected for the diffracted beams. For simplicity we do not show the diffracted beams in Fig. 3.4. The experiment is performed by recording a series of point diffraction

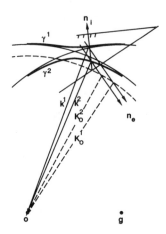

*Figure 3.4.* Dispersion surfaces for a wedge-shaped sample. The entrance surface normal is $n_i$ and the exit face normal is $n_e$. Each diffracted beam becomes split—here in two-beam theory the direct beam is split into $K_0^1$ and $K_0^2$, as shown.

patterns while rotating the crystal around the axis normal to a plane section of the dispersion surface of interest. The component of the incident-beam direction along the rotation axis is held constant. Thus a plane section (normal to the rotation axis) of the dispersion surfaces is obtained. Experimental results are shown in Fig. 3.5.

Parts of the dispersion surfaces can also be observed in conventional point diffraction patterns, in the fine structure of HOLZ lines, or in Kikuchi line patterns. These effects are further discussed in Section 3.8 and demonstrated experimentally in Fig. 3.11.

## 3.5. DEGENERATE AND NONDEGENERATE PERTURBATION THEORY

In this section, we derive some important results based on perturbation theory. These will serve as the basis for our analysis of several three-beam cases, of dynamical effects on HOLZ lines, and for the study of absorption phenomena.

The aim of the perturbation method is to find the change in a known system (in this case a set of Bloch waves) due to a small perturbation in the structure matrix. This change in the structure matrix can occur either in the potential $U_{gh}$, or in the beam direction, as defined by the quantities $S_g$ which arise on the diagonal. Perturbation theory has previously been used in HEED by a number of authors (Buxton, 1976; Hirsch *et al.*, 1977; Hussein and Wagenfeld, 1978), who applied it to problems of absorption, weak beam effects, HOLZ effects, and the study of three- or four-beam interactions. This section discusses the theory, and follows the

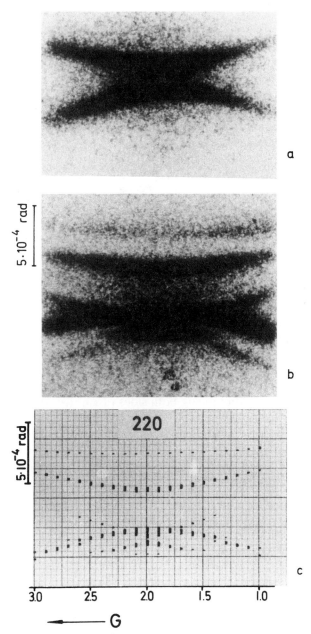

*Figure 3.5.* Fine structure in the (220) reflection from an MgO wedge crystal is shown in (a) as it is rotated under the beam (Lehmpfuhl and Reissland, 1968). A point pattern is formed, and the film is moved as the crystal is rotated about the (220) direction. (b) shows higher resolution (slower rotation) while (c) shows calculated dispersion surfaces for this case, in good agreement. The width of the points is proportional to the strength of the Bloch wave.

formulation outlined in Zuo (1991b). Some applications of perturbation theory are mentioned, but these are mainly treated in other sections.

Our treatment is given in matrix form. We have found this approach to be more straightforward for diffraction than the Rayleigh–Schrödinger perturbation theory of quantum mechanics used by Buxton (1976), which leads to the same results. Our starting equation is Eq. (3.23), which may be written in simplified matrix form as

$$\mathbf{AC} = \mathbf{C}\lambda$$

where $A_{ii} = 2KS_g$ and $A_{ij} = U_{g-h} = U_{gh}$, $\mathbf{C}$ is the eigenvector matrix with $C_{ij} = C_g^j$, and $\lambda$ is the eigenvalue matrix. This is a diagonal matrix with $\lambda_{ii} = 2K\gamma^i$. Let $\mathbf{A} = \mathbf{A}_0 + \mathbf{A}'$, where $\mathbf{A}'$ is a small perturbation resulting from a small change in the potential $U_{gh}$ or the beam direction $S_g$ or both. Then

$$\mathbf{A}_0\mathbf{C} + \mathbf{A}'\mathbf{C} = \mathbf{C}\lambda \tag{3.39}$$

We assume that the eigenvalue and eigenvector matrices $\tau$ and $\Gamma$ corresponding to $\mathbf{A}_0$ are known. (We will use $\tau$ and $\Gamma$ to avoid confusion with $\lambda$ and $C$.) The eigenvector matrix of $\mathbf{A}$ may be written as a combination of the eigenvectors of $\mathbf{A}_0$ and $\mathbf{C} = \Gamma\varepsilon$, where $\varepsilon$ is the coefficient matrix. Multiplying both sides of Eq. (3.39) by $\Gamma^{-1}$ gives

$$\{\Gamma^{-1}\mathbf{A}_0\Gamma + \Gamma^{-1}\mathbf{A}'\Gamma\}\varepsilon = \varepsilon\lambda$$

If $\mathbf{B} = \{\Gamma^{-1}\mathbf{A}_0\Gamma + \Gamma^{-1}\mathbf{A}'\Gamma\}$, then we can write a transformed equation in the basis of the states of $\mathbf{A}_0$,

$$\mathbf{B}\varepsilon = \varepsilon\lambda \tag{3.40}$$

where the matrix $\mathbf{B}$ is defined by

$$B_{ii} = \tau_{ii} + b_{ii} \tag{3.41}$$

and

$$B_{ij} = b_{ij} = \sum_{km} \Gamma_{ik}^{-1} A'_{km}\Gamma_{mj} \tag{3.42}$$

We seek changes in the $i$th Bloch wave and assume that this Bloch wave is nondegenerate. This means that no other Bloch wave has an eigenvalue which is close to that of the $i$th Bloch wave. This condition

# THEORY

may be written as

$$|\lambda_j - \lambda_i| \gg |b_{\max}| \tag{3.43}$$

where $b_{\max}$ is the largest off-diagonal element of **B**. Then the $i$th Bloch wave is said to be nondegenerate. In this case, the change in the $i$th Bloch wave is just

$$\lambda_{ii} = \tau_{ii} + \Delta \tag{3.44}$$

where $\Delta$ is a small quantity. From Eq. (3.40), we then have

$$(\tau_j + b_{jj} - \tau_i - \Delta)\varepsilon_{ji} + \sum_k b_{jk}\varepsilon_{ki} = 0 \tag{3.45}$$

For $j \neq i$, $|\tau_j - \tau_i| \gg |\Delta|$ or $|b_{jj}|$, $\Delta$ and $b_{jj}$ may be neglected, in which case

$$\varepsilon_{ji} \approx b_{ji}/(\tau_i - \tau_j)$$

where $\varepsilon_{ii} \approx 1$ from the normalization condition. On substituting this into Eq. (3.44) and letting $j = i$, Eqs. (3.44) and (3.45) give

$$\lambda_{ii} = \tau_{ii} + b_{ii} + \sum_j |b_{ij}|^2/(\tau_i - \tau_j) \tag{3.46}$$

This result has been derived subject to condition (3.43), known as nondegenerate perturbation theory. It will be used to explain the effects of absorption in Section 3.7. It will also be used in the final stages of structure factor refinement, where the changes in structure factors are small, as discussed in Section 4.2.

Nondegenerate perturbation theory fails if Eq. (3.46) does not converge. This occurs when there are two or more Bloch-wave states belonging to $\mathbf{A}_0$ whose eigenvalues are close to each other, i.e., if condition (3.43) is not satisfied for these Bloch waves. These Bloch-wave states are then said to be degenerate, and the results of nondegenerate perturbation theory cannot be used. For the degenerate case, in the first-order approximation, we may neglect the effects of the nondegenerate states and concentrate only on the effects of the perturbation on the degenerate states. If the degeneracy is $f$-fold, then the first-order wave function may be found from

$$\begin{pmatrix} \tau_i - \lambda + b_{ii} & b_{ij} & \cdots & b_{ik} \\ b_{ji} & \tau_j - \lambda + b_{jj} & \cdots & b_{jk} \\ \vdots & \vdots & \cdots & \vdots \\ b_{ki} & b_{kj} & \cdots & \tau_k - \lambda + b_{kk} \end{pmatrix} \begin{pmatrix} \varepsilon_i \\ \varepsilon_j \\ \vdots \\ \varepsilon_k \end{pmatrix} = 0 \quad (3.47)$$

where $\tau_i, \tau_j, \ldots, \tau_k$ are the eigenvalues of the $f$ degenerate Bloch waves. A general solution to Eq. (3.47) cannot be found unless $f = 2$, that is, unless no more than two Bloch waves are degenerate. In this case

$$\begin{pmatrix} \tau_i - \lambda + b_{ii} & b_{ij} \\ b_{ji} & \tau_j - \lambda + b_{jj} \end{pmatrix} \begin{pmatrix} \varepsilon_i \\ \varepsilon_j \end{pmatrix} = 0 \quad (3.48)$$

This equation can be solved by following the same procedure as used in the two-beam theory of Section 3.3. The result is

$$\lambda^{i,j} = (\tau_i + b_{ii} + \tau_j + b_{jj} \pm \sqrt{(\tau_i + b_{ii} - \tau_j - b_{jj})^2 + 4|b_{ij}|^2})/2 \quad (3.49)$$

and

$$\begin{aligned} \varepsilon_{ii} &= \cos(\beta/2)\exp(i\phi), & \varepsilon_{ji} &= \sin(\beta/2) \\ \varepsilon_{ij} &= -\sin(\beta/2), & \varepsilon_{jj} &= \cos(\beta/2)\exp(-i\phi) \end{aligned} \quad (3.50)$$

where $\phi$ is the phase of $b_{ij}$, and $\beta$ is defined by

$$\cot\beta = (\tau_i + b_{ii} - \tau_j - b_{jj})/2|b_{ij}|$$

Examples of the application of degenerate perturbation theory are given in Sections 3.7 and 3.8.

## 3.6. THREE-BEAM THEORY AND PARTICULAR SOLUTIONS FOR CENTROSYMMETRIC CRYSTALS

In this section we derive the main results of three-beam theory, and specialize to some particular solutions for centrosymmetric crystals. A general discussion of three-beam theory for noncentrosymmetric (acentric) crystals is deferred until the next section.

The three-beam case is the simplest case of many-beam diffraction. The three-beam solution contains within it all the essential features needed to explain such phenomena as the critical voltage effect, dynamical shifts of HOLZ lines, and the principles of structure-factor

# THEORY

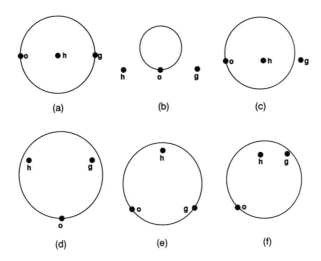

*Figure 3.6.* All possible three-beam geometries. The Laue circle and position of reciprocal lattice points are shown. Symmetry reduction makes many cases easily soluble along certain lines in three-beam CBED patterns.

phase measurement using multiple scattering. None of these effects can be understood using two-beam theory. Since they are among the most interesting and useful applications of high-energy electron diffraction, an understanding of three-beam theory is important.

There are six possible three-beam geometries. These are shown in Fig. 3.6. In each case the scattering is dominated by the interactions between the three beams **0**, **g**, and **h**. The first three cases (a, b, and c) are known as systematic three-beam cases. The rest are nonsystematic cases. If these three beams only are retained and we take **n** antiparallel to **K**, Eq. (3.23) becomes

$$\begin{bmatrix} 0 & U_{-g} & U_{-h} \\ U_g & 2KS_g & U_{g-h} \\ U_h & U_{h-g} & 2KS_h \end{bmatrix} \begin{bmatrix} C_0 \\ C_g \\ C_h \end{bmatrix} = 2K\gamma \begin{bmatrix} C_0 \\ C_g \\ C_h \end{bmatrix} \quad (3.51)$$

This equation was first studied in detail in the important early papers of Fues (1949) and Kambe (1957). The three eigenvalues $\gamma^i$ are given by the roots of the secular equation

$$|2K\gamma^i \mathbf{I} - \mathbf{A}| = 0 \quad (3.52)$$

This gives a cubic equation, whose closed-form solution can be found in mathematical texts but which is too lengthy and complicated to be of much use. Instead, Eq. (3.51) is usually solved using approximations, or for special cases involving symmetry.

In centrosymmetric crystals the structure factors are real, and $U_g = U_{-g}$ if the origin of the reciprocal lattice is taken at the center of the symmetry. This simplifies Eq. (3.51), but a transparent solution is still not possible. In the following we describe two particular solutions to Eq. (3.51) for centrosymmetric crystals. One involves degeneracy in eigenvalues, and the other requires some symmetry.

It was found by Gjønnes and Høier (Gjønnes, 1971) that for centrosymmetric crystals, two of the three eigenvalues of Eq. (3.51) become degenerate at a special point on the dispersion surfaces. This occurs when

$$2KS_g = U_g(U_{gh}^2 - U_h^2)/(U_h U_{gh})$$

and

$$2KS_h = U_h(U_{gh}^2 - U_g^2)/(U_g U_{gh})$$

(3.53)

The degenerate eigenvalue is then given by

$$2K\gamma = -U_g U_h / U_{gh}$$

In the diffraction pattern, this degeneracy causes an intensity minimum. The position of this minimum, as defined by the excitation errors of Eq. (3.53), depends on the values of the three structure factors involved, in particular on the sign of the product $U_{-g} U_h U_{g-h}$. This sign dependence is easily observed in experimental CBED patterns. Thus the sign of the product $U_{-g} U_h U_{g-h}$ is determined by the sign of $S_h$ at the degeneracy point ($S_h$ and $S_g$ have the same sign). The sign of the excitation error can easily be determined, as discussed in Chapter 2. Thus by noting where the intensity minimum occurs, we may determine the sign of the triplet $U_{-g} U_h U_{g-h}$. Since all structure factors $U_g$ have phases of 0° or 180° in centric crystals (corresponding to a sign of plus or minus), this procedure is equivalent to a determination of the sum of the phases of the structure factors. The phase sum is known as a "three-phase structure invariant". [An $n$-phase structure invariant may similarly be defined from the determinantal equation (3.52).] This topic is discussed in more detail in the next section on noncentrosymmetric crystals, for which centrosymmetric crystals form a special case.

In the three-beam geometry of Fig. 3.6a and e, there is a mirror symmetry for centrosymmetric crystals. In this case Eq. (3.51) becomes

$$\begin{bmatrix} 0 & U_g & U_h \\ U_g & 0 & U_h \\ U_h & U_h & 2KS_h \end{bmatrix} \begin{bmatrix} C_0 \\ C_g \\ C_h \end{bmatrix} = 2K\gamma \begin{bmatrix} C_0 \\ C_g \\ C_h \end{bmatrix} \quad (3.54)$$

using the fact that $U_g = U_{-g}$ and $U_{g-h} = U_h$. Because of the symmetry in this equation, there are only two possibilities: (1) $C_0 = C_g$, or (2) $C_0 = -C_g$ and $C_h = 0$. In the first case, Eq. (3.54) is reduced to a $2 \times 2$ matrix, and to a $1 \times 1$ matrix in the second case. These matrices are

$$\begin{pmatrix} U_g & U_h \\ 2U_h & 2KS_h \end{pmatrix} \begin{pmatrix} C_0 \\ C_h \end{pmatrix} = 2K\gamma \begin{pmatrix} C_0 \\ C_h \end{pmatrix} \quad (3.55)$$

and

$$-U_g C_0 = 2K\gamma C_0 \quad (3.56)$$

The two Bloch waves given in Eq. (3.55) are symmetric because $C_0 = C_g$ while the Bloch wave in Eq. (3.56) is antisymmetric. These results may be deduced more elegantly and generally using group theoretical arguments (Kogiso and Takahashi, 1977). Equations (3.55) and (3.56) are easily solved and give

$$2K\gamma^1 = -U_g \quad \text{and} \quad 2K\gamma^{2,3} = \tfrac{1}{2}(2KS_h + U_g \pm \sqrt{(2KS_h - U_g)^2 + 8U_h^2})$$
(3.57)

Eigenvectors and solutions for many other solvable cases involving symmetry (up to twelve beams) can be found in Fukahara (1966). The degeneracy in this symmetric three-beam case occurs when $\gamma^1 = \gamma^3$, at an excitation error

$$2KS_h = (U_h^2 - U_g^2)/U_g$$

This agrees with the result of Eq. (3.53) if we use the fact that $U_{g-h} = U_h$, as assumed in Eq. (3.54).

## 3.7. THREE-BEAM THEORY—NONCENTROSYMMETRIC CRYSTALS AND THE PHASE PROBLEM

Our understanding of the noncentrosymmetric case is much more recent. For acentric crystals, the Fourier coefficients $F_g^X/\Omega$ of $\rho(r)$ are complex, and values of the structure-factor phases $F_g^X$ are required to enable the crystal charge density to be synthesized. These phases depend on the choice of origin in the crystal. Therefore researchers in both X-ray and electron crystallography have for many years sought to find practical solutions to this famous "phase problem." Rather than simply reading-in an entire CBED pattern into a computer and attempting to match it to a set of phases, it is useful to determine first which portions of a CBED pattern are most sensitive to changes in a given structure-factor phase. Our aim in the following is to use the results of three-beam dynamical theory for noncentrosymmetric crystals to find the sensitive region, and to indicate the general form of the intensity pattern. For accurate phase measurement in practice, many-beam calculations will be required.

It is well known that the single-scattering or kinematic theory of diffraction does not allow phases to be measured, nor, since Eq. (3.31) contains only $|U_g|^2$, does the simple two-beam dynamical theory. If absorption is included exactly (or by using perturbation theory), small differences appear between the **g** and −**g** rocking curves in two-beam theory due to the difference between the phases of the structure factors of the real potential and those of the absorption potential. These are responsible for the observed asymmetry in Kikuchi lines (Bird and Wright, 1989) and inner-shell energy-loss spectra (Taftø, 1987). We note here in passing that the phase of the electron structure factor $U_g$ is not equal to that of the X-ray structure factor $F_g^X$. This can be seen from Eqs. (3.4) and (3.5).

For reviews of phase measurement by X-ray diffraction using many-beam effects in acentric crystals, see Chang (1987) and Hummer and Billy (1986). (This has followed somewhat similar lines to the electron diffraction work, with a number of additional complications arising from the vector nature of the electromagnetic wavefield. At the time of writing, the error in phase measurement for low-order reflections in acentric crystals by X-ray diffraction is about 45°. Using electrons it is less than 1°.) For electron diffraction, Høier and Marthinsen, building on earlier work by Kambe and Gjønnes, have provided the basic theory for three-beam diffraction and channeling in acentric crystals in a recent series of papers (Høier and Marthinsen, 1983; Høier *et al.*, 1988; Marthinsen and Høier, 1986; Marthinsen *et al.*, 1988).

The simplest way to expose the phase dependence of three-beam intensities is to use the Bethe potential described in Section 3.3, if one of the three beams, **h**, is weaker than the other two. We write the structure factors as $U_g \exp(i\phi_g)$. In the three-beam case for an acentric crystal, we then have

$$|U_g^{\text{eff}}|^2 = \left|U_g - \frac{U_h U_{g-h}}{2KS_h}\right|^2 = |U_g|^2 \left[\left(1 - \frac{|U_h||U_{g-h}|}{2KS_h |U_g|} \cos \Psi\right)^2 + \left(\frac{|U_h||U_{g-h}|}{2KS_h |U_g|} \sin \Psi\right)^2\right] \quad (3.58)$$

where

$$\Psi = \phi_h + \phi_{-g} + \phi_{g-h} \quad (3.59)$$

is the three-phase invariant, or sum of the phases of the corresponding structure factors. This three-phase invariant is independent of the choice of origin in the crystal, as we expect from the intensities of electron diffraction, since the vectors **h**, −**g**, and **g** − **h** form a closed triangle. Thus the intensities in three-beam diffraction depend on both the amplitudes of the three structure factors and the three-phase invariant involved. The influence of the phases is strongest if

$$|U_\mathbf{h}||U_\mathbf{g-h}|/(2KS_\mathbf{h}|U_\mathbf{g}|) = 1 \quad (3.60)$$

The Bethe approximation is the best approximation for the systematic three-beam cases of Fig. 3.6a and c. In these cases, **g** = 2**h** and $2KS_\mathbf{h} \approx h^2$ near the Bragg condition for **g**. The condition (3.60) may be satisfied by varying the electron high voltage, using the dependence of the structure factor $U_\mathbf{g}$ on accelerating voltage [see Eqs. (3.6) and (3.7)], and this is possible in certain favorable cases. Then the systematics CBED three-beam intensity distribution may be used to measure the three-phase invariant to an accuracy of about 1°, as described in Section 5.1.

The most general three-beam case, however, is the nonsystematics three-beam case shown in Fig. 3.6f. Because of the freedom here in choosing reflections **g** and **h**, the excitation error $2KS_h$ may now be varied in two dimensions over the CBED intensity distribution, and the voltage may also be varied to satisfy condition (3.60). In this way the greatest sensitivity to phases may be obtained. Thus, the nonsystematics three-beam CBED method is the most versatile and general method of phase determination in acentric crystals. Although Eq. (3.58) applies to this case also under certain conditions, the nonsystematic three-beam case is best described by the "Kambe" approximation, as used by Kambe in his classic paper (Kambe, 1957). In the following, we describe the Kambe approximation, and the nonsystematic three-beam case in the language of degenerate perturbation theory given previously in Section 3.5.

For the nonsystematic three-beam diffraction of Fig. 3.9, when the coupling between **g** and **h** is far greater than the coupling between **0** and **g** and **h**, that is, if

$$|U_{gh}| \gg |U_g| \text{ or } |U_h| \tag{3.61}$$

then the structure matrix of the three-beam equation (3.51) may be separated into a primary structure matrix and a perturbation matrix, as follows:

$$\mathbf{A}_0 = \begin{pmatrix} 0 & 0 & 0 \\ 0 & 2KS_g & U_{gh} \\ 0 & U_{hg} & 2KS_h \end{pmatrix} \quad \text{and} \quad \mathbf{A}' = \begin{pmatrix} 0 & U_{-g} & U_{-h} \\ U_g & 0 & 0 \\ U_h & 0 & 0 \end{pmatrix} \tag{3.62}$$

The eigenvalues and eigenvectors of **A** are readily solved using the two-beam solution given in Section 3.3. They are

$$\tau_1 = 0, \quad \tau_{2,3} = [S_g + S_h \pm \sqrt{(S_g - S_h)^2 + |U_{gh}/K|^2}]/2 \tag{3.63}$$

and

$$\Gamma = \begin{pmatrix} 1 & 0 & 0 \\ 0 & \cos(\beta/2)\exp(i\phi_{gh}) & -\sin(\beta/2) \\ 0 & \sin(\beta/2) & \cos(\beta/2)\exp(-i\phi_{gh}) \end{pmatrix} \tag{3.64}$$

where $\beta$ is defined by $\cot \beta = (S_g - S_h)/|U_{gh}|$. Using these results and Eq. (3.40), we obtain the three-beam equation, with the states of $\mathbf{A}_0$ as the basis:

$$\begin{pmatrix} \tau_1 & b_{12} & b_{13} \\ b_{21} & \tau_2 & 0 \\ b_{31} & 0 & \tau_3 \end{pmatrix} \begin{pmatrix} \varepsilon_1 \\ \varepsilon_2 \\ \varepsilon_3 \end{pmatrix} = \lambda \begin{pmatrix} \varepsilon_1 \\ \varepsilon_2 \\ \varepsilon_3 \end{pmatrix} \tag{3.65}$$

Here

$$b_{12} = b_{21}^* = U_{-g}\cos(\beta/2)\exp(i\phi_{gh}) + U_{-h}\sin(\beta/2)$$
$$b_{13} = b_{31}^* = -U_{-g}\sin(\beta/2) + U_{-h}\cos(\beta/2)\exp(-i\phi_{gh}) \tag{3.66}$$

## THEORY

In the region $\tau_2 \approx 0$, $\tau_3 \gg |b_{12}|$ or $|b_{13}|$, to first order $\varepsilon_3 \approx 0$. Then Eqs. (3.65) are reduced to the twofold degenerate perturbation equation

$$\begin{pmatrix} \tau_1 & b_{12} \\ b_{21} & \tau_2 \end{pmatrix} \begin{pmatrix} \varepsilon_1 \\ \varepsilon_2 \end{pmatrix} = \lambda \begin{pmatrix} \varepsilon_1 \\ \varepsilon_2 \end{pmatrix} \quad (3.67)$$

The solutions of this type of equation are given by Eqs. (3.49) and (3.50). Similar results may be obtained in the region $\tau_3 \approx 0$. Combining these results with Eqs. (3.27), (3.63), (3.64), (3.66), and the equation $\mathbf{C} = \Gamma \varepsilon$, and assuming that $\tau_2 \approx 0$ or $\tau_3 \approx 0$, we obtain the intensity expression in the regions defined by $\tau_2 \approx 0$ and $\tau_3 \approx 0$ (Zuo et al., 1989b):

$$I_h(S_g) = \frac{(2KS_g)^2}{(2KS_g)^2 + |U_{h-g}|^2} \sin^2\left\{\frac{\pi t}{K}|U_h^{\text{eff}}|\right\} \quad (3.68)$$

where

$$|U_h^{\text{eff}}|^2 = |U_h|^2 \frac{(2KS_g)^2}{|U_{h-g}|^2 + (2KS_g)^2} \left[\left(1 - \frac{|U_g||U_{h-g}|}{2KS_g |U_h|} \cos \Psi\right)^2 \right.$$

$$\left. + \left(\frac{|U_g||U_{h-g}|}{2KS_g |U_h|} \sin \Psi\right)^2\right] \quad (3.69)$$

The expression for the intensity of reflection **g** is obtained by interchanging **g** and **h** in Eqs. (3.68) and (3.69). The effective potential Eq. (3.69) is the same as the Bethe potential in Eq. (3.58) if $2KS_g \gg |U_{gh}|$, which is the condition for the Bethe approximation. Figure 3.7 shows how the ratio $|U_h^{\text{eff}}|/|U_h|$ depends on the phase invariant. The effective structure factor is most sensitive to the phase invariant near the minimum shown.

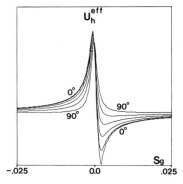

Figure 3.7. Variation of the effective potential $U_h^{\text{eff}}$ (Kambe) with excitation error for equal increments in the phase invariant between 0° and 90°. This quantity is related to the intensity in three-beam CBED patterns along AB ($S_g < 0$) and B'A' ($S_g > 0$) (see Fig. 3.8). Asymptotes are $|U_h|$.

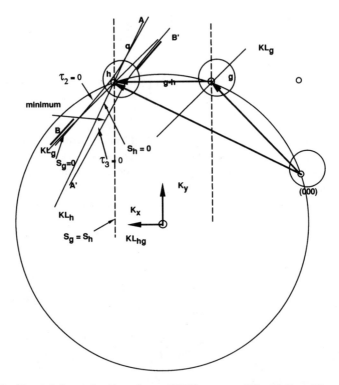

*Figure 3.8.* General form of a three-beam CBED pattern. Kikuchi lines $KL_h$ and $KL_g$ (along which the excitation errors $S_h$ and $S_g$ are zero) are shown. The intensity distribution has extrema on the hyperbolas where $\tau_2 = 0$ and $\tau_3 = 0$, as shown. The intensity approaches two-beam form at A and A′ and fades toward B and B′. For the CdS example discussed in the text, $\mathbf{g} = (4, 1, -2)$, $\mathbf{h} = (4, 1, -4)$, and $\mathbf{h} - \mathbf{g} = (0, 0, -2)$. The position of the minimum is shown for $\Psi = 0$.

These results may be used to draw some qualitative conclusions about the main features of nonsystematic three-beam CBED patterns. These features are shown in Fig. 3.8, which shows three CBED disks in a general nonsystematic pattern in which the Bragg condition is satisfied at the center of each disk $\mathbf{g}$ and $\mathbf{h}$. Kikuchi lines are shown at $KL_g$ and $KL_h$. These lines run normal to the respective $\mathbf{g}$ vectors. We note the line $KL_g$ in disk $\mathbf{h}$, which arises due to multiple scattering. Lines $KL_{h-g}$ and $KL_{g-h}$ are the Kikuchi bands belonging to reflection $\mathbf{g} - \mathbf{h}$. Along these lines $S_g = S_h$. The Kikuchi line of maximum intensity outside the disks (due to inelastic scattering) continues inside the disks (due mainly to elastic scattering) as the locus of the Bragg condition, but is severely perturbed near the center of the disks. Instead of following the geometric locus

THEORY

defined by the Bragg condition ($S_g = 0$), the lines separate, as shown in Fig. 3.8. The resulting gap between the lines at the Bragg condition has been studied for many years since the early work on Kikuchi line patterns of Shinohara (1932). In X-ray diffraction it is known as the Renninger effect. It is this gap which is measured in the intersecting Kikuchi line (IKL) and intersecting HOLZ line (IHL) methods described in Section 5.3. Figure 3.8 shows the hyperbola $\tau_2 = 0$ and $\tau_3 = 0$. According to Eq. (3.68), the locus of maximum intensity follows these two hyperbolas. On the hyperbolas, the incident beam is constrained to move such that

$$2KS_g = \frac{|U_{gh}|^2}{2KS_h} \tag{3.70}$$

We call $\tau_2 = 0$ the upper hyperbola and $\tau_3 = 0$ the lower. On the upper hyperbola, both $S_g$ and $S_h$ are negative. They are positive on the lower hyperbola, thus Eq. (3.70) gives both hyperbolas. The change in excitation error across the gap shown is found from the roots of Eq. (3.63) to be

$$\Delta S_h = |U_{g-h}|/K \tag{3.71}$$

along the line $S_g = S_h$, that is, $K_{h-g}$, in Fig. 3.7. This equation is the basis of the IKL and IHL methods, since it offers a simple method of measuring $U_{g-h}$.

The intensities along the hyperbola also vary. The intensity of beam **h** decreases as the hyperbola approaches the line $S_g = 0$, due to the leading term in Eq. (3.68). Similar conclusions may be drawn about the intensity of beam **g**. A maximum or minimum effective potential (3.69) results in an intensity maxima or minima. According to Eq. (3.69), the maxima and minima occur at the coordinates (Zuo et al., 1989b)

$$S_h \approx \frac{|U_{g-h}|}{4K \cos \Psi} \left\{ \frac{|U_h|}{|U_g|} - \frac{|U_g|}{|U_h|} \pm \left[ \left( \frac{|U_h|}{|U_g|} - \frac{|U_g|}{|U_h|} \right)^2 + 4 \cos^2 \Psi \right]^{1/2} \right\} \tag{3.72}$$

with $S_g$ given by Eq. (3.70). The plus sign in Eq. (3.72) gives the maximum, the minus sign gives the minimum. For centrosymmetric crystals, $\Psi = 0$ or $\pi$. Then the minima occur at

$$S_g = \pm \frac{|U_{g-h}||U_g|}{2K|U_h|} \quad \text{and} \quad S_h = \pm \frac{|U_{g-h}||U_h|}{2K|U_g|} \tag{3.73}$$

These results were derived under the Kambe "strong coupling" approximation where $|U_{g-h}| \gg |U_g|$ or $|U_h|$. Under this condition, Eq. (3.73) agrees with (3.53), the exact solutions of Gjønnes and Høier for centrosymmetric crystals. The position of this minimum immediately tells us whether the three-phase invariant $\Psi$ equals $0$ or $\pi$ in a centrosymmetric crystal. In Fig. 3.8, the minimum occurs on the lower hyperbola A′B′ if $\Psi = 0$, and on the upper hyperbola AB if $\Psi = \pi$. This is also

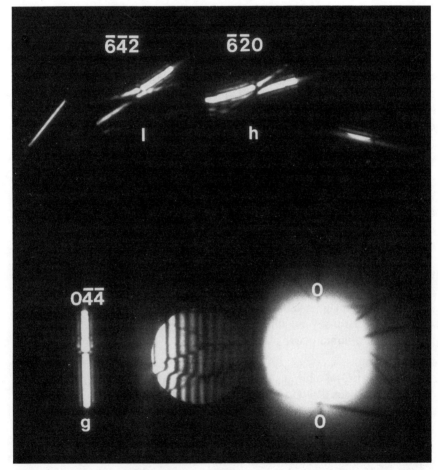

*Figure 3.9.* Nonsystematics three-beam CBED pattern from silicon recorded at 100 kV. The two beams at the Bragg condition $\mathbf{l} = (-6, -4, -2)$ and $\mathbf{h} = (-6, -2, 0)$ differ by the $(0, -2, -2)$ reflection containing vertical bands. The gaps in the intensity along the Bragg hyperbolas in reflections **l** and **g** can be seen, as shown in Fig. 3.8.

illustrated in Fig. 3.9, where the $(-6, -2, 0)$ and $(-6, -4, -2)$ reflections in silicon are strongly excited at 100 kV. These reflections are coupled (differ) by the (022) reciprocal lattice vector. The arrow indicates the minimum of intensity, which occurs for a positive excitation error on the lower hyperbola, showing that the sum of the phases $\Psi$ is zero for these reflections. This is consistent with the fact that the phases of the $(-6, -2, 0)$, $(-6, -4, -2)$ and (022) in silicon are all zero.

For noncentrosymmetric crystals, the measurement of the excitation error at the minimum position may give the values of the phase invariant from Eq. (3.72). The dependence of the nonsystematic three-beam intensity on the phase invariant and the flexibility one has in choosing different reflections **g** and **h** (and thus different three-phase invariants) make it an ideal method for measuring phase invariants. It has been used to study the phase of the (002) reflection in CdS by Zuo et al. (1989b), based on a comparison of experimental and simulated many-beam intensities. Efforts to automate this process (by defining a chi-square goodness-of-fit index) are described in Section 4.2.

The preceding theory applies to elastic Bragg scattering. Yet similar gaps are seen in Kikuchi line patterns. The assumption is made that the inelastic scattering responsible for Kikuchi lines is generated continuously throughout the crystal, and so can be described by integrating Eq. (3.68) over thickness. This affects the last term in Eq. (3.68), but does not alter the expression for the gap.

In summary, for a centric crystal, it may be possible to determine whether the three-phase invariant is 0 or $\pi$ by direct inspection of the position of the minimum in nonsystematic three-beam CBED patterns (Hurley and Moodie, 1980). For acentric crystals, the value of a phase invariant may be estimated from the position of the intensity minimum along the Bragg lines, or may be measured more accurately by comparing computer-simulated CBED patterns with experimental patterns.

## 3.8. DYNAMIC HOLZ INTENSITIES AND POSITIONS. DISPERSION SURFACES FOR HOLZ LINES. HOW THE BRAGG LAW DEPENDS ON LOCAL COMPOSITION

Owing to their potential usefulness for the measurement of strains, accelerating voltage, and composition, we now extend the discussion of Section 2.3 on HOLZ lines to the dynamical case. The study in Section 2.3 was based on the simple Bragg law, but in Section 3.3 we have already seen that this may not be accurate if more than two beams are excited. In this section we quantify this effect, and also provide a geometrical account of the dynamical displacement of HOLZ lines showing how, in principle, the HOLZ Bragg condition depends on structure factors for other excited beams, and therefore on local composition. This effect is suggested schematically in Fig. 3.10a. The experimental evidence for these effects and their applications are summarized in Section 5.4, where suggestions are also made for procedures which minimize these errors due to multiple scattering.

The problem of "dynamical HOLZ line shifts" has been studied by many authors. The earliest relevant work was actually devoted to the problem of the anomalies which were observed in the position and intensities of Kikuchi lines (Gjønnes and Watanabe, 1966; Menzel-Kopp, 1962; Pfister, 1953). These had been analyzed using three-beam considerations as long ago as in 1932 (Shinohara, 1932). A complete analysis using explicit three-beam expressions was given by the Norwegian group (Gjønnes and Høier, 1969; Høier, 1969; Høier, 1972), and these results can also be applied directly to the problem of HOLZ line shifts. Buxton (1976) analyzed the problem by treating the HOLZ reflections as a weak perturbation of the ZOLZ reflections in zone-axis orientations. More recent work has concentrated on the problem of correction schemes for strain measurement (Bithell and Stobbs, 1989; Ecob et al., 1982; Fraser et al., 1985; Jones et al., 1977; Lin et al., 1989; Zuo, 1991a). For a recent three-beam analysis of HOLZ lines, see Britton and Stobbs (1987). We shall assume a parallel-sided slab of crystal whose normal is approximately antiparallel to the beam. In the following, we first derive an expression for the intensity of a HOLZ line, based on the perturbation theory described in Section 3.5, and then draw some conclusions about the HOLZ-line positions based on this theory.

For simplicity, we consider the case where there are no strong interactions among the HOLZ reflections, and we concentrate on the effects on a HOLZ line of interactions in the ZOLZ. Thus we consider the case where, in the ZOLZ, there are $n$ beams $(0, g_1, \ldots, g_{n-1})$ interacting with each other, while in the HOLZ there is one beam $\mathbf{h}$. The interactions between this HOLZ reflection $\mathbf{h}$ and other HOLZ reflections are weak and may be neglected. The interaction between the HOLZ reflection $\mathbf{h}$ and the ZOLZ reflections $\mathbf{g}$ are also assumed to be weaker than the interactions between the ZOLZ reflections, that is,

$$|U_{g_ig_j}| \gg |U_{g_ih}| \tag{3.74}$$

These weak couplings between the ZOLZ and the HOLZ may be treated

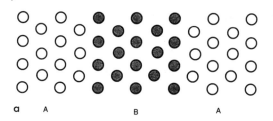
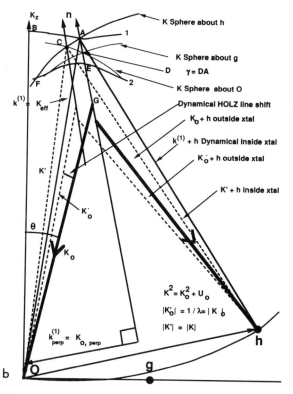

*Figure 3.10.* (a) An unphysical crystal containing different types of atoms on the same lattice. There is a small difference in the directions into which HOLZ beams would be diffracted between region A and region B. (b) The origin of dynamical shifts on HOLZ lines. Dashed rays (such as $\mathbf{K}_0'$) are kinematic "construction lines," defining a kinematic Bragg condition at C. Actual HOLZ-line intensity for $\mathbf{h}$ is a minimum in direction $\mathbf{K}_0$, in the central disk (at $\theta$ to the zone axis), because ZOLZ potential opens gap DA, shifting Bragg condition from C to A. Medium lines are dynamical waves inside the crystal. Thick lines are dynamical beams outside the crystal. The surface normal is $\mathbf{n}$. The angle between $\mathbf{K}_0$ and $\mathbf{K}_0'$ is the dynamical HOLZ-line shift. $\mathbf{K}_{\text{eff}}$ is the wavevector which would be derived, using the simple Bragg law, from an experimental measurement of $\theta$, given $|\mathbf{h}|$.

as a perturbation, that is, we may define, as in Section 3.5,

$$\mathbf{A}_0 = \begin{pmatrix} 0 & U_{-g_1} & \cdots & U_{-g_{n-1}} & 0 \\ U_{g_1} & 2KS_{g_1} & \cdots & U_{g_1-g_{n-1}} & 0 \\ \vdots & \vdots & \cdots & \vdots & \vdots \\ U_{g_{n-1}} & U_{g_{n-1}-g_1} & \cdots & 2KS_{g_{n-1}} & 0 \\ 0 & 0 & \cdots & 0 & 2KS_h \end{pmatrix}$$

$$\text{and} \quad \mathbf{A}' = \begin{pmatrix} 0 & 0 & \cdots & 0 & U_{-h} \\ 0 & 0 & \cdots & 0 & U_{g_1-h} \\ \vdots & \vdots & \cdots & \vdots & \vdots \\ 0 & 0 & \cdots & 0 & U_{g_{n-1}-h} \\ U_h & U_{h-g_1} & \cdots & U_{h-g_{n-1}} & 0 \end{pmatrix} \quad (3.75)$$

The solutions of the unperturbed $\mathbf{A}_0$ give the first $n$ eigenvalues and eigenvectors, together with one eigenvalue and one eigenvector due to the HOLZ reflection:

$$\tau_{n+1} = 2KS_h$$

$$\Gamma_{i,n+1} = \Gamma_{n+1,i} = 0 \quad \text{for } i = 1, \ldots, n \quad \text{and} \quad \Gamma_{n+1,n+1} = 1 \quad (3.76)$$

The eigenvalues and eigenvectors of $\mathbf{A}_0$ may be found numerically or analytically. With these results, we obtain a new equation of type (3.40). Near the intersection of $\tau_{n+1} = 2KS_h$ and one of the dispersion surfaces of the ZOLZ we have a twofold degenerate perturbation problem, since the minimum gap in the ZOLZ dispersion surfaces is about $|U_g|$, which is far larger than the perturbation, according to relation (3.74). The important interaction is that between the HOLZ reflection and the dispersion surface of the ZOLZ closest to the sphere of radius $K$ drawn about the origin. This follows because the smaller $\gamma$, the larger the excitation $c_i$, according to

$$c_i^* = C_0^i = \sum_{\mathbf{g}} U_{-\mathbf{g}} C_{\mathbf{g}}/2K\gamma^i$$

When $\gamma$ is close to zero, $C_g \sim 0$ and $C_0 \sim 1$. This can be stated as a useful rule of thumb: the branch of the dispersion surface which is nearest to the $K$-sphere drawn about the origin contributes most to the intensity. Assuming that the $i$th eigenvalue of the ZOLZ is closest to the

## THEORY

incident-beam $K$-sphere, the strongest HOLZ line is then described by

$$\begin{pmatrix} \tau_i & b_{i,n+1} \\ b_{n+1,i} & 2KS_h \end{pmatrix} \begin{pmatrix} \varepsilon_i \\ \varepsilon_{n+1} \end{pmatrix} = \lambda \begin{pmatrix} \varepsilon_i \\ \varepsilon_{n+1} \end{pmatrix} \quad (3.77)$$

where a small correction term $b_{ii}$ has been neglected. According to Eqs. (3.49) and (3.50), the solutions of Eq. (3.77) are

$$\lambda^{i,n+1} = (\tau_i + 2KS_h \pm \sqrt{(\tau_i - 2KS_h)^2 + 4|b_{i,n+1}|^2})/2 \quad (3.78)$$

and

$$\begin{aligned} \varepsilon_{ii} &= \cos(\beta/2)\exp(i\phi), & \varepsilon_{n+1,i} &= \sin(\beta/2) \\ \varepsilon_{i,n+1} &= -\sin(\beta/2), & \varepsilon_{n+1,n+1} &= \cos(\beta/2)\exp(-i\phi) \end{aligned} \quad (3.79)$$

where $\phi$ is the phase of $b_{i,n+1}$ and $\beta$ is defined by

$$\cot \beta = (\tau_i - 2KS_h)/(2|b_{i,n+1}|)$$

For the other terms, $\lambda_j = \tau_j$, $\varepsilon_{jj} = 1$, and $\varepsilon_{jk} = 0$ ($j \neq i, n+1, k \neq j$). Using $C = \Gamma \varepsilon$, we thus obtain

$$C_0^i = \varepsilon_{ii}\Gamma_0^i, \quad C_h^i = \varepsilon_{i,n+1} \quad \text{and} \quad C_0^{n+1} = \varepsilon_{i,n+1}\Gamma_0^i, \quad C_h^{n+1} = \varepsilon_{n+1,n+1} \quad (3.80)$$

and $C_h^j = 0$ for all other values of $j$. By employing the more familiar notation of $C$ and $\gamma$ to replace $\Gamma$ and $\tau$ for the eigenvectors and eigenvalues of the ZOLZ, we obtain from the above results and Eq. (3.27) (Buxton, 1976; Zuo, 1991b)

$$I_h = |C_0^i|^2 \frac{|U_h^{\text{eff}}|^2}{|U_h^{\text{eff}}|^2 + K^2(S_h - \gamma^i)} \sin^2 \frac{\pi t}{K} \sqrt{|U_h^{\text{eff}}|^2 + K^2(S_h - \gamma^i)} \quad (3.81)$$

where

$$U_h^{\text{eff}} = \sum_{k=1}^{n} C_k^{i*} U_{g_{k-1}-h} \quad (3.82)$$

One other case can be readily solved—where there are two HOLZ reflections interacting strongly with each other but all ZOLZ reflections are weak, except the incident beam. Then the three-beam results of the last section may be used. The situation becomes rather complicated when there are also many strong beams in the ZOLZ, but four-beam results have been given for this case (Zuo, 1991b).

The intensity expression (3.81) for a HOLZ line enables us to draw some conclusions about HOLZ-line positions. We expect the HOLZ intensity for beam **h** to follow the trajectory defined by Zuo (1991a):

$$S_h - \gamma^i = 0 \tag{3.83a}$$

This may simply be derived by using the condition for the *dynamical* Bragg condition for the strongest Bloch wave,

$$K^2 - (\mathbf{k} + \mathbf{h})^2 \approx K^2 - (\mathbf{K} + \mathbf{h})^2 - 2K\gamma$$
$$= 2K(S_h - \gamma) = 0 \tag{3.83b}$$

where **k** is the Bloch-wave wavevector AO in the ZOLZ in Fig. 3.10b. [The kinematic expression for the HOLZ line trajectory, Eq. (2.11), was based on the condition $\gamma = 0$, and should be compared with the above.] We now discuss the geometric interpretation which these results give to dynamical HOLZ-line shifts.

Figure 3.10b shows the scattering geometry for a HOLZ line **h** crossing the (000) ZOLZ disk at the Bragg condition for **h**. Because of the large number of wavevectors, we depart temporarily from the definitions of wavevectors given at the beginning of the book. The first-order reflection **g** in the ZOLZ is slightly off the Bragg condition. (The HOLZ line **h** and the first-order Bragg condition **g** would therefore lie at the extreme left-hand edges of the central and first-order CBED disks if the center of the axial disk corresponded to the zone axis in an experimental pattern.) The scale of the diagram has been distorted for clarity—on a diagram drawn to scale it is not possible to show all relevant detail. (Typical dimensions at 100 kV might actually be $|K| \approx 27$, $|g| \approx 0.5$, the distances AD $\approx 0.005$ and DG $\approx 0.01$, all in reciprocal angstroms, and scattering angles all less than 10°.) We shall use the expression "$K$-sphere" to mean a sphere of radius $|K|$, and "dynamical Bragg condition" to mean the dominant minimum in an experimental HOLZ-line pattern. The figure shows several sets of wavevectors. The boldest lines show the true dynamical wavevectors $\mathbf{K}_0$ and $\mathbf{K}_0 + \mathbf{h}$ outside the crystal at the dynamical Bragg condition. Here $\mathbf{K}_0$ makes an angle $\theta$ with the zone axis, which can be measured. This dynamical Bragg condition is defined by the intersection of a $K$-sphere about **h** with dispersion surface 1, which we assume to be dominant.

Continuous lines show Bloch-wave wavevectors excited inside the crystal. These originate on the dispersion surfaces, and (in two dimensions) their components $\mathbf{k}_{\text{perp}}^{(i)}$ in the direction normal to the surface normal **n** must equal that of $\mathbf{K}_0$. A second Bloch-wave wavevector $\mathbf{k}^{(2)}$

(not shown) might therefore originate at E. Although the various ZOLZ Bloch wavevectors $\mathbf{k}^{(i)}$ travel in different directions inside the crystal, since they have the same component $\mathbf{k}_{\text{perp}}^{(i)}$ normal to $\mathbf{n}$ they combine to form a single beam on leaving the crystal. For the HOLZ reflections, each beam remains split into its Bloch-wave components after leaving the crystal.

Short-dashed lines indicate kinematic wavevectors $\mathbf{K}'$, based on the true accelerating voltage. We assume initially that this true accelerating voltage is known, so that $\mathbf{K}'$ (inside the crystal) and $\mathbf{K}_0'$ (outside the crystal) can be determined using Eqs. (2.1) and (2.2). These wavevectors are then used solely as construction lines to provide a first approximation to the shape of the dynamical dispersion surfaces AB and EF, as described in Section 3.4. No such wavevectors are actually excited inside the crystal in these directions at the dynamical Bragg condition. Exploring the intensity along a line in the central disk (i.e., varying $\theta$) corresponds to moving the wavevectors which meet at A (the dynamical Bragg condition) to C and F, where a secondary minimum might occur due to branch 2 of the dispersion surfaces. The Ewald sphere, of radius $|K'|$, drawn about C, therefore passes through $\mathbf{h}$ at the Bragg condition. Then the projection of $\mathbf{K}'$ onto $\mathbf{h}$ must equal half the length of $\mathbf{h}$, as required by the simple Bragg law [Eq. (2.11)].

In order to find the true diffracted wavevectors ($\mathbf{K}_0 + \mathbf{h}$) leaving the crystal, we need first to find the allowed dynamical wavevectors $\mathbf{k}^{(i)}$ inside the crystal. This requires both a differential equation (the Schrödinger equation) and a boundary condition. Equation (3.16) provides solutions to the differential equation, giving the allowed Bloch-wave labeling wavevectors $\mathbf{k}^{(i)}$ inside the crystal. The boundary condition requires that the components of the incident and diffracted wavevectors in the plane of the surface $\mathbf{K}_{o,\text{perp}}$ and $\mathbf{k}_{\text{perp}}^{(1)}$ be equal. Figure 3.10b is an attempt to summarize these boundary conditions geometrically, while also imposing the energy and momentum conservation conditions. The figure tells us little about the intensities of the Bragg beams—we assume here that there is a minimum of intensity in the central disk whenever the $K$-sphere about $\mathbf{h}$ crosses a dispersion surface. This then defines an angle $\theta$ and hence a point in our one-dimensional central disk. (In fact, the branch of the dispersion surface which is nearest to the $K$-sphere drawn about the $\mathbf{g}$ vector of interest contributes most to the intensity in beam $\mathbf{g}$. That is, the scattering kinematics deviate as little as possible from that allowed in "vacuum.")

The effect of "swtiching on" the lattice potential is therefore to open up a gap DA near D, so that the true dispersion surfaces (on which all wavevectors indide the crystal must commence) becomes curved, as shown at AB and EF. If the HOLZ line is treated as a weak perturbation, then the Bragg point of interest moves from C to A. Weaker lines (fine structure) may also be seen in the outer ring of HOLZ disks at an incident-beam direction at which the $K$-sphere intersects branch 2 at F. This fine structure is shown experimentally in Fig. 3.11. In this sense, the fine structure in HOLZ disks can give a map of the ZOLZ dispersion surface. (It should be distinguished from the "splitting" fine structure on HOLZ lines which occurs due to strain in crystals. This occurs on a finer scale and is discussed in Section 8.5.)

The length of the diffracted wavevector $\mathbf{K}_0$ is fixed by Eq. (2.2); its direction is therefore also now fixed (in two dimensions) by the requirement that it originates on the surface normal as shown. There is therefore a small difference between the directions of the wavevectors $\mathbf{K}_0$ and $\mathbf{K}_0'$, which is the dynamical HOLZ-line shift. Changes in composition, causing changes in structure factors and hence in $\gamma$, will alter the shape of branch 1 and the distance DA. In two-beam theory this distance is just $1/\xi_g$. If the gap DA widens, for example, the figure shows that A moves to the left, and the angle between $\mathbf{K}_0$ and $\mathbf{K}_0'$ will increase. Thus the direction $\Theta$ in which the minimum of intensity for HOLZ line **h** occurs in the central disk depends on structure factor **g** for the ZOLZ. Use of the simple Bragg law [Eq. (2.4)] corresponds to the use of wavevectors $\mathbf{K}_0'$ rather than $\mathbf{K}_0$, but this, if based on the true accelerating voltage, would not agree with the measured angle at which the HOLZ minimum occurs.

Equation (3.83) gives a dynamically corrected HOLZ-line equation

$$K_y = -\frac{g_x}{g_y} K_x + \frac{g_z}{g_y}\left(K_z - \frac{K\gamma}{g_z}\right) - \frac{g^2}{2g_y} \qquad (3.84)$$

using the same notation as in Eq. (2.15). This dynamical expression differs from the kinematic expression [Eq. (2.15)] only by the term in $\gamma$. Over a small region of the dispersion surface, we might assume that the dispersion $\gamma$ is approximately constant. Then the effects of dynamical dispersion may be thought of as a correction to the accelerating voltage, and accommodated by a change in the term $K_z$ in Eq. (3.84). However, this correction to the high voltage differs from zone to zone because of the weighting $g_z$.

The strongest experimental evidence for HOLZ-line shifts comes from the work of Lin *et al.* (1987), who found variations of several kilovolts when using Eq. (2.4) to determine the microscope accelerating voltage from indexed HOLZ lines taken from different zone axes of the same silicon crystal. For example, they found $E_0 = 195.6\,\text{kV}$ at the [112] pole (based on IHL), but $E_0 = 198.4$ kV at the [356] pole. Using complete many-beam computations to match the line positions, the calculated values become 198.1 kV and 198.5 kV for the same axes. (The additional correction for $U_0$ has not been included. This is about 20 V for silicon.) The reason for these changes is that, in the zone-axis center, $\gamma$ is positive, thus the effective $K_z$ value is lower, and this can be simulated by lowering the high voltage. The value of $\gamma$ decreases as one moves away from the center of a high-symmetry zone axis.

In summary, the following conclusions may be drawn:

1. Dynamical effects will always result in the accelerating voltage being underestimated near a center of a zone axis, if Eq. (2.4) is used.
2. The dynamical correction is least at high-index zone axes, or may be greatly reduced by avoiding zone axes altogether, where $\gamma$ is small (Zuo, 1991a).
3. The dynamical correction increases with atomic number (for a similar projected density of atoms).
4. The correction will be least (other things being equal) in smaller unit cell crystals in which the reciprocal lattice is sparse.
5. Since branch 1 is relatively flat at the zone center, it has frequently been assumed that the correction is orientation-independent over the central disk.
6. The correction is least from the highest HOLZ layer. For example, it was found (Lin *et al.*, 1989) that, at the [113] pole of silicon, the voltage used to match the FOLZ was 0.6 kV lower that required to match the TOLZ, using Eq. (2.4).
7. Errors of several kilovolts are likely in measurements of accelerating voltage based on Eq. (2.4). If the perturbation correction given below is used, this error is reduced to perhaps 200 V. Automated refinement, using many profiles across the lines matched to dynamical calculations, can reduce the error to as little as 14 V (Zuo, 1991a).

An expression for the dynamical correction to the accelerating voltage has been given (Lin *et al.*, 1989; Zuo, 1991a) based on Bethe's perturbation method described in Section 3.4. This gives the correction (increase) to the "kinematic" voltage $E'_0$, obtained by applying Eq. (2.4) to a HOLZ pattern with known lattice spacings, as

$$\Delta E_0 = \frac{300 K^2 \gamma^{(1)}}{(1 + 1.956 \times 10^{-6} E_0) nH} \qquad (3.85)$$

where $nH$ is the height of the HOLZ layer in reciprocal angstroms, and $\gamma$ and $K$ are measured in similar units. The true microscope voltage is approximately $E_0' + \Delta E_0$. We may now use the results of Section 3.3 on the Bethe potential to approximate the value of $\gamma$ which is closest to the incident-beam $K$-sphere. Away from the Bragg condition, where $S_g$ is large, we may expand Eq. (3.7) to first order, so that

$$\gamma^{(1)} \approx S_0^{\text{eff}} = -\frac{1}{2K} \sum_{g \neq 0} \frac{|U_g|^2}{2KS_g} \tag{3.86a}$$

*Figure 3.11.* (a) HOLZ lines in the central disk of silicon at 100 kV. [111] orientation at 183 °C. (b) Kinematic calculation of the pattern in Fig. 3.11a for the voltage which gives the best fit (98.5 kV). The true voltage is 100.0 kV. (c) A section of the dispersion surfaces for Si [111] at 100 kV from the zone center to $K_t = (-2, -2, 4)/4$. (d) HOLZ line in outer ring (dark field) corresponding to Fig. 3.11a for the $(9, -9, 1)$ reflection, showing fine structure. The lines are labeled to correspond with the ZOLZ dispersion surfaces shown in Fig. 3.11c. (Figure 3.10b illustrates this effect.)

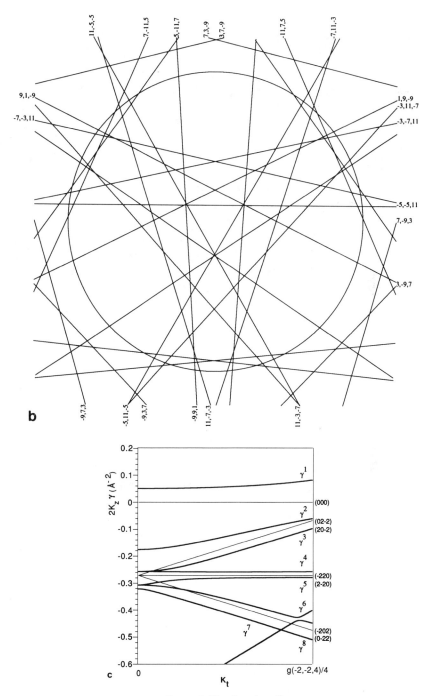

Figure 3.11. (continued)

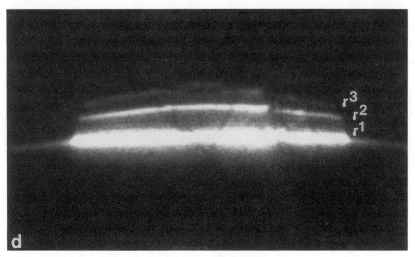

*Figure 3.11. (continued)*

Near the zone center we have $2KS_g = -g^2$, and thus

$$\gamma^{(1)} \approx \frac{1}{2K} \sum_{g \neq 0} \frac{|U_g|^2}{|g|^2} \tag{3.86b}$$

where the sum is over structure factors $U_g$ in the ZOLZ.

Figure 3.11a shows a silicon [111] zone-axis HOLZ pattern at 100 kV, recorded at a temperature of $-183\,°C$. Figure 3.11b shows a simulated HOLZ pattern based on the kinematic approximation of Eq. (2.15). The high voltage has been varied to match the experimental HOLZ pattern of Fig. 3.11a near the center of the disk. This kinematic matching gives a high voltage of 98.5 kV. Figure 3.11c shows the calculated dispersion surfaces of the Si [111] zone axis from $\mathbf{K}_t = 0$ to $\mathbf{K}_t = (-2, -2, 4)/4$, obtained using the Bloch-wave method described in Section 3.2. The bold lines are the dispersion surfaces, and the thin lines are the kinematical $K$-spheres. Figure 3.11d shows the fine structure in the CBED disk for the $(9, -9, 1)$ HOLZ reflection. This HOLZ reflection has a very weak interaction with the other HOLZ reflections, thus the fine structure shown is produced by the ZOLZ dispersion surfaces, in the manner discussed above. The fine structure in Fig. 3.11d is thus a direct image of the various branches of the dispersion surfaces shown in Fig. 3.11c, as indicated by the labeling on the figure.

From Fig. 3.11b, a measurement of the high voltage using the kinematic approximation yields the value 98.5 kV. This is 1.5 kV less than the actual value of 100 kV, obtained by an independent method. From Fig. 3.11c we see that $2K\gamma$ is about 0.05 Å$^{-2}$ near the [111] zone-axis center and, for the Si [111] zone axis, $g_z$ is about 1/9.4 Å$^{-1}$ for the FOLZ. At 100 kV, $K = 1/0.037$ Å$^{-1}$. From Eq. (3.84), the effective wavelength is therefore

$$\lambda^{\text{eff}} = 1 \bigg/ \left( K_z - \frac{K\gamma}{g_z} \right) = 0.037325 \text{ Å}$$

The wavelength at 98.5 kV is 0.03732 Å. This is therefore in good agreement with the predicted value of the effective wavelength.

A different type of fine structure has also been observed on the satellite reflections produced by semiconductor multilayers (Gong and Schapink, 1991). This is discussed in Section 8.5.

## 3.9. ABSORPTION AND ITS EFFECTS

In this section we will see how the effects of inelastic scattering influence the elastic scattering which is detected in electron microdiffraction experiments. The theoretical discussion in this section therefore refers to experiments in which an ideal energy filter is used which excludes all inelastic scattering from the microdiffraction pattern. We will see that inelastic scattering causes two main effects: first, an asymmetry is introduced into the zero-order disk around the Bragg condition, and second, differences arise between the intensities of the **g** and −**g** disks at the Bragg condition in noncentrosymmetric crystals. [This has corresponding effects on energy loss spectra (Taftø, 1987).] To understand these effects, we will use the results of the perturbation theory given in Section 3.5.

We have introduced the concept of an optical potential $V(\mathbf{r}) = V^c(\mathbf{r}) + iV'(\mathbf{r})$ in Section 3.2 [Eq. (3.17)], with $V^c(\mathbf{r})$ the real crystal potential (describing the interaction of the incident electron with the crystal electrons and the nuclei) and $V'(\mathbf{r})$ a second real potential which accounts for depletion of the elastic wavefield by inelastic scattering. The term "absorption" is used to describe this depletion, since the probability that the electron is scattered back into the original state is very small. The use of such a phenomenological absorption potential in HEED has been justified theoretically (Yoshioka, 1957). For high-energy electrons, there are three important inelastic scattering mechanisms: (1) inelastic scattering resulting from the excitation of crystal electrons, (2) excitation of plasmons, and (3) excitation of phonons. The plasmon scattering contributes to the mean absorption ($V'_0$) only, and the contributions of electron scattering decrease rapidly as the scattering angle increases. The contribution to $V'_g$ ($g \neq 0$) comes mostly from phonon scattering (Hall and Hirsch, 1965; Radi, 1970; Yoshioka and Kainuma, 1962). The mean absorption describes an overall attenuation of the incident electrons. Figure 3.12 shows calculated absorption coefficients for copper according to Humphreys and Hirsch (1968). The dotted lines show the phonon scattering contribution to the absorption. The total absorption is seen to deviate only slightly from the phonon scattering at very low angles. These absorption coefficients have been calculated in the past for a number of crystals by several authors [see Reimer (1984) for a review]. As discussed in Section 3.1, the mean atomic vibration amplitude $\langle u_i^2 \rangle$ is related to the Debye–Waller factor $\exp(-Bs^2) = \exp(-Bg^2/4)$ by

$$B_i = 8\pi^2 \langle u_i^2 \rangle \tag{3.87a}$$

In expressions for the intensity of a diffracted beam, the factor $\exp(-2Bs^2)$ appears.

The absorption coefficients for phonon scattering may be calculated analytically most simply using an Einstein model. This model assumes that the atoms in the crystal vibrate independently of each other, in

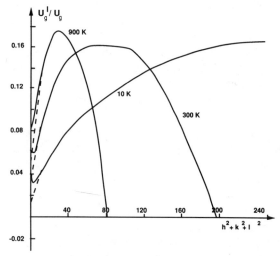

*Figure 3.12.* The absorption potential for copper at three temperatures as a function of $h^2 + k^2 + l^2$ for 100-kV electrons. Dotted curve is contribution from phonon scattering alone, continuous line is phonon plus single electron excitation. Except at very small scattering angles, phonon scattering dominates (Humphreys and Hirsch, 1968).

contrast to more accurate models based on discrete vibrational modes. Thus the thermal diffuse scattering calculated using the Einstein model cannot predict the thermal diffuse streaks which are often observed in experimental patterns and which can be related to the phonon dispersion (Kitamura, 1966). A more realistic model is the Debye model. However, it has also been shown (Hall and Hirsch, 1965) that for gold, at least, the Debye model and Einstein model give similar absorption coefficients. It is not clear if the same conclusion holds for more complex crystals, however the incorporation of lattice vibration models greatly complicates the theory. A parametric fit for the absorption coefficients [such as $V'_g/V_g = (A + Bg)g$] has also been proposed, and values of $A$ and $B$ have been given for some crystals (Voss et al., 1980). In summary, calculations based on the Einstein model, using the Debye–Waller factor as input, appear to be a useful first approximation and act as a basis for further refinement. The Debye–Waller factor can be found from experimental X-ray or neutron diffraction results, from theoretical calculations (Sears and Shelley, 1991), or from the values listed in the *International Tables for Crystallography*. The simplest procedure is then to treat the absorption coefficients as adjustable parameters in a refinement, on an equal footing with the other structure factors, and sample thickness, as described in Section 4.2.

Calculations for total absorption coefficients, including plasmon excitation, single electron excitation, and phonon scattering, are given in Radi (1970) and Humphreys and Hirsch (1968) for a range of crystal structures. Otherwise, measurements and calculations for particular crystals are scattered throughout the literature [see Reimer (1984) for a summary]. A comparison of measured and experimental values for Al, Cu, Au, Si, Ge, MgO, and NaCl is given by Weickenmeier and Kohl (1991), in which a Fortran program is offered on request. A Fortran program for computing the phonon scattering contribution to the absorption coefficients is also described in Bird and King (1990), together with tabulated values for Al, Cu, Ag, Au, C, and Ga. It is commonly assumed that other electronic processes make a significant contribution only for $\mathbf{g} = 0$. These programs can provide values of the absorption coefficients $V'_\mathbf{g}$, and hence $U'_\mathbf{g}$ through Eq. (3.6) (with primes added to $V_\mathbf{g}$ and $U_\mathbf{g}$). They require the Debye–Waller factor for the crystal of interest, and are based on an Einstein model.

Equations (3.2b), (3.3), and (3.4) yield the Fourier coefficients of the elastic portion of the optical potential for a crystal (in volts):

$$V_g = \frac{h^2}{2\pi\gamma m_0 e\Omega} \sum_i f_i^B(g) \exp(-2\pi i\mathbf{g} \cdot \mathbf{r}_i) \exp(-B_i g^2/4) \qquad (3.87b)$$

where $B_i$ is the Debye–Waller factor as defined following Eq. (3.4). An atomic absorptive form factor $f'_i(g)$ can be defined similarly by

$$V'_g = \frac{h^2}{2\pi\gamma m_0 e\Omega} \sum_i f'_i(g) \exp(-2\pi i \mathbf{g} \cdot \mathbf{r}_i) \exp(-B_i g^2/4) \quad (3.87c)$$

where $V'_g$ is the imaginary part of the Fourier coefficient of the optical potential. The "atomic" absorption coefficient $f'_i(g)$ is then given by Hall and Hirsch (1965):

$$f'(g) = \frac{1}{K} \int f^B(\mathbf{q}) f^B(\mathbf{q} - \mathbf{g})[\exp(-B g^2/4) - \exp\{-B[q^2 - (\mathbf{q} - \mathbf{g})^2]/4\}] d^2\mathbf{q}$$

$$(3.87d)$$

The integral may be evaluated on a grid of $B$ and $g$ values (Bird and King, 1990), or analytically if the atomic scattering factor is first expanded as a sum of Gaussians (Weickenmeier and Kohl, 1991).

Absorption has two effects on all crystals in electron diffraction. The first is the average absorption, which gives rise to a mean complex wavevector

$$K_z = K_{0z} + U_0/2K_{0z} + iU'_0/2K_{0z} \quad (3.88)$$

The imaginary part yields a damping term

$$\exp[-4\pi(2meV'_0/h^2)t/2K_{0z}] = \exp(-t/\lambda) \quad (3.89)$$

which multiplies all the diffracted beam intensities, where

$$\lambda = h^2 K_{0z}/(4\pi m e V'_0) \quad (3.90)$$

is the equivalent mean free path of the incident electron.

A second effect of absorption has become known as the anomalous transmission effect, by analogy with the Borrmann effect in X-ray diffraction. This is due to the term $V'_g$. The effect in electron diffraction was first observed by Honjo and Mihama (1954), who used a wedged-shaped crystal to produce spot spitting (doublets) in diffraction patterns [similar to Lehmpfuhl and Reissland (1968)], so that each spot of the doublet can be related to one of the dispersion surfaces. They then detected a difference in intensity between the two spots. In CBED, this anomalous transmission effect is revealed by the intensity asymmetry around the Bragg condition in the transmitted (zero-order) disk, instead of the symmetry which we expect from the two-beam intensity expression of Eq. (3.31). The two-beam theory has been applied to electron diffraction with absorption, and comparisons made with experiment in the work of Hashimoto et al. (1962). In following, we shall apply the perturbation theory presented in Section 3.5 to the problem of absorption in the two-beam case.

In high-energy electron diffraction, the absorption potential is much smaller than the crystal potential, typically less than one-tenth of the crystal potential. Thus, absorption may be treated as a perturbation. This perturbation is nondegenerate. Therefore, from Section 3.5,

$$\mathbf{A}_0 = \begin{pmatrix} 0 & U_{-g} \\ U_g & 2KS_g \end{pmatrix} \quad \text{and} \quad \mathbf{A}' = \begin{pmatrix} 0 & iU'_{-g} \\ iU'_g & 0 \end{pmatrix} \quad (3.91)$$

The solutions to $\mathbf{A}_0$ are given in Eqs. (3.29) and (3.30). Equation (3.42) yields

$$b_{11} = \sum_{km} \Gamma^*_{k1} A'_{km} \Gamma_{m1}$$
$$= \Gamma^*_{11} A'_{12} \Gamma_{21} + \Gamma^*_{21} A'_{21} \Gamma_{11}$$
$$= \cos(\beta/2) \exp(i\varphi_g) iU'_{-g} \sin(\beta/2) + \sin(\beta/2) iU'_g \cos(\beta/2) \exp(i\varphi_g)$$
$$= i |U'_g| \sin \beta \cos(\varphi_g - \varphi'_g) \quad (3.92)$$

and similarly

$$b_{22} = -i |U'_g| \sin \beta \cos(\varphi_g - \varphi'_g)$$

and

$$b_{12} = -b^*_{21} = i |U'_g| \{\cos^2(\beta/2) \exp[i(\varphi_g - \varphi'_g)] - \sin^2(\beta/2) \exp[i(\varphi_g - \varphi'_g)]\} \exp(i\varphi_g) \quad (3.93)$$

Here $\varphi_g$ and $\varphi'_g$ are the phases of $U_g$ and $U'_g$.

For centrosymmetric crystals, $\varphi_g = \varphi'_g = 0$ or $\pi$, thus

$$b_{11} = -b_{22} = i |U'_g| \sin \beta = \frac{i |U'_g|}{\sqrt{1 + \omega^2}} \quad \text{and}$$

$$b_{12} = -b^*_{21} = i |U'_g| \cos \beta = \frac{iU'_g \omega}{\sqrt{1 + \omega^2}} \quad (3.94)$$

The terms $b_{11}$ and $b_{22}$ are thus the first-order corrections to the eigenvalues 1 and 2 of Eq. (3.29), while $b_{12}/(\tau_1 - \tau_2)$ and $b_{21}/(\tau_2 - \tau_1)$ are the corrections to the eigenvectors resulting from the introduction of an absorption potential. Near the Bragg condition $\omega$ is small, so that the corrections to the eigenvectors may be neglected. Then the amplitudes for two-beam diffraction with absorption are given by

$$\phi_0(t) = \cos^2(\beta/2)\exp(-iXt) + \sin^2(\beta/2)\exp(iXt)$$
$$\phi_g(t) = -\cos(\beta/2)\sin(\beta/2)\{\exp(-iXt) - \exp(iXt)\} \quad (3.95)$$

where

$$X = \frac{\pi |U_g|}{K}\sqrt{1+\omega^2} + i\frac{\pi |U'_g|}{K}\frac{1}{\sqrt{1+\omega^2}}$$

The mean absorption effect has been neglected here. Rocking curves for **0** and **g** based on Eq. (3.95) with $|U'_g|/|U_g| = 0$ and 0.1 are plotted in Fig. 3.13. It is seen from this figure that the absorption potential causes an intensity asymmetry around the Bragg angle in the rocking curve of the (0, 0, 0) reflection, the intensity at $-S_g$ being lower than the intensity at $+S_g$. This asymmetry arises because Bloch wave 1 is strongly excited on the side where $S_g$ is less than zero, while Bloch wave 2 is strongly excited on the side where $S_g$ is larger than zero. According to Eq. (3.94) Bloch wave 1 is strongly absorbed, while Bloch wave 2 is less absorbed. As discussed in many texts [such as work by Hirsch et al., 1977)], this is attributed to the fact that, in real space, Bloch wave 1 has maxima of intensity located on the atomic sites (resulting in a higher probability of inelastic scattering by localized processes), while Bloch wave 2 has maxima which fall between the atoms. (We are here making the

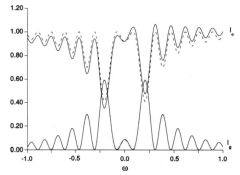

Figure 3.13. Intensity of beams $I_0$ and $I_g$ according to two-beam theory, with absorption, as a function of deviation parameter $\omega = s_g \xi_g$ around the Bragg condition $\omega = 0$. The thickness is $t = 1.9\ \xi_g$. The full lines show $U'_g/U_g = 0.1$ and $U_0 = 0$, while the dashed lines show $U'_g/U_g = 0$ and $U'_0 = 0$.

independent Bloch-wave approximation. The total wavefield in the crystal actually contains contributions from both Bloch waves.)

For noncentrosymmetric crystals, $\varphi_g \neq \varphi'_g$ in general, and the corrections to the eigenvectors cannot be neglected. In the kinematic approximation,

$$I_g \propto |U_g + iU'_g|^2 = |U_g|^2 + |U'_g|^2 + 2|U_g||U'_g|\sin(\varphi_g - \varphi'_g) \quad (3.96)$$

Because of the sine function, there is a difference between the intensities of the **g** and −**g** reflections; this difference is

$$I = I_g - I_{-g} \propto 4|U_g||U'_g|\sin(\varphi_g - \varphi'_g) \quad (3.97)$$

In X-ray diffraction this effect is called *anomalous dispersion* (Karle, 1989). Its effects in electron diffraction have been discussed recently by Bird (1990) and Taftø (1983). To observe this effect of absorption in acentric crystals, we again use two-beam theory to provide an example. From Section 3.5, the correction to the eigenvalues for an acentric crystal is seen to have the same effect as in a centric crystal, but with a new effective absorption potential $|U'_g|\cos(\varphi_g - \varphi'_g)$. Thus in the following we may concentrate on the effects of the correction to the eigenvector. To simplify matters, we consider the intensity of reflection **g** at the Bragg condition, where $\cos^2(\beta/2) = \sin^2(\beta/2) = 0.5$ and $\tau_1 - \tau_2 = 2|U_g|$. Then

$$\varepsilon_{12} = \varepsilon^*_{21} = -\frac{|U'_g|}{2|U_g|}\sin(\varphi_g - \varphi'_g)\exp(i\varphi_g) \quad (3.98)$$

The perturbed eigenvector matrix may now be written in the form

$$\mathbf{C} = \Gamma(1 + \Delta)$$

where $\Delta$ is a small off-diagonal matrix with $\Delta_{11} = \Delta_{22} = 0$, $\Delta_{12} = \varepsilon_{12}$, and $\Delta_{21} = \varepsilon_{21}$. The inverse of matrix **C** is approximately

$$\mathbf{C}^{-1} \approx (1 - \Delta)\Gamma^+$$

With these results, and results of the two-beam solutions from Section 3.4, we find that

$$\Delta \phi_g(t, S_g = 0) = i\frac{|U'_g|}{|U_g|}\sin(\varphi_g - \varphi'_g)\exp(i\varphi_g)\sin\left(\frac{\pi|U_g|t}{K}\right) \quad (3.99)$$

due to the correction in the eigenvectors. If the effect of absorption in the wave amplitude $\phi_g$ is neglected, then

$$\phi_g(t, S_g = 0) = i \exp(i\varphi_g) \sin\left(\frac{\pi |U_g| t}{K}\right) \quad (3.100)$$

Thus the intensity at the Bragg condition is approximately

$$I_g(t, S_g = 0) \approx \left[1 + 2\frac{|U_g'|}{|U_g|}\sin(\varphi_g - \varphi_g')\right]\sin^2\left(\frac{\pi |U_g| t}{K}\right) \quad (3.101)$$

From this equation, we would expect the same intensity asymmetry between $I_\mathbf{g}$ and $I_{-\mathbf{g}}$ at the Bragg condition as predicted by the kinematic approximation. The intensity away from the Bragg condition is given by rather complicated expressions.

In summary, the effects of the inelastic scattering of electrons on elastic scattering may be described by a complex optical potential composed (in real space) of a real crystal potential and an imaginary absorption potential. In the two-beam case, this absorption potential causes two anomalous dispersion effects. One is the difference between the intensities of the central beam at orientations $+S_g$ and $-S_g$. The other is an intensity difference between the **g** and $-\mathbf{g}$ reflections at their respective Bragg conditions in acentric crystals.

Absorption coefficients may be measured using the intensity distribution in the zero-order disk of a CBED pattern at the Bragg condition. The phases of the Fourier coefficients of the absorption potential may be measured by comparing the intensities of the **g** and $-\mathbf{g}$ reflections near the Bragg condition. An example of these measurements is given in Section 4.2, and Section 5.1 describes in more detail the measurement of the real and imaginary parts of the Fourier coefficients of the optical potential.

## 3.10. CHANNELING, BOUND STATES, AND ATOMIC STRINGS

Channeling is the tendency of a beam of high-energy charged particles to run along the paths of lowest potential energy in a crystal. It may be of two types. If the beam is aligned with a major zone axis it is known as axial or two-dimensional chaneling. For a beam running normal to a systematics line of reflections it is known as planar or one-dimensional channeling. For positively charged particles this path runs along the empty tunnels between the atom strings. For electrons it runs along the nuclei. Channeling results in an increase in the range of positively charged particles, which channeling theories seek to explain. For electrons and positrons it is a quantum-mechanical phenomena, while for ion beams it may be described using classical concepts.

For electron beams in the kilovolt energy range, the experimental geometry and the theory of particle channeling thus become very similar to those of CBED. The channeling literature, however, is usually concerned with rather thick samples by the standards of CBED work, in which absorption and inelastic scattering effects dominate. Channeling patterns are therefore usually Kikuchi patterns from very thick crystals, and so require a complete analysis of dynamical thermal diffuse scattering for their study (Gjønnes and Taftø, 1976). Since the main theoretical approach used to analyze the elastic portion of electron channeling has also proven useful in analyzing CBED patterns (Buxton and Tremewan, 1980; Vincent *et al.*, 1984), we give it here briefly in order to relate it to the previous many-beam treatment.

When a collimated, high-energy beam of charged particles strikes a crystal, a very large number of elastic and inelastic processes result. These may include the ionization of atoms, the excitation of valence band electrons, Rutherford scattering, Bragg diffraction, nuclear reactions, X-ray emission, Auger electron production, and phonon excitation, among other processes. For an amorphous (glassy) target we do not expect that the rates of these processes should depend on the direction of the incident beam with respect to the target. The finding that they do for crystalline targets is also known as channeling. Thus channeling may also be studied by measuring the intensity of these secondary emissions as a function of the incident-beam direction. From such studies it is possible to determine the crystallographic site of dopant atoms [see Spence (1992b) for a review].

The elastic contribution to electron channeling patterns is normally obtained by expressing the solution to Eq. (3.14) in the form (Howie, 1966)

$$\Psi(\mathbf{r}) = \sum_j \alpha^{(j)} \exp(2\pi i k_z^{(j)} z) B^{(j)}(x, y) \qquad (3.102)$$

This separates the energetic forward free-particle motion of the electron (now described by a plane wave) from its transverse motion, described by the lateral eigenfunction $B^{(j)}(x, y)$. This separation is valid if all the important reciprocal lattice vectors lie in a plane normal to the beam, that is, if the variation of crystal potential in the beam direction can be neglected and the "projection" or ZOLZ approximation made. HOLZ effects are ignored. The Bloch-wave excitation amplitude $\alpha^{(j)}$ is determined by matching the wave in the crystal to the incident wave at the boundary. From Eq. (3.14) we then obtain

$$\frac{-h^2}{8\pi^2 m}\left[\frac{\partial^2 B^{(j)}}{\partial x^2}+\frac{\partial^2 B^{(j)}}{\partial y^2}\right]-|e|V_p(x,y)B^{(j)}=\frac{h^2}{2m}[K_0^2-\mathbf{k}_z^{(j)2}]B^{(j)} \quad (3.103)$$

where $V_p(x, y)$ is the (positive) crystal potential (in volts) averaged in the $z$ direction. This averaging is known as the continuum model in channeling theory, or the projection or ZOLZ approximation in electron diffraction. The eigenvalues $\varepsilon^{(j)}$ corresponding to the transverse eigenstates $B^{(j)}$ are thus

$$\varepsilon^{(j)}(\mathbf{K}_t) = \frac{h^2}{2m}(K_0^2 - k_z^{(j)2})$$

$$\approx -\frac{h^2 K}{m}(k_z^{(j)} - K) - |e|V_0$$

$$\approx -\frac{h^2 K}{m}\left(\gamma^{(j)} - \frac{K_t^2}{2K}\right) - |e|V_0 \quad (3.104)$$

Here $K_t = (K^2 - K_z^2)^{1/2}$ specifies the incident-beam direction. In the axial orientation $K_t = 0$. In addition, in the projection approximation

$$\gamma^{(j)}(\mathbf{K}_t) = \mathbf{k}_z^{(j)} - \mathbf{K}_z \quad (3.105)$$

for the case where the surface normal is parallel to the zone axis, which is taken to be the $z$ axis.

Since $\varepsilon^{(j)}$ is an eigenvalue, with the units of energy, we may imagine that the transverse motion of the electron is described by bound (or free) states $B^{(j)}$ within the (negative) transverse crystal potential energy well given by $-|e|V_p(x, y)$. For bound states $\varepsilon^{(j)}$ is also negative, while for free states it lies above the maximum value of the crystal potential energy. In order to establish a relation with the dispersion surfaces of dynamical electron diffraction theory, we note that these are given by the curves $k_z^{(j)}(K_t)$, with the uppermost dispersion surface corresponding to the deepest bound state.

# 4

# The Measurement of Low-Order Structure Factors and Thickness

In this chapter we discuss the accurate measurement of structure-factor amplitudes and specimen thickness by CBED. We assume that a simple, inorganic, small unit-cell crystal of known structure is to be analyzed. Then the refinement of a small number of low-order "bonding" reflections will allow a detailed three-dimensional map of the crystal charge density to be made. These low-order structure factors are refined together with other diffraction parameters such as the specimen thickness, the incident-beam directions, and the absorption coefficients. In general, these parameters are treated on an equal footing with the structure factors. The refined structure factors include the effect of Debye–Waller factors, so they are temperature dependent (see Section 3.1). (The refinement of lattice parameters is treated in Sections 5.4 and 3.6.) Earlier reviews of the subject can be found in Cowley (1967, 1969), Goodman (1978), and Spence and Carpenter (1986). A recent review of structure factor measurement by electron diffraction can be found in Spence (1992c). Sections 4.5 and 4.6 discuss the significance of the measured charge density and of the zero-order Fourier coefficient of potential.

A considerable amount of work has been conducted on the analysis of crystal structures using "point" or "ring" electron diffraction patterns, based mainly on the kinematic theory (with some dynamical corrections). Although outside the scope of a book on microdiffraction, we mention it here briefly, since many of the methods used in this work may also have application to the analysis of HOLZ lines in sparse zones (see Sections 6.2 and 3.8). This work has been mainly concerned with clay minerals (Zvyagin, 1967), organic films, polycrystalline samples, and the detection of hydrogen (Vainshtein, 1964b). Recent work based on the dynamical theory can be found in Avilov *et al.* (1989), devoted to the analysis of crystal of LiF, PbSe, BiOCl, PbTe, and $Bi_2Se_3$. All this work has been reviewed in English in Cowley *et al.* (1992).

This chapter concerns mainly many-beam computational methods for structure-factor refinement, but they are introduced with a description of the two-beam method for thickness determination, since this exposes the important principles of any CBED refinement. The critical voltage effect, being a three-beam effect, is treated in the next chapter, together with the accurate measurement of structure-factor phases, which also requires three-beam interactions. Thus the arrangement of these topics is somewhat arbitrary and has been made according to the complexity of the underlying theory. For materials scientists, the most useful section will be 4.1 on sample-thickness measurement, which includes a worked example.

## 4.1. TWO-BEAM METHODS FOR THICKNESS DETERMINATION

Sample thickness may be determined by a variety of methods in TEM, including the use of the projected width of inclined stacking fault images and plasmon loss spectra (Egerton, 1986). A cleaved sample provides the most accurate thickness calibration, and 90° wedges of MgO and GaAs have frequently been used to assist the quantitative interpretation of lattice images. The popular CBED method (Kelly *et al.*, 1975) is based on Eq. (3.31), and we apply it here to the pattern from MgO shown in Fig. 4.1 as an exercise. This shows the (220) and central CBED disks recorded in the systematics orientation at 120 kV. We write Eq. (3.33) in the form

$$\left(\frac{S_i}{n_i}\right)^2 + \left(\frac{1}{n_i}\right)^2 \left(\frac{1}{\xi_g}\right)^2 = \frac{1}{t^2} \tag{4.1}$$

with $S_i$ the excitation error at the $i$th minimum and $t$ the effective specimen thickness along the beam direction. A plot of $(S_i/n_i)^2$ against $(1/n_i)^2$ therefore gives $(1/t)^2$ as the intercept, and hence the thickness. The slope gives $(1/\xi_g)^2$, and hence $\xi_g$, from which $|U_g|$ is obtained. To obtain the values of $S_i$ at the minima, the pattern must be calibrated (i.e., the beam direction must be found at each point) using Eq. (2.10). Since the rocking curve is symmetrical (see Chapter 7) the required values of $\Delta X$ were obtained by taking half the distance between corresponding minima on either side of the Bragg condition, which itself happens to occur at a minimum in this case, as shown at A and B in Fig. 3.1. The distance $X$ was obtained by measuring the distance between the edges of the CBED disks corresponding to twice the Bragg angle. A table of

# MEASUREMENT OF LOW-ORDER STRUCTURE FACTORS AND THICKNESS

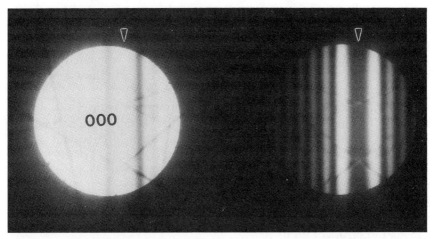

*Figure 4.1.* MgO (220) systematics at 120 kV. The arrows indicate the Bragg conditions. The sample thickness is determined from measurements of the excitation error at the positions of the fringe intensity minima.

values yielded the data plotted in Figure 4.2. We note that $i = 1$ corresponds to the first minimum either inside or outside the Bragg condition. However, an important problem arises of how to determine the first value of $n$ ($i = 1$). For a sample thickness $t < \xi_g$, we require $n_1 = 1$; for $\xi_g < t < 2\xi_g$, $n_1 = 2$, and so on. Thus $n_1$ is the first integer larger than $t/\xi_g$. In practice, one tries several initial values until a good straight-line fit is obtained for a reasonable value of $\xi_g$. As shown, a reasonable straight line was found for $n_1 = 1$ whose intercept and slope give

$$t = 104.4 \pm 1 \text{ nm} \quad \text{and} \quad \xi_g = 74.9 \pm 1 \text{ nm} \quad \text{(at 120 kV)}$$

There are a number of difficulties with this method; however, for

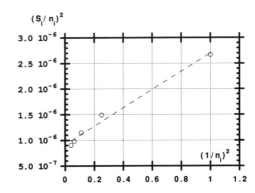

*Figure 4.2.* Graph of $(S_i/n_i)^2$ against $(1/n_i)^2$ for thickness determination, obtained from Fig. 4.1.

small unit-cell crystals an accuracy of 2% in thickness has been claimed. In fact, it will be seen in the next section that we have just performed the most elementary form of structure-factor refinement by CBED. This measured value for the extinction distance may be compared with the two-beam value obtainable from the scattering factors in the *International Tables for Crystallography,* which give 69 nm at 120 kV.

This method is frequently used in conjunction with X-ray microanalysis. The most important limitation of the method results from the excitation of other beams, which become increasingly important as the reciprocal lattice becomes more dense. Obviously the fine HOLZ lines in the ZOLZ disk in Fig. 4.1 cannot be reproduced, but it is important also to appreciate that the excitation of other orders in the ZOLZ systematic row can produce subtle changes in the intensity distribution of the disk at the Bragg condition, including shifts in the positions of the minima. This can lead to errors of up to 10%.

Several authors (Allen, Castro-Fernandez, Glazer, Blake, etc.) have analyzed the method, mainly to emphasize the effects of neglecting absorption or many-beam effects [for references to all this work and a review, see Ecob (1986)]. Many-beam effects are minimized by using a large reciprocal lattice vector and a small unit cell. The problem of finding $n_1$ can also be resolved by conducting thickness measurements at several thicknesses, and by ensuring that the value of $\xi_g$ obtained is independent of thickness. Greater accuracy can also be obtained by using the maxima [from Eq. (3.34)] in addition to the minima.

If highly accurate values of thickness are required, it is recommended that the methods described in the next two sections be used. Here it is shown that one can determine thicknesses to within a few angstroms. To obtain greater accuracy, it is necessary to observe the forbidden termination reflections which occur in crystals containing incomplete unit cells at their surface.

## 4.2. AUTOMATED DYNAMICAL LEAST-SQUARES REFINEMENT METHODS

An obvious improvement over the preceding analysis can be obtained by fitting many-beam computations based on Eqs. (3.22) and (3.24) to the CBED disks in order to measure structure factors or other crystallographic parameters. This has been attempted by many researchers over the past twenty years [a review of early work can be found in Cowley (1969)]. Typical of the early work comparing dynamical calculations with CBED experiments are the papers of Goodman and Lehmpfuhl (1967) on MgO and, more recently, Voss et al. (1980) on silicon. See also Cowley (1967), Gjønnes et al. (1988), Goodman (1978), Høier et al. (1988), Ishizuka and Taftø (1984), Kreutle and Meyer-Ehmson (1969), Marthinsen and Høier (1988), Shishido and Tanaka (1976), Smart and Humphreys (1978), Smith and Lehmpfuhl (1975), Steeds and Mansfield (1984), Terasaki et al. (1979), and Zuo et al. (1988).

A brute-force computational approach will be extremely inefficient [but see Bird (1992)], unless the two- and three-beam results given previously are used as a guide to find which regions of the pattern are most sensitive to changes in the structure factors. This can be done, for example, by differentiating the two-beam result with respect to each of the parameters $t$, $U_g$, $\lambda$, and $S_g$. Thus we have seen that the best thickness is the steepest part of the thickness fringe, the best orientation for refinement of $U_g$ is the Bragg condition for reflection **g** (see Fig. 3.1), while the three-beam degeneracy point gives greatest sensitivity for phase measurement (see Section 5.1 for acentric crystals). A general fitting program, such as "Refine/CB" (described below), will treat both centric and acentric crystals, and an example of its use for acentric crystals is given in Section 5.1. For a given crystal, the adjustable parameters include $t$, $U_g$, and choice of orientation. We assume initially that the low-order structure factors of a small unit-cell crystal are to be refined, and that the Debye–Waller factors and atomic coordinates are known accurately. It will be seen, however, that the automated refinement technique to be described can readily be generalized to refine other parameters, such as atomic coordinates, strain, and Debye–Waller factors.

First, let us consider in more detail the experimental conditions which provide the greatest sensitivity to changes in structure factors. In Eq. (3.31) the position of minima well away from the Bragg condition ($S_g$ large) are more sensitive to changes in thickness than to changes in structure factor, due to the factor of $n^2$ in Eq. (3.33). For example, when $n = 3$, a 0.1% change in $t$ is equivalent to a 1% change in $|U_g|$. The position of these minima are therefore commonly used to determine sample thickness, as described in Section 4.1. Structure factors, however, are best determined from the ratios of the intensity maxima B/A shown in Fig. 3.1, at the Bragg condition for the reflection of interest and at one of the optimum thicknesses $t_n$. In practice the value of $t_n$ will be chosen to maximize the ratio of elastic to inelastic scattering. For $S_g = 0$, from Eq. (3.31), $I_g$ is seen to be equally sensitive to $t$ and $U_g$, so that the optimum thicknesses $t_n = \xi_g(n + 1/4) = (n + 1/4)K/U_g$ should be used, where the slope of the thickness fringes is greatest. Thus, due to the sine function in this equation, $I_g$ can be very sensitive to changes in structure factor. For example, when $t|U_g|/K = 5/4$, a 1% change in $U_g$ results in a 6% change in the B/A ratio.

We have seen (in Fig. 2.3) that it is possible to obtain two Bragg conditions in a single systematics CBED recording, and more in two dimensions (Fig. 2.4). For accurate work, zero-loss energy-filtered data should be collected along the entire systematics line, in addition to photographic recordings which will be needed to determine the approximate beam direction. This process must be repeated with each order to be refined successively at the Bragg condition. The accelerating voltage can be found using dynamical simulations for the HOLZ-line patterns in a reference crystal such as silicon. (The gun bias and other microscope settings should not then be altered before inserting the crystal of interest.) Similar dynamical simulations of the HOLZ lines (assuming a reasonable value for the ionicities) will accurately calibrate the three-dimensional beam direction. Absorption parameters are most accurately refined using the variation of intensity within the central disk. The outer fringes of a CBED disk can be used to refine thickness. Figure 4.3 shows experimental data from MgO systematics recorded both on film, and as recorded using a photomultiplier and serial energy-loss filter tuned to the elastic peak. Figure 4.4 compares these results with dynamical calculations. As shown in Fig. 9.2, energy filtering is essential for accurate results.

Since different parts of the rocking curve show different sensitivities to different parameters, the entire CBED refinement process is best handled by minimizing a weighted "goodness of fit" index $\chi$ (Marthinsen et al., 1990; Zuo and Spence, 1991), as in the Reitvelt method used in neutron diffraction. The refinement is thus accomplished by finding a minimum in $\chi^2$ as a function of the adjustable parameters. We define

$$\chi^2 = \sum_i \frac{f_i \cdot (cI_i^{\text{theory}} - I_i^{\text{exp}})^2}{\sigma_i^2} \qquad (4.2)$$

The experimental CBED intensities are given by $I_i^{\text{exp}}$ and the calculated points [from Eq. (3.27)] by $I_i^{\text{theory}}$. The $f_i$ is a weight coefficient, which can be adjusted to increase the importance of certain contributions to $\chi^2$ which are sensitive to particular parameters. Here $\sigma_i^2$ is the variance of the $i$th point, which can be measured from successive experiments or by using $\sigma_i^2 = I_i^{\text{exp}}$, assuming Poisson statistics. Further, $c$ is a normalization coefficient, which can be found by either normalizing the theory and experiment at a particular point, or by taking the first-order derivative

# MEASUREMENT OF LOW-ORDER STRUCTURE FACTORS AND THICKNESS

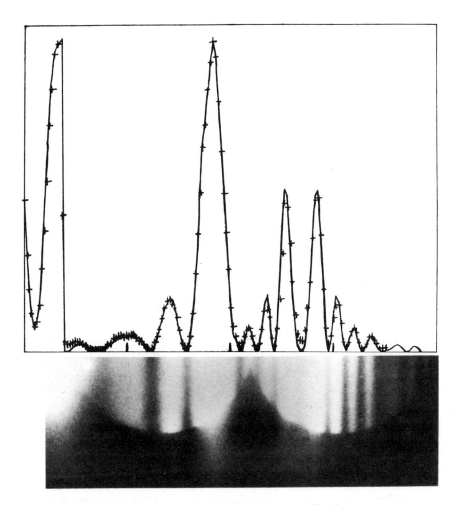

*Figure 4.3.* Three types of CBED data. A comparison of the [111] systematics at 120 kV for MgO recorded on film (below); with elastic energy filtering (crosses); and calculations refined for best fit (continuous line).

of $\chi^2$. This gives

$$c = \frac{\sum_i \frac{f_i}{\sigma_i^2} I_i^{\text{exp}} I_i^{\text{theory}}}{\sum_i \frac{f_i}{\sigma_i^2} I_i^{\text{theory}} I_i^{\text{theory}}} \quad (4.3)$$

For a typical initial refinement in a centric crystal with, say, three

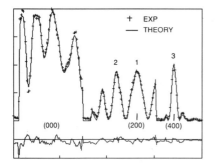

*Figure 4.4.* Final refinement of MgO [h00] systematics at 120 kV: dynamical calculations (continuous line) and experimental data (crosses). The Bragg conditions are indicated and their difference shown below.

CBED disks $(0, g, h)$, one therefore performs a search in five-parameter space if absorption and thickness are included. (The zero-order absorption coefficient is not refined since it causes only a uniform exponential attenuation of all beams.) It will be appreciated that a separate diagonalization of the structure matrix is required for every data point, since each point in the rocking curve corresponds to a different incident-beam direction. This process must then be repeated for every set of parameters. The use of perturbation methods to increase computing speed is discussed below.

There is, however, no optimization method which can at present guarantee finding a global minimum in $\chi^2$, and thus the best fit. Our experience has been, after testing several algorithms, that the "simplex" method is the most robust, if not the fastest. It has been consistently more successful in our work than other methods for finding the global minimum.

A FORTRAN program (called REFINE/CB) has been developed for this refinement and is available from the authors, or as part of the "XTALS" X-ray and neutron crystallography package (Hall, 1991). This is essentially a combination of the Bloch-wave program based on Eq. (3.27) (listed in Appendix 5.2) with the Simplex least-squares nonlinear optimization program. The program is described in Zuo and Spence (1991). A simplified flow chart is shown in Fig. 4.5. The optimization routine (subroutine Aomeba) is discussed fully elsewhere (Press *et al.*, 1986). The results for MgO shown in Fig. 4.4 were obtained with this program. As a practical example, we give here the experimental conditions for this work, in which the aim was to refine the low-order structure factors of MgO.

Data was collected at 120 kV (nominal) from an MgO smoke crystal on a Philips EM400 electron microscope using a 100-nm probe and double-tilt stage to reduce contamination. The path of a typical scan (A–A') is shown in Fig. 2.3, in which the pattern was deflected under

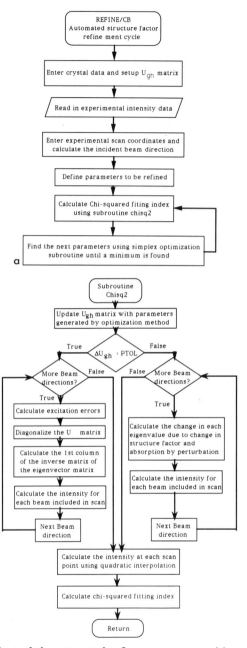

*Figure 4.5.* Flow chart of the automated refinement program: (a) main program, (b) subroutine to evaluate $\chi^2$.

computer control over the entrance aperture of a Gatan model 607 serial ELS spectrometer. An energy window of 5 eV was used, centered on the elastic peak. The ELS entrance aperture size was 1 mm (giving an angular resolution of 0.15 mrad) and the camera length used as 6500 mm. The LaB6 electron source employed provided sufficient stability for the collection of 200 points, each with a dwell time of 100 ms. (Stability is checked by comparing counts at the same point before and after the scan.) The zero disk contained 68 points.

(1) The incident-beam direction was measured from Kikuchi-line features in the CBED pattern recorded on film, as described in Chapter 2. Excitation errors were assigned to points along the scan line (for disk **g**) using Eq. (2.11). Figure 4.6 shows the method used to define the scan coordinates.

(2) The initial thickness refinement was performed using 9 systematic beams. Structure factors for neutral atoms were used, together with absorption coefficients from the "atom" subroutine of Bird and King (1990). The starting values were therefore

$$U(200) = 0.0592 \qquad U'(200) = 0.00076317$$
$$U(400) = 0.025334 \qquad U'(400) = 0.00058483$$

Point 1 (see Fig. 4.4) was used for normalization. The initial thickness refinement was carried out by a separate subroutine specially written for thickness determination. This uses the golden section method

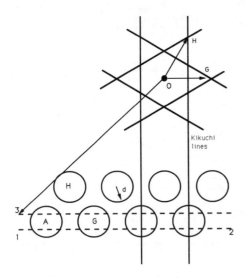

*Figure 4.6.* Method used to determine the beam direction for the MgO refinement. The dimensionless coordinates of points 1, 2, 3, and G must be given as vectors using AG and AH as unit vectors and 0 as origin. The indices of the zone axis 0 and distance $d$ are also needed. The scan line 12 is assumed to lie along AG.

for one-dimensional optimization. A simple method was used first to find an interval containing the minimum in $\chi^2$. This method requires an initial value and a search range. An initial guess of 100 nm for thickness and a 10-nm search range was used. Five function evaluations were required to find the interval (70 nm, 80 nm, and 100 nm) such that $\chi^2(80\,\text{nm}) < \chi^2(70\,\text{nm})$, $\chi^2(80\,\text{nm}) < \chi^2(100\,\text{nm})$, and 10 evaluations to find the minimum, where the thickness was found to be 80 nm. (A function evaluation consists of a set of Bloch-wave calculations for each point in the central disk—in this case 68 points.) Each calculation requires a matrix diagonalization.

(3) The incident-beam direction was varied manually within the error range of measurement (see step 1 above) until the lowest $\chi^2$ was found, and until no systematic error features (such as an asymmetric peak) appeared in the difference plot.

(4) An initial structure-factor refinement was performed using 9 systematic beams, starting with the structure factors listed in step 2 above, and the thickness obtained in step 2. Thus five parameters were adjusted: $U(200)$, $U'(200)$, $U(400)$, $U'(400)$, and thickness. Point 1 was used for normalization. It took 126 function evaluations for subroutine Amoeba to reach the minimum. This took 16 min and 43 s processor time on a VAX station 3200 computer. The results were

$$U(200) = 0.058259 \qquad U'(200) = 0.0016306$$
$$U(400) = 0.025347 \qquad U'(400) = 0.00057725$$
$$\text{Thickness} = 81.023\,\text{nm}$$

(5) The final refinement was performed using 33 beams, including HOLZ beams, and the same five parameters. A weight window was used with $f = 1.0$ for each point on the rocking curve except points 2 and 3 near the Bragg conditions of interest, where the refinement is most sensitive. Here $f = 20.0$. Point 1 was again used as the normalization point. The values obtained in the last step were used as the initial values. It took 84 function evaluations and a processor time of about 6 hours to reach the minimum value of $\chi^2$. The final results were, at 120 kV,

$$U(200) = 0.058563 \qquad U'(200) = 0.0016407$$
$$U(400) = 0.024864 \qquad U'(400) = 0.00056942$$
$$\text{Thickness} = 80.693\,\text{nm}$$

Error are discussed in the next section. Figure 4.7 shows contours of equal $\chi^2$ in the neighborhood of the minimum, plotted as a function of

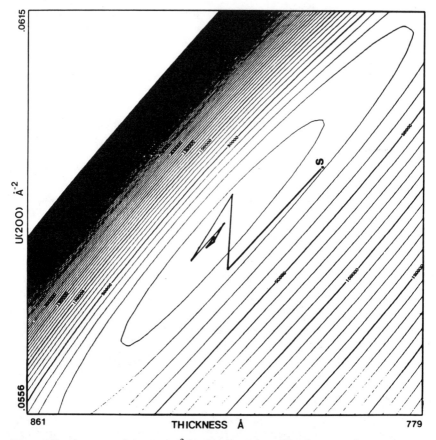

*Figure 4.7.* Contours of constant $\chi^2$ in two dimensions. For demonstration purposes, a refinement has been completed using only thickness and one structure factor as adjustable parameters. The path taken by the Simplex algorithm to the minimum is shown by the continuous line from S. The contour increment is $\Delta\chi^2 = 50$.

$U(200)$ and thickness. The path chosen by the Simplex algorithm is also shown.

(6) Step 5 was repeated using the values from step 5 as the initial values. This step is required by the Simplex method to ensure that a true minimum is found, rather than a saddle point. The results are shown in Fig. 4.4.

The procedure used for the subtraction of the background due to inelastic scattering is most important. This background consists of inelastic phonon scattering, plasmon scattering, and losses due to electronic excitations. The use of an electron energy loss spectrometer to collect only elastically scattered electrons (plus those which have excited phonons) considerably improves the accuracy of the refinement by eliminating the direct plasmon and single electron excitation contribution. The removal of most of the Kikuchi pattern in two-dimensional elastically filtered CBED patterns, however, suggests that the dominant contribution to the background is the (large-angle) phonon scattering of (large-loss) plasmon loss electrons. If filtering is not possible, densitometer scans taken just outside the CBED disks can also be used to estimate this phonon/plasmon diffuse scattering in the systematic case. This must be subtracted. Note that the calculated intensities were convoluted with the measured detector response rather than attempting deconvolution

The use of the perturbation methods described in Chapter 3 in the Refine/CB algorithm can reduce computing times dramatically. Most of the computing time (6 hours) in the preceding refinement was devoted to the final search, for which the diagonalization of a $33 \times 33$ complex matrix is required for each of the 68 data points, and this must be repeated for each of the 4 structure-factor parameters. For bonding and atomic position parameter studies, the structure matrix can be broken up into the sum of a zero-order matrix (including excitation errors) $\mathbf{A}_0$ and a smaller pertubation matrix $\mathbf{A}'$, containing the required changes in the structure factors due to bonding. The essential requirement for an absorption potential, however, complicates this somewhat, since the matrix of eigenvectors is then not unitary. (For an acentric crystal with absorption it is neither symmetric nor Hermitian.) We assume that the eigenvectors are unchanged (Hirsch et al., 1977). The changes in the eigenvalues due to a small change $\Delta U_\mathbf{g}$ in structure factors can be obtained from Eqs. (3.46) and (3.42). If we retain only the first-order terms $b_{ii}$ in Eq. (3.46), the change in the eigenvalues becomes

$$2K\Delta\gamma^i = \sum_{gh} C_{ig}^{-1} \Delta U_{gh} C_{hi} \tag{4.4}$$

where $C_{hi}$ is the eigenvector matrix of $\mathbf{A}_0$ and $C_{ig}^{-1}$ is the inverse matrix of the eigenvector matrix. To implement the perturbation method in the refinement program, we use a switch mechanism with a control parameter PTOL (see Fig. 4.5). If the percent change in the structure factor ($|\Delta U_{gh}|/|U_{gh}|$) is smaller than PTOL, then the perturbation method is used, otherwise the diagonalization is used. The program will update the $\mathbf{A}_0$ matrix, its eigenvector matrix, and the inverse of the eigenvector matrix, if two successive cycles of the optimization routine result in a smaller $\chi^2$ and the percent change in the structure factor is larger than

PTOL. We have found that it is better to give a rather small value (<0.05) to PTOL. The test run shows that the times are about equal for perturbation and diagonalization for a nine-beam refinement, however the time for a 33-beam refinement with pertubation is only about 0.5 hour, which is significantly lower than the 6 hours spent on the 33-beam refinement using diagonalization. For a noncentrosymmetric crystal, the neglect of changes in the eigenvectors may lead to significant error, as discussed in Section 3.9. This procedure is therefore not recommended for the refinement of acentric crystals.

In summary, the automated refinement method makes it possible to refine many parameters, such as bonding reflections, position parameters, Debye–Waller factors, or lattice spacings for strain measurement (see Section 5.4). However, the limitations of this technique have yet to be fully determined, and difficulties will certainly be experienced with larger unit-cell crystals unless the wide-angle methods described in Chapter 6 can be used, at some cost in probe size.

Through the weight factor, it should also be possible to separate to some extent the effects of the parameters on the data, since each affects different groups of reflections. For example, the effects of bonding (which influence low-order reflections) might be disentangled from the position parameters which, together with the Debye–Waller factors, affect mainly the high-order reflections.

The application of this algorithm to a noncentrosymmetric crystal (BeO) for the purpose of measuring structure-factor phases is described in Section 5.1. For recent developments see Zyo (1992b).

## 4.3. ERROR PROPAGATION IN REFINEMENT BY CBED

Error analysis is all-important in structure-factor refinement, since the effects sought are so small (Maslen, 1988). The accuracy of measurements conducted by the CBED method depends on the errors in all the parameters—thickness $t$, $U_g$ (the assumed, nonrefined values), accelerating voltage $E_0$, electron counting noise, absorption coefficients $U'_g$, and errors in the calibration of the incident-beam direction. In this section we investigate how errors in each of these parameters propagate through the dynamical scattering theory to affect the intensities. By differentiating the two-beam expression, we find that, including all sources of error,

$$\left(\frac{\Delta U_g}{U_g}\right)^2 = C_1\left(\frac{\Delta l_g}{l_g}\right)^2 + \left(\frac{\Delta t}{t}\right)^2 + 0.25\left(\frac{\Delta V}{V}\right)^2 + C_2\left(\frac{\Delta U'_g}{U'_g}\right)^2 \quad (4.5)$$

where the values $C_1 \approx 0.0$ and $C_2 = 0.01$ were obtained, for example, from computational trials for the (004) reflection in GaAs. Equation (4.5) is still useful if one or more parameters are found from other sources. In a least-squares refinement, the standard deviation $\sigma_{a_k}^2$ in any parameter is the root sum square of the products of the standard deviation of each data point $\sigma_i^2$ multiplied by the effect which that data point has on the determination of the coefficient $a_k$ (Bevington, 1969). Hence

$$\sigma_{a_k}^2 = \sum_{i=1}^{n} \sigma_i^2 \left(\frac{\partial a_k}{\partial I_i}\right)^2 \qquad (4.6)$$

If a minimum in $\chi^2$ is found and a parabolic expansion around the minimum can be introduced then the standard deviation is found to be (Wolberg, 1967)

$$\sigma_{a_k}^2 \approx \frac{\chi^2}{n-p} C_{kk}^{-1} \qquad (4.7)$$

Here $C_{kk}^{-1}$ is the $k$th diagonal term of the inverse matrix of $C$ which is defined by

$$C_{kl} = \sum_{i=1}^{n} \frac{f_i}{\sigma_i^2} \left(\frac{\partial I_i}{\partial a_k}\right)\left(\frac{\partial I_i}{\partial a_l}\right) = C_{ik} \qquad (4.8)$$

Equations (4.7) and (4.8) may be used to estimate the standard deviation in each parameter. The derivatives in Eq. (4.8) can be calculated by first-order perturbation theory (Zuo and Spence 1991). For the MgO 200 and 400 refinement we find, at 120 kV and room temperature,

$U(200) = 0.05847 \pm 0.00025 \qquad U(400) = 0.02484 \pm 0.00048$

$U'(200) = 0.00158 \pm 0.00014 \qquad U'(400) = 0.00059 \pm 0.00014$

$t = 808.0 \pm 5.4$ angstroms

The definition of $\chi^2$ requires the variance $\sigma_i$ in the denominator of Eq. (4.2) (Bevington, 1969). In this case, we have used $\sigma_i^2 = I_i^{\exp}$. This assumes that the statistics of the electron counting noise are Poisson. Experiments to test this assumption suggest that deviations from Poisson statistics are encountered at large count rates when using a photomultiplier and scintillator (Zuo and Spence, 1991).

## 4.4. MEASUREMENT OF DEBYE–WALLER FACTORS BY CBED

An analysis of the effects of temperature on extinction distances in electron diffraction is given by Howie and Valdre (1967). From Eqs. (3.4), (3.17), and (3.87) we see that the temperature dependence of electron structure factors for an isotropic, centric crystal containing only one type of atom can be written as

$$U_g(\text{tot}) = U_g \exp(-B\mathbf{g}^2/4) + U'_g(\text{phonon}) \qquad (4.9)$$

where $B = B(T)$ is the Debye–Waller factor. By measuring the intensity of several reflections (particularly high orders) as a function of $\mathbf{g}$ it is possible to measure $B$ using electron diffraction data. Measurements may also be conducted at several temperatures. In one case, for example (Goodman, 1971), values of both $U'_g$ and $B$ were determined separately for MgO. In such a centric crystal, the Debye–Waller factor affects the real (elastic) potential, and therefore the ratio A/B in Fig. 3.1a and, most sensitively, the intensity of the high-order reflections. We have seen, however, that the effects of absorption for beam $\mathbf{g}$ are best measured from the asymmetry of the zero-order disk with $\mathbf{g}$ at the Bragg condition. The two effects may therefore be separated in centric crystals. The influence of bonding effects may be disentangled from the Debye–Waller effect by matching reflections as a function of $\mathbf{g}$, high-order reflections being more sensitive to the latter and low orders to the former. Both the critical voltage effect and the intersecting HOLZ (or Kikuchi line) methods have also been used to measure Debye–Waller factors. These depend sensitively on temperature, since the IKL gap [Eq. (5.11)] and the critical voltage [Eq. (5.10)] both depend on the real part of the structure factor in centric crystals. Alternatively, temperature may be used to fine-tune patterns near the critical voltage (Sellar *et al.*, 1980). In recent work, both critical voltages and Kikuchi line splittings for Si, Ge, Al, Cu, and Fe have all been measured as a function of temperature (Matsumura *et al.*, 1989). These workers find that the anharmonic contribution to the temperature factor in the metals is readily detectable above 300 K but is small for the semiconductors. They derive values of structure factors in good agreement with those obtained by X-ray and neutron diffraction. There would appear to be considerable scope for more studies of this type, provided that adequate CBED facilities and energy filtering can be fitted to high and medium voltage electron microscopes.

The use of HOLZ lines in CBED patterns to measure thermal expansion coefficients is described by Angelini and Bentley (1984).

## 4.5. WHAT IS MEASURED. CHARGE DENSITY AND THE TOTAL CRYSTAL ENERGY

In this section we discuss the relationship between the charge density in a crystal (which may be measured by CBED) and other quantum-mechanical quantities which determine their properties. The most important of these is the total energy of the crystal, which will be related to the charge density. In the process, we will gain some understanding of the origin of the optical potential used to describe diffraction from crystals in Chapter 3.

An electrostatic energy $E_H$ may be defined for a crystal such that

$$E_H = \frac{e^2}{2} \int \int \frac{\rho(\mathbf{r}_i)\rho(\mathbf{r}_j)}{4\pi\varepsilon_0 |\mathbf{r}_i - \mathbf{r}_j|} d\mathbf{r}_i \, d\mathbf{r}_j \qquad (4.10)$$

This could in principle be calculated from the results of a structure-factor refinement. Here $\rho(\mathbf{r})$ excludes the nuclear contribution but includes bonding effects. The interpretation given to $E_H$ above is the classical electrostatic work needed to assemble the charge distribution $\rho(\mathbf{r})$ from infinitely separated point charges, including their self-energy. Another important energy, the cohesive energy of a crystal, is defined as the difference between the true total energy of the crystal $E_{cr}$ and that of its constituent atoms $E_{at}$ at infinite separation and zero temperature.

We now consider the electron diffraction experiment. The Schrödinger equation for the system of the beam electron (with interaction Hamiltonian $H_b$ and total energy $E_b$) and crystal (Hamiltonian $H_c$ and total energy $E_{cr}$) is (in SI units)

$$\left(-\frac{h^2}{8\pi^2 m}\nabla^2 + H_c + H_b\right)\Psi = (E_b + E_{cr})\Psi \qquad (4.11)$$

where

$$H_c = -\sum_{i=1}^{ZN}\frac{h^2}{8\pi^2 m}\nabla_i^2 - \sum_{i=1}^{ZN}\sum_{a=1}^{N}\frac{Ze^2}{4\pi\varepsilon_0|\mathbf{R}_a - \mathbf{r}_i|} + \frac{1}{2}\sum_{i\neq j}^{ZN}\sum^{ZN}\frac{e^2}{4\pi\varepsilon_0 r_{ij}}$$

$$+ \frac{1}{2}\sum_{a\neq b}^{N}\sum^{N}\frac{Z^2 e^2}{4\pi\varepsilon_0 |R_{ab}|} \qquad (4.12)$$

for a crystal containing $N$ identical atoms, each of $Z$ electrons. We have ignored inelastic scattering. Here $r_{ij} = |\mathbf{r}_i - \mathbf{r}_j|$ are crystal electron coordinates, while $R_{ab} = |\mathbf{R}_a - \mathbf{R}_b|$ are nuclear coordinates. If the beam electron coordinate is $\mathbf{r}$, then

$$H_b = \sum_j \frac{e^2}{4\pi\varepsilon_0 |\mathbf{r} - \mathbf{r}_j|} - \sum_a \frac{Ze^2}{4\pi\varepsilon_0 |\mathbf{r} - \mathbf{R}_a|} \tag{4.13}$$

It has been shown (Rez, 1978) that the exchange energy between the beam and crystal electrons is negligible. Thus, because of its high energy, the beam electron can be distinguished from the crystal electrons, and the total wavefunction satisfying the eigenvalue Eq. (4.11) can therefore be written as a product:

$$\Psi(\mathbf{r}, \mathbf{r}_1, \mathbf{r}_2, \ldots, \mathbf{r}_n) = \psi_c(\mathbf{r}_1, \mathbf{r}_2, \ldots, \mathbf{r}_n)\psi_b(\mathbf{r}) \tag{4.14}$$

where $\psi_c$ is an antisymmetric linear combination of one-electron wavefunctions.

For the crystal electrons,

$$H_c\psi_c = E_{cr}\psi_c \tag{4.15}$$

Equations (4.14), (4.15), and (4.11) can be employed to show that the beam electron wavefunction alone now satisfies a Schrödinger equation:

$$\frac{1}{4\pi^2}\nabla^2\psi_b + \frac{2m}{h^2}\left(\int \psi_c^* H_b \psi_c \, d\tau\right)\psi_b = K^2\psi_b \tag{4.16}$$

where $K^2 = (2m/h^2)E_b$. From Eq. (3.14), we see that $V(\mathbf{r})$, the potential seen by the beam electron whose structure factors $V_g = h^2 U_g/2m |e|$ we measure, is the matrix element in Eq. (4.16) (divided by $e$). Inelastic scattering gives rise to an additional imaginary potential. We now wish to relate $V(\mathbf{r})$ to $E_{cr}$.

If $\psi_c$ is written as a Slater determinant in terms of one-electron wavefunctions $\phi_j(\mathbf{r}_j)\alpha_j$ (with spin part $\alpha_j$), then (Raimes, 1961)

$$eV(\mathbf{r}) = \int \psi_c^* H_b \psi_c \, d\tau = \sum_j \int \frac{e^2 |\phi_j(\mathbf{r}')|^2}{4\pi\varepsilon_0 |\mathbf{r} - \mathbf{r}'|} d\mathbf{r}' - \sum_i \frac{Ze^2}{4\pi\varepsilon_0 |\mathbf{r} - \mathbf{R}_i|}$$

$$= e\int \frac{e\rho(\mathbf{r}')}{4\pi\varepsilon_0 |\mathbf{r} - \mathbf{r}'|} d\mathbf{r}' - \sum_i \frac{Ze^2}{4\pi\varepsilon_0 |\mathbf{r} - \mathbf{r}_i|} \tag{4.17}$$

This equation has the form of the integral solution of Poisson's equation and expresses $V(\mathbf{r})$, the potential seen by the beam electron, as the sum of the potential due to a (dimensionless) electronic charge $\rho(\mathbf{r}) = \psi_c \psi_c^* = \Sigma_j |\phi_j(\mathbf{r})|^2$ and a nuclear potential. This defines $\rho(\mathbf{r})$, as obtainable by electron or X-ray diffraction studies, in terms of the many-electron crystal wavefunction $\psi_c$. Unfortunately, a method has not yet been devised which allows $E_{cr}$ to be written simply in terms of $\rho(\mathbf{r})$.

The total crystal ground state energy is, in an obvious notation,

$$E_{cr} = \int \psi_c^* H_c \psi_c \, d\tau$$
$$= E_{kin} + E_{e/nuc} + E_{e/e} + E_{nuc/nuc} + E_{ex} \quad (4.18)$$

where

$$E_{e/e} = \frac{1}{2} \int \int \frac{e^2 \rho(\mathbf{r}_j) \rho(\mathbf{r}_i)}{4\pi\varepsilon_0 |\mathbf{r}_i - \mathbf{r}_j|} d\mathbf{r}_i = E_H \quad (4.19)$$

using the same determinant of one-electron wavefunctions. Thus the quantity which can be measured by electron diffraction is the electrostatic, Coulomb, or Hartree energy $E_H$. In addition, the core potential

$$E_{el\,nuc} = \sum_a \int \frac{Ze^2 \rho(\mathbf{r}_j)}{4\pi\varepsilon_0 |\mathbf{R}_a - \mathbf{r}_j|} d\mathbf{r}_j \quad (4.20)$$

might be measured. The kinetic energy of the crystal electrons $E_{kin}$ is related to their potential energy through the virial theorem, while the nuclear interaction term $E_{nuc/nuc}$ may be evaluated by an Ewald lattice sum. Thus it is the exchange and correlation energy $E_{ex}$ which is not accessible in electron diffraction experiments, since this depends in a complicated way on cross terms between the complex one-electron crystal electron wavefunctions $\phi(\mathbf{r}_j)$. It has, however, been shown to be a unique universal functional of $\rho(\mathbf{r})$ (Hohenberg and Kohn, 1964).

The electrostatic energy $E_H$ may be expressed in terms of X-ray structure factors $f(s)$. For example, for a diatomic molecule it is proportional to (Spackman and Maslen, 1986)

$$E_H' = \int_0^\infty [Z_a - f_a(\mathbf{S})][Z_b - f_b(\mathbf{s})] J_0(\mathbf{sR}) \, ds \quad (4.21)$$

The magnitude of the missing term $E_{ex}$ depends on material; however, as an example, for diamond $E_{coh} = 7.4 \, \text{eV}$ per atom while $E_{ex} = 2 \, \text{eV}$ (Chadi, 1989). We conclude that $E_{ex}$ cannot be neglected and that $E_{coh}$

cannot easily be determined from electron diffraction data. For a fuller discussion of the calculation of total energy in terms of charge density, see Lundqvist and March (1983).

It has also been shown that the corrections to $V(\mathbf{r})$, the potential seen by the beam electron, due to virtual inelastic scattering (other than phonon scattering) are negligible (Smart and Humphreys, 1978).*

Accurate measurements of the charge-density distribution in crystals using either electron diffraction or X-ray diffraction require careful interpretation, since the distribution measured depends on temperature. A difference density is usually plotted, being the difference between the measured density and that calculated from neutral atoms placed on lattice sites. (Since they provide an arbitrary reference, we suggest that the neutral atom densities be obtained from the scattering factors in the *International Tables for Crystallography*. In this way comparisons can be made between the results of different groups.) The effect of the Debye–Waller factor (even at $T = 0$ K) causes the structure factors of this difference density to fall off rather rapidly. It is important to appreciate how small the adjustment of the valence electrons is to the energy-lowering process of crystal formation—typically this involves changes to less than $10^{-4}$ of the total charge density. Thus contour maps of charge density for neutral atoms placed on lattice sites are indistinguishable from those of the fully bonded and relaxed crystal. Solid state effects can only be seen in difference maps and affect only low-order structure factors. The statistical significance of these difference maps are discussed by Maslen (1988).

## 4.6. THE SIGNIFICANCE OF THE MEAN INNER POTENTIAL $V_0$

We now consider the interpretation of the zero-order Fourier coefficient of crystal potential $V_0$ and derive a simple relationship between this and the diamagnetic susceptibility. In a monatomic crystal consisting of atoms on lattice sites $i$, with unit cell volume $\Omega$, the Fourier coefficients $V_g$ of potential in SI units are, from Eqs. (3.3) and (3.4) (with $\mathbf{g} = 2\mathbf{s}$),

$$V_g = \frac{1}{\Omega} \sum_i f^e(g) \exp(-2\pi i \mathbf{g} \cdot \mathbf{r}_i)$$

$$= \left(\frac{|e|}{4\pi^2 \varepsilon_0 \Omega}\right) \sum_i \frac{[Z - f^x(\mathbf{g})]}{g^2} \exp(-2\pi i \mathbf{g} \cdot \mathbf{r}_i) \qquad (4.22)$$

---

* For this reason, Poisson's equation (4.17) may be used.

Since we are interested in the value of this expression for $\mathbf{g}=0$, the Debye–Waller factor has been neglected. Now for spherically symmetric atoms, the X-ray scattering factor is

$$f^x(\mathbf{g}) = 4\pi \int \rho(r)\left(\frac{\sin 2\pi gr}{2\pi gr}\right) r^2\, dr \qquad (4.23)$$

For $V_0$, we must consider the limiting form of $f^x(\mathbf{g})$ for small $\mathbf{g}$. Expanding $\sin x / x$, we have

$$\frac{Z - f^x(g)}{g^2} = T_1$$

where

$$T_1 = \frac{1}{g^2}\left[Z - 4\pi \int \rho(r) r^2\, dr + \frac{4\pi}{3}\int \rho(r) r^2 (4\pi^2 g^2 r^2)\, dr - \cdots\right]$$

The second term is $Z$, the number of atomic electrons. Hence

$$\begin{aligned} T_1 &= \frac{4\pi^2}{3}\int \rho(r) r^2 (4\pi r^2)\, dr \\ &= \frac{4\pi^2}{3}\int \rho(r, \theta, \varphi) r^2\, d\tau \\ &= \frac{4\pi^2}{3}\int \phi r^2 \phi^*\, d\tau \\ &= \frac{4\pi^2}{3}\langle r^2 \rangle \end{aligned} \qquad (4.24)$$

where $\phi$ are the wavefunctions of the atomic orbitals. Thus

$$V_0 = \frac{|e|\, n_0}{3\Omega \varepsilon_0}\langle r^2 \rangle = \frac{|e|\, N \langle r^2 \rangle}{3\varepsilon_0} \qquad (4.25)$$

where $n_0$ is the number of atoms per cell, $N$ is the number per unit volume, and $\langle r^2 \rangle$ is the mean-square radius of the atom.

Now the diamagnetic susceptibility per unit volume is, in dimensionless SI units, given by Langevin's atomic theory for gases as

$$\chi = -\frac{\mu_0 N Z e^2}{6m}\langle r^2 \rangle \qquad (4.26)$$

Hence we might expect an approximate relationship in crystals of the form

$$\chi = -\left(\frac{\varepsilon_0\mu_0 Z |e|}{2m}\right)V_0 \qquad (4.27)$$

This result will not be exact, since Eq. (4.26) was derived for gases, and because we have represented the crystal as a simple superposition of single atoms. An exact relationship can, however, be derived between $f(0)$ and $\chi$. In the old system (first Born approximation) the result is

$$f^B(0) = [4\pi me^2 Z/3h^2]\langle r^2 \rangle$$

The first and most extensive comparison of $V_0$ values (derived from $\chi$) with the values measured by reflection electron diffraction was given by Miyake (1940). A further discussion can be found in Ibers (1958).

Values of $V_0$ have been measured by many workers, most commonly using an electron biprism in an interference experiment. It may also be measured from RHEED experiments. For a summary of measurements, see Spence (1992a,c). Interferometry measurements provide values of the product $V_0 t$, so that thickness must be known accurately. This problem is avoided in the reflection geometry. In transmission work, the highest accuracy has been obtained by using cleaved crystalline wedges, for which the thickness is known accurately at each point (Gajdardziska-Josifovska et al., 1992).

In conclusion, we see that $V_0$ has two important interpretations—first as a measure of diamagnetic susceptibility, and second as a measure of the "size" of an atom. It is thus the most sensitive of all the structure factors to the state of ionicity of atoms in a crystal and, because of the terms in $r^2$ and $r^4$ in Eqs. (4.24), depends strongly on the distribution of outer valence electrons. For example, a simple expression for $V_0$ in terms of the Gaussian expansion for atomic scattering factors is given by Eq. (3.13). For MgO, this atomic estimate (17.6 V) differs considerably from the experimental value of 13.6 V. The difference is due to the redistribution of charge in the solid state, and to the sensitivity that Eq. (4.24) provides for this effect. Methods for evaluating $V_0$ in terms of either a multipole expansion of the crystal charge density or using an infinite sum of X-ray structure factors are described in Becker and Coppens (1990).

# 5

# Applications of Three- and Many-Beam Theory

In this chapter we describe several experimental applications of the ideas outlined in Section 3.7 on three-beam theory. The previous chapter dealt mainly with techniques which depend on the matching of many-beam computed microdiffraction patterns to experimental patterns. In this chapter we are concerned with extracting simple rules and general predictions from the three-beam expression for the intensity distribution in CBED patterns, and in using these rules to extract information from experimental patterns. The determination of structure-factor signs (in centric crystals), the critical voltage effect, and the determination of crystal polarity can all be understood using this approach. Phase measurement in acentric crystals and dynamical shifts of HOLZ lines are also treated, since these topics may also be understood using three-beam theory. A more quantitative many-beam analysis will be required for accurate measurements.

## 5.1. FINDING ATOMIC POSITIONS. STRUCTURE-FACTOR PHASE MEASUREMENT. ENANTIAMORPHS

Attempts to determine unknown crystal structures directly by electron diffraction have a long history. Early researchers, inspired by the success of the X-ray method, attempted to use kinematic scattering theory, especially in polycrystalline materials (Vainshtein, 1964), where it was hoped that dynamical effects would "average out." Slow but steady progress has been made in the analysis of extremely thin organic films, using a combination of imaging and diffraction methods (Unwin and Henderson, 1975). (By indicating atomic positions directly, images thus "solve" the phase problem of X-ray crystallography.) In favorable cases where thin films crystallize, these structures have now been solved in three dimensions to a resolution of a few angstroms. Important problems (among many) include radiation damage and bending of the films.

For inorganic crystals, considerable preliminary information may be collected in order to constrain the search for possible trial structures. CBED patterns may then be computed for each of these structures for comparison with experimental patterns. This preliminary information includes the crystal space group, the cell constants and angles, and the number and type of atoms in the unit cell. Using CBED, point diffraction patterns, and X-ray microanalysis, much of this information can be obvtained, although with rather low accuracy. A serious problem arises with small crystals in the determination of the number of atoms in the cell, since there appears to be no way to determine the density of, say, a 10-nm crystalline precipitate. Other useful constraints are obtained from a table of interatomic bond lengths which occur commonly in nature, and from atomic resolution lattice images of the same structure.

The most comprehensive attack on the problem of structure determination by CBED is described by Vincent *et al.* (1984), who succesfully determined the structure of AuGeAs precipitates at Au—Ge—In contacts on GaAs substrates. Their method was as follows. Isostructural crystals were grown, and both these and the precipitates of interest were studied. The space group was determined to be centrosymmetric, monoclinic, C2/c (number 15), using the methods described in Chapter 7. Cell constants and angles were measured from spot patterns with an accuracy of about 0.5%, and cell constant ratios to higher accuracy (about 0.1%) using HOLZ lines (see Section 5.4). The $NiP_2$ structure, which has similar cell constants, angles, and space group, was used as an early trial structure. From the space group, stoichiometry, and likely tetrahedral coordination, it was concluded that either all atoms occupy separate fourfold sites, or two species randomly occupy the 8(f) sites, with the third element on a fourfold site. The structure is then defined by the three position parameters $\mathbf{r} = (x, y, z)$ defining the 8(f) site containing Ge and As atoms.

Many possible structures were then eliminated using bond-length arguments. Dispersion surfaces were then calculated for different values of $\mathbf{r}$, from which thickness-independent conclusions could be drawn about the width and visibility of HOLZ lines. The intensity of HOLZ reflections were then graded on a three-point scale and compared with calculations. (In thin crystals, because of their long extinction distances, the intensity of these reflections is relatively insensitive to thickness, as discussed in Section 5.4.) Reflections in widely differing projections were used, providing a severe three-dimensional test of the model. Voltage-dependent features were sought in the zone-axis orientation, at which dispersion surfaces cross. The structure was finally confirmed to be isostructural with $NiP_2$ and $PdP_2$.

Approaches to generalizing the preceding approach have also been discussed (Vincent et al., 1984). Here, a conditional projected potential is introduced which involves a sum over Fourier coefficients in only one HOLZ layer. This is then used in an equation similar to (3.103) describing channeling effects. The transverse eigenstates can then be treated as atomic-like states, following the LCAO method of band theory (Buxton and Tremewan, 1980), and labeled accordingly as $1s$, $2p$, etc. according to symmetry. These states may be either bound or free, depending on how $\varepsilon$ in Eq. (3.104) compares with the maximum in the interatomic potential. The aim of this work is to relate the intensities of HOLZ lines to modified structure factors, perturbed by the strong ZOLZ dynamical interactions. The choice of zone axes needed to distinguish trial structures with different numbers of free parameters are discussed in detail. The value of using HOLZ intensities for structure analysis emerges clearly from this work, as further discussed (with additional examples) in Section 6.2 on large-angle methods for the HOLZ.

In noncentrosymmetric (acentric) crystals, the determination of atomic coordinates requires a knowledge of the phases of structure factors. We therefore devote the remainder of this section to the practical methods which have been developed for measuring the phases of structure factors accurately by electron diffraction. Section 3.7 contains the theoretical background to this section. We have seen that, in noncentrosymmetric crystals, the Fourier coefficients $F_g^X/\Omega$ of $\rho(\mathbf{r})$ are complex, and values of the structure factor phases $F_g^X$ are therefore required to enable the crystal charge density to be synthesized. As discussed in Section 3.1, the phase of the electron structure factor $U_g^c$ is not equal to that of the X-ray structure factor $F_g^X$, but values of $F_g^X$ can be determined from measurements of $U_g^c$ and $U_g'$. In the presence of absorption, the coefficients $U_g^c$ and $U_g'$ in acentric crystals each become the complex Fourier coefficients of real, real-space potentials, so that $U_g^c = U_{-g}^{c*}$ and $U_g' = U_{-g}'^{*}$. In general, it therefore becomes necessary to refine four numbers for each structure factor. However, the real and imaginary parts of $U_g^c$ and $U_g'$ may become mixed when entered as input data to a dynamical computer program, which usually represents $U_g^c + iU_g'$ by a single complex number. In order to refine $U_g^c$ and $U_g'$, it will therefore be found necessary in general to use rocking curve data from both the $\mathbf{g}$ and $-\mathbf{g}$ reflections. Let $U_g^c(R)$ and $U_g^c(lm)$ be the real and imaginary parts of the Fourier coefficients of the elastic real-space potential in an acentric crystal, with similar primed quantities for the absorption potential. Then the Fourier coefficients of the total complex optical potential obtained from a refinement program are

$$X(\mathbf{g}) = [U_\mathbf{g}^c(\mathrm{R}) + iU_\mathbf{g}^c(\mathrm{lm})] + i[U_\mathbf{g}'(\mathrm{R}) + iU_\mathbf{g}'(\mathrm{lm})]$$
$$= [U_\mathbf{g}^c(\mathrm{R}) - U_\mathbf{g}'(\mathrm{lm})] + i[U_\mathbf{g}^c(\mathrm{lm}) + U_\mathbf{g}'(\mathrm{R})]$$

if the complex value of $X(-\mathbf{g})$ can also be determined, it is possible by addition and subtraction to recover the required values of the complex $U_\mathbf{g}^c$ and $U_\mathbf{g}'$.

We now briefly summarize the results of Section 3.7. It is clear that the single-scattering or kinematic theory of diffraction does not allow phases to be measured, nor, since Eq. (3.31) contains only $|U_\mathbf{g}^c|^2$, does the two-beam dynamical theory. In 1957, however, Kambe showed that the lowest-order elastic scattering theory which provides observable information on phases is three (Kambe, 1957). Now the phases of $U_\mathbf{g}^c$ depend on the choice of origin in the crystal. However, Kambe was also able to show that the dynamical three-beam intensity in centric crystals depends only on a certain sum of the three relevant structure-factor phases, the three-phase structure invariant defined in Eq. (3.59). If this is reckoned according to that equation, so that the reciprocal lattice vectors form a closed triangle, then this quantity is independent of choice of origin in the crystal. Following Kambe's discovery, Gjønnes and Høier (1971) showed that, in the general nonsystematics three-beam geometry shown in Fig. 3.8, two special points existed in centric crystals (labeled "q" and "minimum" in Fig. 3.8) at which the intensity was zero due to a degeneracy in the eigenvalues. The position of zero intensity (or of a minimum in $n$-beam theory) indicates immediately whether the three-phase invariant is 0 or $\pi$ for centric crystals. The intensity along the hyperbolae in Fig. 3.8 is given by Eq. (3.68). The position of the minimum depends on accelerating voltage. Thus, for a centric crystal, it may be possible to determine whether the three-phase invariant is 0 or $\pi$ by inspecting the position of the minimum in three-beam CBED patterns (Gjønnes and Høier, 1971; Hurley and Moodie, 1980).

For noncentrosymmetric crystals, experimental phase measurements have been reported by Bird et al. (1987), Ichimiya and Uyeda (1977), Zuo et al. (1989b), Zuo et al. (1989c), and others, whose work we now discuss. We commence with the three-beam nonsystematics CBED method, since this appears to be the most versatile and general method of phase determination. It depends on an analysis of the intensity along the hyperbolas AA' and BB' in Fig. 3.8, given by Eq. (3.68). The hyperbolas are asymptotic to the Bragg lines $S_g = 0$ and $S_h = 0$. The prefactor in Eq.

(3.68) causes the intensity in disk **g** along both the hyperbolas to decline toward B and B' as $S_g$ becomes large and $S_h$ small, while that in disk **h** declines for large $S_h$ and small $S_g$. Most important for phase analysis are the positions and magnitudes of the maxima and minima along the hyperbolas. The positions of these maxima and minima are given by Eq. (3.72). We see that both depend on the three-phase invariant $\Psi$. The minimum occurs on the lower hyperbola (BB' in disk **h** in Fig. 8.2) if $\cos \Psi$ is positive, and on the upper branch AA' if $\cos \Psi$ is negative.

Figure 5.1 shows a series of three-beam calculations for the $(-1, 7, -5)$ CBED disk in GaAs at 120 kV for various thicknesses and phases $\Psi$. The minimum indicated is seen to vary in width with both thickness and phase. The other reflection excited is $(-2, 6, -4)$. These calculations were based on Eq. (3.68), in which the sensitivity of the intensity to phase enters through the effective potential $U_h^{\text{eff}}$. (Figure 3.7 shows the variation of this thickness-independent quantity with $S_g$ for various phases $\Psi$.) We see, as confirmed in Fig. 5.1, that the minimum

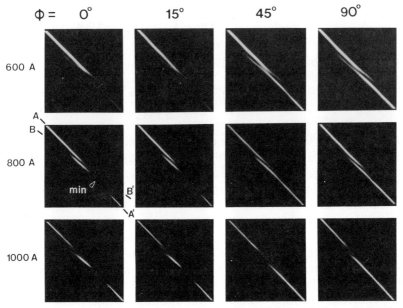

*Figure 5.1.* The variation of three-beam patterns with thickness and phase. Each column corresponds to the same value of the three-phase invariant, while the thickness changes with each row. These are three-beam dynamical calculations for the $(-1, 7, -5)$ reflection in GaAs at 120 kV. The other reflections are (000) and $(-2, 6, -4)$. The values $U(-1, 7, -5) = 0.0139$, $U(-2, 6, -4) = 0.0250$, and $U(1, 1, -1) = 0.0512$ were used.

varies from deep (for $\Psi = 0$) to nonexistent for $\Psi = 90°$. Thus the sensitivity of the CBED intensity to changes in phase varies along the hyperbolas and is greatest at the minimum. The intensity well away from the minimum should therefore be used to normalize the experimental intensity to the calculations.

There have now been several published applications of these principles to experimental phase measurements in the transmission geometry. The first attempt to measure low-order structure-factor phases experimentally by dynamical electron diffraction appears to be that of Ichimiya and Uyeda (1977). Here (0002) thickness fringes from a 60° wedge in CdS were compared with calculations. As a result of the many-beam interactions (as we have seen in Section 3.7), the intensity is sensitively dependent on structure-factor phase. The authors give an error of 0.04 for their measurement of $\tan \phi(0002) = -0.54$.

The second measurement, also for CdS (Bird *et al.*, 1987; Bird and James, 1988), depends on a measurement of the distance between the maximum and minimum intensity on the three-beam hyperbolas. From Eq. (3.72) this is

$$d = S_\mathbf{h}^{\max} - S_\mathbf{h}^{\min} = \frac{|U_{\mathbf{g-h}}|}{2\kappa \cos \Psi} \left[ \left( \left| \frac{U_\mathbf{g}}{U_\mathbf{h}} \right| - \left| \frac{U_\mathbf{h}}{U_\mathbf{g}} \right| \right)^2 + 4 \cos^2 \Psi \right]^{1/2} \quad (5.1)$$

The phase may then be found since

$$|\cos \Psi| = \left| \frac{U_g}{U_h} \right| - \left| \frac{U_h}{U_g} \right| \Big/ 2 \sqrt{\frac{d^2 K^2}{|U_{gh}|^2} - 1} \quad (5.2)$$

The sign of $\cos \Psi$ is indicated by the positions of the maxima and minima. Distances measured on film must first be converted to excitation errors. (Bird *et al.* actually give an approximate result in place of Eq. (5.1), based on first-order corrections to kinematic theory.) Using this method, Bird *et al.* measure the $(-3, 7, -5)$, $(-1, 7, -9)$, $(-2, 0, 4)$ phase triplet in InP to an accuracy of $\pm 15°$.

A third paper (Zuo *et al.*, 1989b) also uses the nonsystematics three-beam geometry. For CdS, these workers used many-beam calculations to match the intensity along lines across the $(4, 1, -4)$ hyperbolas. The $(4, 1, -2)$ reflection was also excited (the geometry is shown in Fig. 3.8). The result of their analysis gives the sum of the phases of these electron structure factors, together with that of the (00-2), to be $49.6 \pm 5°$. Data were recorded on film and read into a computer for comparison with dynamical calculations. Allowance must be made for the logarithmic response of the film (Valentine, 1966). The background was accounted for by convolving the computed intensity with a model distribution for inelastic scattering. When it was assumed that two of these phases were known exactly, the error in the remaining (00-2) X-ray structure factor was found to be ±0.75°.

The fourth case concerns measurements made in the systematics orientation, based on a similar principle to that of the critical voltage method but extended to noncentrosymmetric crystals (Zuo et al., 1989c). These appear to be the most accurate phase measurements yet published by any method. The principle of the method is as follows. As in the critical volatge method, we consider a second-order reflection g at the Bragg angle in an acentric crystal. For a certain range of accelerating voltage, the perturbation to the rocking curve for the second-order reflection **g** due to the unavoidable weak excitation of the first-order reflection **h** is shown below to be very sensitive to the sum of the phases of the two relevant structure factors. (Excitation of the third-order beam is much weaker.) By comparing the second-order beam's rocking curve with the results of many-beam calculations (including nonsystematic interactions for increased accuracy), the phase sum may be found. We use the "Bethe potential" correction [Eq. (3.58)] to analyze the method. With structure-factor phases $\phi_g$, the three-phase invariant is

$$\Psi = -\phi_g + \phi_h + \phi_{g-h} = -\phi(00\text{-}4) + \phi(00\text{-}2) + \phi(00\text{-}2)$$
$$= 2\phi(00\text{-}2) - \phi(00\text{-}4) \tag{5.3}$$

Unlike the nonsystematics three-beam case, only two phases are now involved. The rocking-curve intensity observed in a CBED disk is given by Eq. (3.31) with $U_g$ replaced by the effective structure factor of Eq. (3.58), or

$$|U_g^{\text{eff}}(B)|^2 = |U_g|^2\left[1 - \frac{|U_h||U_{g-h}|}{\kappa S_h |U_g|}\cos\Psi + \left(\frac{|U_h||U_{g-h}|}{2\kappa S_h |U_g|}\right)^2\right] \tag{5.4}$$

which depends on $\Psi$. A plot of $|U_g^{\text{eff}}/U_g|$ from Eq. (5.4) for various values of the phase $\Psi$ is similar to Fig. 3.7, if the abscissa on that figure is relabeled $S_h$ and the curves allowed to extend to infinity for $S_h = 0$. We restrict attention to the region near $S_g = 0$, where the Bethe approximation is most accurate (i.e., for $|S_g| \ll |S_h|$ and $|S_h| \gg |U_{\text{max}}|/2K$, with $U_{\text{max}}$ the largest value of $U$). The quantity $|U_g^{\text{eff}}|$ is most sensitive to changes in $\Psi$ near its minimum:

$$\left(\frac{|U_g^{\text{eff}}(B)|}{|U_g|}\right)_{\text{min}} = |\sin\Psi| \quad \text{for } 2\kappa S_h = \frac{|U_h||U_{g-h}|}{|U_g|\cos\Psi} \tag{5.5}$$

Thus to obtain sensitivity to phase in disk **g** near $S_g = 0$, we require $S_h$ to satisfy the (materials-dependent) constant of Eq. (5.5). We now use a similar argument to that used for the critical voltage. If $S_g = 0$, then the Ewald sphere geometry requires $2\kappa S_h = \mathbf{h}^2$ in Eq. (5.5). Using the definition of $U_g$ to solve for $\gamma$, and the expression $E_0 = m_0 c^2 (\gamma - 1)/|e|$ for accelerating voltage, we find the accelerating voltage in the form

$$E_A = \frac{m_0 c^2}{|e|} \left[ \left( \frac{h^2 h_p^2 |V_g| \cos \Psi}{|V_h| |V_{g-h}| 2 m_0 |e|} \right) - 1 \right] \tag{5.6}$$

where $h_p$ is Planck's constant and **h** is a reciprocal lattice vector. This equation reduces to an approximate form of the critical voltage formula for $\Psi = 0$ or $\pi$ (for centric crystals). The critical voltage corresponds to a choice of $E_0$ which makes $|U_g^{\text{eff}}(B)| = 0$. Here we have extended the theory to acentric crystals and found the voltage $E_0$ at which, for a given phase, $U_g^{\text{eff}}$ is a minimum near $S_g = 0$, and therefore most sensitive to phase. Since, however, the excitation error $S_h$ is used as a variable in CBED experiments, the choice of accelerating voltage is no longer critical. Unlike the critical voltage method (which is restricted to centric crystals), this method also allows independent refinement of several reflections, rather than giving a relationship between structure-factor amplitudes.

As an example, for CdS, with **g** = (00-4) and **h** = (00-2), $|U_h|$ = 0.0577, $|U_g|$ = 0.0142, and $\Psi = 55.332°$ for neutral atoms (room temperature, $E_0 = 120$ kV). Then at $S_g = 0$, a one-degree change in $\Psi$ leads to a 1% relative change in $U(00-4)^{\text{eff}}$. This produces a readily detectable change in $I_g$. Equation (5.4) also shows how insensitive the (00-4) intensity is to $|U(00-4)|$ and how sensitive it is to $|U(00-2)|$. Thus (and in view of the relatively large size of **g**), the use of scattering factors for neutral atoms (rather than ions) is a good approximation for $|U(00-4)|$.

In analyzing these results, different scans were compared taken parallel to the systematics line. From these it was found that the variation in the experimental intensity ratios was about 6%. In view of the insensitivity of the (00-4) intensity to the $|U(00-4)|$ structure factor, it was not included in the refinement parameters. The approximation was made that the phases of the absorption structure factors were equal to those of the elastic potential (see Bird et al., 1989). Debye–Waller factors and the position parameter $u = 0.37717(8)$ were obtained from recent X-ray work. The final refinement therefore included (1) the phase invariant $\Psi$, (2) a starting value of $|U(00-2)|$ obtained from the *International Tables for Crystallography*, (3) absorption coefficients, and (4) specimen thickness. The high voltage and incident-beam direction were found as described in Section 2.3. All beams of appreciable intensity were included in the calculations, including nonsystematics. Absorption coefficients $U'_g$ were obtained by matching the asymmetry in the central disk, with each of the reflections satisfied in turn. The refinement of the remaining parameters $\Psi$ and $|U(00\text{-}2)|$ was based on the sensitivity of the (00-2) rocking curve (near its center) to $|U(00\text{-}2)|$ and that of the (00-4) curve to $\Psi$. Errors were estimated from Eq. (4.5). The 6% error in the experimental data gives $C_1 = 0.02$. Errors in thickness and $E_0$ were both 0.5%. A 10% error in $U'(00\text{-}2)$ causes a 3.3% change in $R'$ giving $C_2 = 0.002$. Errors in other parameters are negligible. Thus the error in $|U(00\text{-}2)|$ is found to be about 1%. The refinement of $\Psi$ is completed last, using the (00-4) rocking curve. For the best fit it was found that

$$\Psi = 54.4° \pm 0.9°$$

The error is obtained from a quadrature sum of the phase changes due to errors in intensity measurement, the measured $|U_g|$, absorption factors, and thickness. These were 0.8°, 0.3°, 0.3°, and 0.2°, respectively. If we assume $\phi(00\text{-}4) = 2.94°$ (known), then Eq. (3.11) gives the corresponding error in the deduced X-ray structure-factor phase $\phi^X(00\text{-}2)$ as ±0.069°.

It is possible to convert these errors in the measurement of structure-factor phases to changes in atomic-position parameter, if a degree of ionicity can be assumed. For CdS (a one-parameter structure) the preceding error makes it possible to determine the dimensionless atomic-position parameter to within about ±0.0005. These ionicity and atomic coordinate effects might be disentangled from a series of patterns emphasizing different orders. For example, we find that the (004) structure-factor phase is very sensitive to the position parameter $u$, while the phase of the (002) depends more strongly on bonding.

Our final example of structure-factor phase measurement concerns the (002) reflection in the noncentrosymmetric BeO crystal structure (Zuo et al., 1992). BeO is toxic, and has the wurtzite structure, space group P6$_3$mc, with cell parameters $a = 2.6979(2)$ and $c = 4.3772(2)$ Å (Downs et al., 1985). For the ideal wurtzite structure, $c/a = 1.633$ and the $z$ parameter is 0.375. BeO is compressed along the $c$ axis relative to this ideal; we have used $z = 0.3775$. We choose an origin at the lighter Be atom. The polarity of the crystal is defined as follows: a vector drawn from Be to O in the wurtzite structure is defined as the positive [001] direction.

This example shows how the automated refinement techniques described in Section 4.2 may be applied to acentric structures to determine individual phases (with respect to a specified origin) rather than phase sums. We will also see how the same values of the structure factors are obtained after analyzing crystals of different thicknesses. In this work the origin was taken at the Be atom site. Zero-loss filtered data was recorded at 80 kV using a serial ELS system fitted with a photomultiplier. Line scans were taken along the [00$h$] systematics near the [−130] zone, as shown in Figs. 5.2a and 5.2b. Although all the data was used for the refinements, by adjusting the weight coefficient and observing the effect on the goodness of fit, it was found that: 1. Ratios such as $A/B$ shown in Fig. 3.1 were most sensitive to $|U(002)|$; 2. The (002) and (004) profile intensity ratio was most sensitive to $\phi(002)$; 3. The internal variation of the (000) disk is most sensitive to $|U'(002)|$; 4. The (000) and (002) relative intensities were most senitive to $\phi'(002)$; 5. Thickness is most sensitive to the outer (002) and (004) fringes. A refinement of $\phi'(002)$ based on the (002) and (00-2) reflections recorded simultaneously was also attempted (Bird et al., 1989), but was found to give larger $\chi^2$. As shown in Fig. 5.2a, it was possible to obtain both the (002) and (004) Bragg conditions in the central disk, and this condition was used to refine the $U(200)$ amplitude $|U(200)|$, the $U(200)$ phase, and the amplitude of the absorption coefficient $|U'(002)|$. Figure 5.2b shows similar data taken from a different thickness of the same crystal. The phase of $U'(200)$ could also be refined using an axial orientation, in which both the (002) and (00-2) reflections were obtained at their Bragg conditions in a single CBED recording. Here, as discussed in association with Eq. (3.97), use is made of the fact that the difference in the intensity of the **g** and -**g** reflections in an acentric crystal is sensitive to the difference between the phases of the structure factors of the elastic and absorption potentials.

The data in Fig. 5.2a and 5.2b are shown compared with the results of many-beam Bloch-wave calculations. The calibration of the incident-beam coordinates and the determination of the center of the (000) disk are important in this work, and these must be treated as refinement parameters in the CB/REFINE algorithm. Initial estimates were based on micrographs (two-dimensional data) taken with the scans, since these allow both components of $K_t$ to be determined approximately. In particular, the $x$ component of $K_t$, the center of the (000) disk, and the thickness were refined together initially (refinement 1). About 20 search steps were needed to minimize $\chi^2$. Using these values, a further seven-beam refinement was made of thickness, $|U(002)|$, $\phi(002)$, $|U'(002)|$, and $\phi'(002)$ (refinement 2). These two refinements were then repeated until $\chi^2 = 5.04$. At this point we have $t = 709.1$ Å, $|U(00.2)| = 0.039230$, $\phi(00.2) = -0.89719$ rad, $|U'(002)| = 0.000755$, $\phi'(002) = -0.62$ rad, and $|U'(00.4)| = 0.000159$. (All $U_g$ values are in Å$^{-2}$.) A further

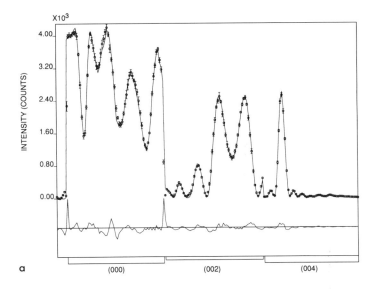

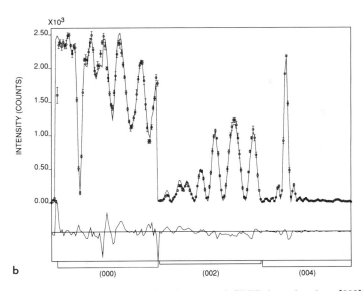

*Figure 5.2.* (a) Zero-loss experimental and computed CBED intensity along [002] in the central disk for BeO at 80 kV. The (002) and (004) Bragg conditions are indicated. The orientation is near the [130] zone axis. Refined sample thickness $t = 70.929$ nm. The plot below shows the difference between calculation and experiment. (b) Similar to Fig. 5.1a, but recorded from a different thickness, found after refinement to be $t = 106.03$ nm.

reduction in $\chi^2$ was then obtained by measuring the y component of $K_t$ from micrographs, in order to more accurately include off-systematics reflections. The refinement was then continued (refinement 3) using 124 beam calculations (to include all reflections visible on the micrographs), and treating the complex $U(00.2)$, $U'(00.2)$, $|U'(004)|$, and $t$ as adjustable parameters. Refinement 3 was then restarted, to confirm that a global minimum had been found. (This procedure is essential with the Simplex algorithm.) This final refinement was carried out using 124 dynamically interacting beams, among which the 19 strong beams with small excitation errors were treated by exact matrix diagonalization, while the remainder of the weak beams were treated using the Bethe perturbation potential. For pupposes of checking other computer programs against ours, it may be useful to specify the beam direction for this computation, since it provides a severe test for sign errors and agrees well with experiment. The final refinement gives the components $K_t$ of a wavevector drawn in the ZOLZ from the center of the $[-1, 3, 0]$ zone to the first point of the scan in Fig. 5.2a to be $-1.282$ g $(00.2) - 0.01$ g $(31.0)$, while that of the last point was $1.755$ g $(00.2) - 0.01$ g $(31.0)$. $(-0.02961, -0.0987, -2.5644)$ (origin at the Be atom). The Debye–Waller factors used in this work were $B_{Be} = 0.355$ and $B_{Ox} = 0.28$.

The acentric nature of the crystal was seen in the different heights of the (002) and (00-2) Bragg peaks obtained simultaneously in this pattern. Thus they are correctly normalized since they both occur within the (000) disk. A comparison with computations then allowed the polarity of the crystal to be determined, and the absolute indexing of the pattern to be found. The refinement of this pattern followed similar lines to that of Fig. 5.2a, using now $\phi'(002)$ and thickness as refinement parameters together with the values of $U(002)$ and $|U'(002)|$ obtained previously. The final values found at 80 kV, with $\chi^2 = 5.04$, were

$$|U(002)| = 0.039592 \pm 0.00014 \text{ Å}^{-2}$$
$$\phi(002) = -0.88478 \pm 0.017 \text{ rad}$$
$$|U'(002)| = 0.00073 \pm 0.00006 \text{ Å}^{-2}$$
$$\phi'(002) = -1.1 \pm 0.5 \text{ rad}$$
$$|U'(004)| = 0.0002 \pm 0.0001 \text{ Å}^{-2}$$
$$t = 711.6 \pm 1.6 \text{ Å}$$

Finally, the last refinement was repeated with a slightly different starting point, and the minimum thus confirmed to be unique. The polarity of the crystal was confirmed by repeating the analysis with every g replaced by −g in the computations. It is interesting to note that this reversal of polarity was found to increase $\chi^2$ to about 20.

Considerable confidence in the method is given by comparing these values with those refined from Fig. 5.2b, which shows data from the same crystal at a different thickness. The refinements from this figure give

$$|U(002)| = 0.03982 \pm 0.00013 \text{ Å}^{-2}$$

$$\phi(002) = -0.87867 \pm 0.017 \text{ rad}$$

$$|U'(002)| = 0.00090 \pm 0.00007 \text{ Å}^{-2}$$

$$\phi'(002) = -0.4 \pm 0.5 \text{ rad}$$

$$|U'(004)| = 0.00004 \pm 0.0001 \text{ Å}^{-2}$$

$$t = 1059.7 \pm 2.0 \text{ Å}$$

These fall within the errors of the previous analysis. The corresponding values of the X-ray structure-factor phase is obtained from Eq. (3.11) as $\phi^X(00.2) = -1.190 \pm 0.0009$ rad, corresponding to an accuracy of better than one-tenth of a degree. This value of $\phi^X$ agrees closely (but is about fifty times more accurate) than that which can be computed from X-ray measurements (Zuo et al., 1992).

Again, the error in $\phi(002)$ might be converted into an error in atomic coordinates. In this work, since it had been measured previously, the one adjustable parameter which defines this structure (the z coordinate of one of the atoms) was not treated as a refinement parameter. In other cases this may be necessary.

Reversing the sign of the phase of the plane wave representing the incident electron beam would reverse the sign of $\phi(002)$, and this must be considered (together with the choice of origin) when comparing this work with results from other researchers.

A different group of techniques has been developed based on the straightforward measurement of the intensity of many high-order reflections, in both the ZOLZ and HOLZ (Kolby and Taftø, 1991; Ma et al., 1991; Tomokiyo and Kuroiwa, 1990). (Some of these are also briefly described in Sections 6.1 and 6.2.) In particular, the slight rotation of the oxygen octahedra which occurs during the phase transition in $SrTiO_3$ has been studied by Tanaka and Tsuda (1990), who use direct measurement of HOLZ reflections and comparison with dynamical computations. By using a refinement technique similar to that described in Section 4.2, they are able to determine this rotation to within an accuracy of about 0.2°.

The "handedness" of a crystal can be determined by analysis of CBED intensities. This corresponds to the difference between a right-handed and a left-handed screw, or the difference between an individual's right and left hand. Space groups which differ in this way are known as enantiomorphs and are mirror images of one another. We note that two enantiomorphs may have either the same (e.g., 1) or different symmetry (as in the case of quartz). Applications of CBED for this purpose are described by Goodman and Johnson (1977) for quartz, and by Tanaka and Terauchi (1985a) for MnSi. In both cases the absolute orientation can be determined. The topic is also discussed by Vincent et al. (1986).

The effect of enantiomorphism on dynamical intensities has been analyzed by Marthinsen and Høier (1989). It can be understood by noting that any odd number of mirror operations reverses handedness, and that three mirrors is equivalent to a center of inversion. The effect of replacing $\mathbf{r}$ by $-\mathbf{r}$ in the general expression for dynamical intensity may be understood by writing Eq. (3.27) as

$$I_h(t) = \sum_i |C_0^{i*} C_h^i| + 2 \sum_{i>j} |C_0^{i*} C_h^i C_0^j C_h^{j*}| \cos[+2\pi(\gamma_i - \gamma_j)t + \alpha_{ij}] \quad (5.7)$$

where $\alpha_{ij}$ is the phase of $C_0^{i*} C_h^i C_0^j C_h^{j*}$. The replacement of $\mathbf{r}$ by $-\mathbf{r}$ leaves the eigenvalues unaffected but conjugates all the eigenvectors. This reverses the sign of $\alpha_{ij}$, which can lead to large intensity differences between the two crystals in CBED patterns along Bragg lines.

In three-beam theory, the effect of enantiomorphism is to reverse the phase of the three-phase invariant. From Eq. (3.69) we see that this does not lead to an observable effect (within either the Bethe or Kambe approximations). In the wurtzite structure the application of a mirror operation (or a twofold axis) reverses the polarity of the structure, and we have seen above with BeO that the use of the incorrect polarity gave a much poorer refinement.

The effect of reversing the phase of the incident plane wave is to reverse both the sign of $\alpha_{ij}$ and the plus sign before the term $2\pi(\gamma_i - \gamma_j)$ in Eq. (5.7), so that the same intensity distribution is predicted from the same crystal, regardless of plane-wave sign convention.

## 5.2. CRITICAL VOLTAGES

The critical voltage (CV) method is the most accurate method of structure-factor measurement by electron diffraction for centric crystals. What is measured is the ratio of the magnitude of the first-order structure factor to that of the second-order reflection. The method requires an electron microscope whose accelerating voltage may be varied continuously over a large range, typically from 100 kV to 1 MeV. The field has been reviewed by Fox and Fisher (1988).

In its simplest form, the critical voltage effect consists in the observation of a minimum of intensity in a second-order reflection at the Bragg condition for a particular accelerating voltage. This voltage depends sensitively on the ratio of the first- to second-order structure factors, and so may be used to measure this ratio with high precision from measurements of the critical accelerating voltage. If the second-order reflection is assumed to depend mainly on the known atomic coordinates, the first-order reflection, which is more sensitive to bonding effects, may be found with high precision. Accuracy is often ultimately limited by knowledge of the Debye–Waller factor for the high-order reflections. Introductory reviews of the theory may be found in Cowley (1981b), Humprheys (1979), and Reimer (1984).

The effect may be understood either by employing three-beam theory, derived using the Bethe potentials (Uyeda, 1968), or from a (related) two-Bloch-wave picture (Lally et al., 1972). We commence with the Bethe approximation, as described in Section 3.3 It is assumed that Eq. (3.31) gives the two-beam result for the intensity of a strong second-order beam $\mathbf{g}$ near the Bragg condition. We now consider the perturbation to its intensity due to the unavoidable excitation of a weaker first-order reflection $\mathbf{h} = \mathbf{g}/2$. Bethe's eigenvalue perturbation expression (3.37) may now be used in this two-beam expression to account for the effect of beam $\mathbf{h}$. If the analysis is limited to these three beams, only the term with $\mathbf{k} = \mathbf{h}$ and $\mathbf{g} - \mathbf{k} = \mathbf{h}$ contributes in Eq. (3.37). (We note that $\mathbf{g} - \mathbf{h} = \mathbf{h}$.) Since $U_\mathbf{g}^{\text{eff}}$ depends on $\gamma = m/m_0$ and hence the accelerating voltage $E_0$, an absence of intensity in $I_\mathbf{g}(S_\mathbf{g}, t)$ (called an extinction) occurs *for all thicknesses* (in this three-beam approximation) at the critical voltage $E_c$ when $U_\mathbf{g}^{\text{eff}} = 0$. (A strong minimum is seen in the many-beam case.) We may thus set Eq. (3.37) equal to zero and solve the resulting equation for $\gamma = m/m_0$ with $S_\mathbf{g} = 0$ (which fixes $S_\mathbf{k}$ by the Ewald sphere geometry). We may then use the expression

$$E = \frac{m_0 c^2 (\gamma - 1)}{|e|} \tag{5.8}$$

for the accelerating voltage in order to find $E_c$. The result is (with $\mathbf{g} = 2\mathbf{h}$)

$$E_c = \frac{m_0 c^2}{|e|} \left( \frac{h^2 g^2 |e| V_{2h}}{2 m_0 |e|^2 V_h^2} - 1 \right) \tag{4.9}$$

This expression gives an approximate value of the critical voltage $E_c$ in terms of the two lowest-order structure factors. The extension of these equations to the noncentrosymmetric case was discussed in Section 5.1.

Alternatively, we may use Eq. (3.57), which gives the exact solution of the symmetric three-beam case shown in Fig. 3.6a. The critical voltage effects occurs when $\gamma^1 = \gamma^3$, at

$$\mathbf{h}^2 = (U_\mathbf{h}^2 - U_{2\mathbf{h}}^2)/U_{2\mathbf{h}}$$

This is satisfied at the critical voltage

$$E_c = \frac{m_0 c^2}{|e|} \left[ \frac{h^2 g^2 |e| V_{2h}}{2m_0 |e|^2 (V_h^2 - V_{2h}^2)} - 1 \right] \qquad (5.10)$$

Equation (5.9) agrees with Eq. (5.10) for $V_h^2 \gg V_{2h}^2$, which is usually the case. A critical voltage does not occur if $V_h^2 < V_{2h}^2$. At the critical voltage, $U^{\text{eff}} = 0$ and the two-beam intensity expression (3.38) exactly cancels. In the general many-beam case, this destructive interference between two Bloch waves for one particular beam occurs when two eigenvalues $\gamma^i$ of the structure matrix $\mathbf{A}$ become degenerate (equal) for a particular voltage $E_0$. The symmetries of the two Bloch waves also interchange as $E_0$ passes through $E_c$. The problem has been analyzed in detail in the three-beam approximation for acentric crystals [see Section 3.6 and work by Marthinsen et al. (1988)].

The effect of accidental degeneracies in the eigenvalues has also been observed in the axial orientation by Matsuhata and Steeds (1987) at more conveniently lower voltages [see also the paper by Buxton and Loveluck (1977)]. The effect may consist of a bright spot in the central disk of a zone-axis CBED pattern, or of a reversal of symmetry in the other orders. Five- and seven-beam closed-form solutions (reducible by symmetry) are used to analyze these results, together with many-beam calculations. Corresponding effects are found in the HOLZ reflections.

It was realized at an early stage that, since Eq. (3.37) contains the product $|K| S_\mathbf{g}$, minima can be expected in CBED patterns for particular excitation errors (i.e., at certain points in the pattern) at voltages other than $E_c$ (which determines $(|K|)$. An error in choosing $E_c$ may be compensated for by changing $S_\mathbf{k}$, i.e., by looking at a different point in the CBED pattern. This is the basis of the intersecting Kikuchi line (IKL) method described below, which might therefore be said to provide "critical voltage at any voltage" (Taftø and Gjønnes, 1985). It is also the basis of the phase-determination method outlined in Section 5.1 and is used in the "CBED critical voltage" method of Sellar et al. (1980). In this work the temperature of the sample (and hence the Debye–Waller factor) was used for fine-tuning the critical voltage condition. In general, degeneracies may be identified for certain combinations of excitation error and accelerating voltage by solving, in closed form, the few-beam dispersion equations, as described in Section 3.6. These solutions have now been given for many orientations of high symmetry (Fukahara, 1966). Intensity minima can therefore be found in many orientations of high symmetry.

The first observations of the critical voltage effect were made by Nagata and Fukuhara (1967), Uyeda (1968), and Watenabe et al. (1968). The accuracy of the method is usually better than 1% and may be as high as 0.6% in $V_g$. It has also been used to measure Debye temperatures and order parameters in alloys (Lally et al., 1972). Accuracy is ultimately limited by the Debye–Waller factors used and the contrast of the intensity minimum.

In practice, by varying $E_0$ the experimentalist seeks either the disappearance of the Kikuchi line associated with the satisfied Bragg beam (Thomas, et al., 1974), or the central maximum of a dark-field bend contour (Lally et al., 1972), or that of the central maximum of the rocking curve displayed in a CBED disk in the CBED-CV method (Sellar et al., 1980). Many-beam dynamical calculations are then performed, in which the two lowest-order structure factors are adjusted to give a minimum of intensity in the second-order (satisfied) reflection for the observed value of $E_c$. Absorption coefficients must be known.

Spence (1992a,c) summarizes measurements of structure factors made by the critical voltage (and other) methods. Readers must be cautioned against comparing values recorded at different temperatures. Errors may be as small as 0.4% in the measured $f_j$ values for Si (111) (Hewat and Humphreys, 1974), which corresponds to an error of 0.11% in the corresponding X-ray scattering factor recovered using Eq. (3.10). This may be compared with the best accuracy obtainable by the X-ray diffraction Pendellosung method for the same reflection, which is 0.07%. For the determination of Debye temperatures, short-range order parameters, and scattering factors in metal alloys, see work of Shirley and Fisher (1979).

## 5.3. THE INTERSECTING KIKUCHI-LINE AND HOLZ-LINE METHODS

The critical voltage method has the important advantage of being a "null" method, in which a minimum of intensity is sought. Most electron diffraction methods depend on the direct measurement of electron intensities and their comparison with calculations. The intersecting Kikuchi line (IKL) method is unusual in that it allows structure factors to be determined simply from the distance between features measured on a photographic plate. This distance is the width of the gap which appears when Kikuchi lines (or HOLZ lines) cross. We shall also refer to the intersecting HOLZ line (IHL) method.

Instead of following the geometric locus defined by the Bragg condition ($S_g = 0$), HOLZ lines separate where they cross, as shown in Fig. 3.8. The resulting gap between the lines at the Bragg condition has been studied for many years since the first work of Shinohara (1932) on Kikuchi-line patterns. In X-ray diffraction it is known as the Renninger effect. It is this gap which is measured in the IKL and related methods.

The situation can be analyzed with the aid of the three-beam dynamical theory for acentric crystals given in Section 3.7. There we saw [Eq. (3.71)] that the change in excitation error along the line $S_g = S_h$ across the gap is

$$\Delta S_h = |U_{g-h}|/K \tag{5.11}$$

Hence the structure factor $U_{g-h}$ can be measured from a measurement of the gap. Figure 3.8 shows the parabolas $S_+ = 0$ and $S_- = 0$ defined by Kambe (along which intensity is greatest), in addition to the asymptotic lines $S_g = 0$ and $S_h = 0$ along which the geometric Bragg condition is satisfied. On the hyperbolas (as discussed in Section 3.7) the incident beam is constrained to move such that

$$S_g = \frac{|U_{g-h}|^2}{4K^2 S_h} \tag{5.12}$$

The intensity variation along the hyperbolas is given by Eq. (3.68) in three-beam theory.

The latter results apply to elastic Bragg scattering. Yet similar gaps are seen in Kikuchi-line patterns. The assumption is made that the inelastic scattering responsible for Kikuchi lines is generated continuously throughout the crystal, and so can be described by integrating Eq. (3.68) over thickness. This affects the last term in Eq. (3.68), but does not alter the expression for the gap.

A deeper understanding of both the critical voltage and the closely related IKL method can be obtained by plotting the relevant many-beam dispersion surfaces for a range of incident-beam directions and accelerating voltages. The eigenvalues $\gamma^i$ then describe the deviation of these surfaces from asymptotic spheres of radius $|K|$ erected about each reciprocal lattice point. It was seen in Section 3.4 that this deviation is greatest near Bragg conditions. The surfaces may touch at certain accelerating voltages and orientations, giving rise to the CV and IKL effects. In three dimensions, the form of these surfaces can be very complicated.

The IKL and IHL techniques are simple to apply, however finding suitable cases showing a clear splitting unperturbed by other interactions at a convenient accelerating voltage is somewhat fortuitous, since a search through all possible diffraction conditions is rarely practical. The method is thus simpler and more flexible (if applicable), but less accurate than the critical voltage method. An instructive case of high symmetry in which only one excitation error is varied is given in Taftø and Gjønnes (1985), applied to SiC. Other applications of the method can be found as follows: The concentration of vanadium atoms at interstitial tetrahedral sites in vanadium oxide above the ordering temperature has been determined using structure-factor measurements based on the IKL technique by Høier and Andersson (1974). Structure factors for Cu and $Cu_3Au$ have been measured using this method by Matsuhata et al. (1984). Structure factors in silicon have also been measured by the IKL technique (Terasaki et al., 1979). The displacement of Kikuchi lines near intersections can also be understood using this approach (Høier, 1972). The resulting shifts observed by Høier are exactly those discussed in the next section on strain measurement.

## 5.4. LOCAL MEASUREMENT OF STRAINS, COMPOSITION, AND ACCELERATING VOLTAGE USING HOLZ LINES

This section summarizes work on the measurement of strains in thin crystals from measurements of the positions of HOLZ lines. The material of Sections 2.3 and 3.8 must be read in conjunction with this section.

It is important to appreciate that, in very thin crystals, the strains measured may have resulted from the thinning process. These thin-film relaxation processes have been studied in detail (Bangert and Charsley, 1989; Perovic et al., 1991; Treacy et al., 1985). They present a much more serious problem for the very thin specimens used for lattice imaging (or nanodiffraction work of the type described in Section 8.3 with overlapping orders) than for CBED studies.

HOLZ reflections involve long extinction distances, so it was widely believed that HOLZ lines in the central disk might be relatively insensitive to thickness changes and local composition variations, and that a purely geometric description of their positions could therefore be used to measure strain, based on the simple Bragg law [Eq. (2.4)]. The value of these lines lies also in the fact that they transfer information on fine lattice spacings into the region near the optic axis, where geometric lens distortions are small and where more accurate measurements can be made. HOLZ lines are also sharp (owing to the curvature of the Ewald sphere), so that rather accurate measurements can be made from them. Since CBED patterns can be obtained from nanometer-sized regions, the usefulness of such an interpretation for local mapping of strains was obvious. These strains may be related to composition variations, using Vegard's law relating composition to local lattice parameter changes under an isotropic expansion or contraction. The difficulties with this approach will be apparent from the discussion of Section 3.6. Approximate and exact methods of dealing with these difficulties are now described.

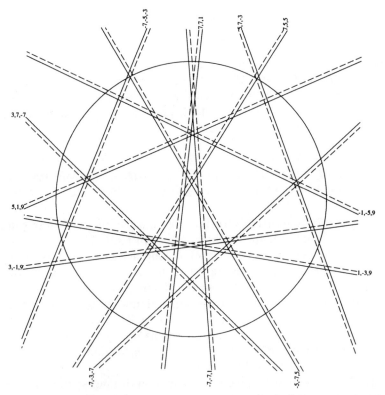

*Figure 5.3.* HOLZ lines for the Si [−2, 2, 1] zone axis, computed for 100 kV (full lines) and 99 kV (dashed lines).

Figure 5.3 shows a calculated pattern of HOLZ lines from the zero-order disk of the $(-2, 2, 1)$ zone axis in silicon, for 100 kV and 99 kV. The program listed in Appendix 5.1 has been used to generate a pattern of lines according to the kinematic Eq. (2.15). From the figure, it is clear that a large and easily measured shift occurs in the HOLZ lines for a change in accelerating voltage of much less that 1 kV. In addition, the two-dimensional nature of the pattern of lines makes angular calibration unnecessary, since the ratio of the distances between intersections can be used. (In the systematics orientation, additional angular calibration would be needed to determine $E_0$ from $d_{hkl}$.) The strain $\Delta a/a$ due to a shift $\Delta \theta$ in a HOLZ reflection **g** may be estimated from

$$-\frac{\Delta a}{a} = \frac{\Delta \theta}{\theta} = \frac{\Delta E_0}{2E_0} = \frac{\Delta g}{g} \approx \frac{2K_0\gamma}{g^2} \qquad (5.13)$$

From this we see that the sensitivity of HOLZ line positions to strain increases with accelerating voltage and angle, so that the highest-order HOLZ lines should be used. HOLZ lines become sharper in thicker crystals. If these are faint, a cooling holder and energy filter will increase their contrast. Any stacking disorder in the beam direction will reduce the contrast of HOLZ lines (see Section 9.2). In an instrument fitted with continuous accelerating voltage controls, strain may be measured directly at the microscope in cubic crystals by restoring a HOLZ line pattern to its reference shape for the unstrained crystal and noting the voltage change $\Delta E_0$ needed to do this. Since a fractional change $\Delta$ in lattice spacing produces a fractional change $\Delta/2$ in accelerating voltage, it is clear that HOLZ lines may provide a rather accurate method of relative strain mapping (Randle et al., 1989). In the general case involving noncubic crystals, it would be necessary to measure all the independent lattice parameters allowed by the symmetry to fully characterize the strain. (The maximum number is six, three cell constants and three angles, however a frequently used assumption is that the lattice expansion is isotropic, so that no symmetry change is involved.

We have seen in Section 3.8 that dynamical shifts can complicate the situation. The most compelling experimental evidence for the importance of dynamical corrections comes from the work of Lin et al. (1987), who found variations of up to 6 kV when using Eq. (2.4) to determine the same microscope accelerating voltage from indexed HOLZ lines taken from different zone axes of the same silicon crystal. We see in general that the effects of composition and lattice strain on HOLZ-line positions become entangled in the presence of strong dynamical interactions. The implications of this work for attempts to map local composition variations based on Vegard's law are as follows:

(1) Simulations (Bithell and Stobbs, 1989; Lin et al., 1989) show that changes in atomic number alone greater than about five will cause significant errors in lattice parameter determination. The neglect of dynamical effects may lead to errors in determining lattice parameters of 0.2%.

(2) Elastic relaxation in the very thin, heterogeneous samples used in TEM may render them unrepresentative of bulk material (Treacy et al., 1985). These thin-film effects must be expected to influence HOLZ-line analysis. They may be minimized by choice of geometry (see Section 8.5), or through the use of thicker samples, which becomes possible if an imaging energy filter is used. In unfavorable cases these effects will be much larger than the dynamical corrections (Kaufman et al., 1986).

(3) An approximate procedure for measuring local strains has been proposed as follows (Lin et al., 1989). The true accelerating voltage $E_0$ is measured using many-beam calculations and a silicon reference sample. The correction $\Delta E_0$ is calculated from Eq. (3.85) using structure factors from the ZOLZ of the sample of interest. This value of $\Delta E_0$ is subtracted from $E_0$ to give an effective "kinematic" voltage $E_0'$, corresponding to the wavevector $K_{\text{eff}}$ shown in Fig. 3.10b. The silicon sample is replaced by the sample of interest, and lattice parameters deduced from measurements of the HOLZ-line positions using the voltage $E_0'$. This procedure assumes that branch (1) of the dispersion surface is dominant. The effect of several will be to cause fine structure to appear within the HOLZ lines, as discussed in Section 3.6 (see Fig. 3.11).

(4) An alternative and preferable approach is to choose an orientation far from a zone axis in which few beams are excited, so that the dynamical correction is greatly reduced. For the most accurate work, the automated refinement procedure described in Section 4.2 may then be applied to measurements of *many* distances between HOLZ line intersections. These may be analyzed using the true accelerating voltage, which must be measured by employing a standard crystal (such as silicon). The cell constants are treated as adjustable parameters in the refinement instead of the structure factors. In $YBa_2Cu_3O_{7-\delta}$, for example, X-ray and neutron diffraction studies have given a reliable relationship between the $b/a$ ratio and oxygen deficiency $\delta$. The ratio $b/a$ was measured by fitting calculations to the distances between many HOLZ intersections (Zuo, 1992a). These were measured accurately by digitizing CBED patterns taken in orientations well away from zone axes. When fitting the HOLZ lines Eq. (3.84) was used. The small dispersion $\gamma$ is assumed constant and estimated from many-beam calculations. A full nonlinear least-squares error analysis was used, as in the Rietvelt method

for neutron diffraction, and a goodness-of-fit index defined, as in Section 4.2. In this way it was possible to determine the accelerating voltage to an accuracy of 14 V, and the cell constants to an accuracy of 0.001 and 0.0006 Å. The oxygen deficiency was found to be 0.1 in one local region.

(5) The later comments concern absolute lattice parameter determinations. Relative changes, however, can be observed with much greater sensitivity, although these may also be influenced by composition variations. If only relative changes in lattice parameter are required, the two-dimensional HOLZ pattern may be fitted to kinematic calculations in some unstrained region and the accelerating voltage used as a free parameter for best fit.

(6) The important question arises as to the "spatial resolution" of this technique. From how small an area can strains be measured using HOLZ lines? This question is discussed in Section 8.3 on CBED from defects, and in the note at the end of Chapter 8. Throughout this section, however, we have been implicitly assuming the validity of the column approximation (Hirsch *et al.*, 1977) and treating the crystal under the probe as uniformly strained. For strain fields which vary slowly on the scale of the probe diameter, it is a reasonable approximation to assume that the method measures the average strain for the crystal under the probe. The failure of this approximation is discussed in Section 8.3.

We conclude that high accuracy in absolute lattice parameter or accelerating voltage determination can be achieved only at the cost of considerable computational effort. The average composition and structure factors must be known, and thin-film relaxation effects pose the greatest problem in certain materials, depending on elastic constants, specimen geometry, and inhomogeneity. Some rather rough conclusions may be drawn, however. At about 100 kV, and ignoring thin-film relaxation and atomic number and density effects: strains and accelerating voltages may be determined to about one part in 100 at best using straightforward application of the Bragg law [Eq. (2.4)]; they may be determined to about one part in 1000 at best using the perturbation correction for composition effects [Eq. (3.85)]; they may be determined to perhaps one part in 10,000 at best using a full dynamical refinement.

The following papers have applied this technique to the stated systems with varying degrees of sophistication in the data anslysis: Si (dynamical match and first full analysis) (Jones *et al.*, 1977), Si/Ge superlattices (Eagelsham *et al.*, 1989; Fraser *et al.*, 1985; Maher *et al.*, 1987), precipitates in superalloys (Ecob *et al.*, 1982; Vincent and Bielicki, 1981), strain in silver halide particles (Vincent *et al.*, 1988). Strains associated with interfaces have also been measured in superalloys (Ecob *et al.*, 1981; Fraser, 1983) and at grain boundaries (Stoter, 1981).

The above discussion has referred to incoherent CBED patterns (as defined in Section 8.1). The strain analysis technique may also be applied to patterns obtained from coherent nanodiffraction probes on instruments fitted with a field-emission gun. Provided that the strain varies slowly on the scale of the electron probe and that the orders do not overlap, the patterns will be the same as the incoherent ones. Thus the "strain mapping" technique may be extended down to the nanometer range (but see the *note added in proof* at end of Chapter 8).

As an example, we consider work (Pike *et al.*, 1991; Pike *et al.*, 1987) on the measurement of strains in Si/Si–Ge multilayers. The multilayers consisted of 45 nm of pure silicon, followed by 9 nm of 85 at% Si alloyed with 15 at% Ge. A modified Vacuum Generators HB501 STEM was used, with a quadrupole imaging energy filter and CCD camera, viewing a scintillator screen. The fine HOLZ lines in the central CBED disk were used for the analysis, and these patterns were obtained from regions of crystal smaller than 1 nm in diameter. (A typical pattern obtained from this system is shown in Fig. 9.4.) Dynamical corrections were incorporated into the analysis using Eq. (3.85), and a sparse (low-symmetry) zone axis used to minimize this effect. For this [012] zone axis at 100 kV, the dynamical shift due to composition change alone between HOLZ lines in silicon and those in 85 at% Si–Ge corresponds to a reduction in the effective accelerating voltage of 100 ($= \Delta E_0$). By fitting the energy-filtered HOLZ lines to dynamically corrected kinematic calculations, these workers were able to measure the tetragonal distortion $\varepsilon_T = 0.99 + 0.09\%$. (The quantity $\varepsilon_T$ is defined as the relative difference between the lattice parameter parallel and perpendicular to the growth direction.) This gives a misfit parameter of $0.56 + 0.05\%$, which may be compared with the X-ray measured value of $0.62 + 0.04\%$. The X-ray value is an average over many layers—the error in the microdiffraction result is limited due to broadening of the HOLZ lines by the finite width of the single layer from which the measurement was taken.

In work on Si–Ge multilayers with a slightly different composition and using a larger probe, evidence for monoclinic forms of the alloy have also been found in thin films (Eagelsham *et al.*, 1989). Unit-cell parameters were measured with an accuracy of about 0.0001 nm in this work, however the form of the dynamical corrections is not indicated. The significance of these changes in symmetry in thin films for "band-gap engineering" of thin-film devices are clear. (Figure 8.16 shows a wide-angle CBED image of the variation in HOLZ-line position across GeSi multilayer interfaces.)

The use of HOLZ lines in CBED patterns to measure thermal expansion coefficients is described by Angelini and Bentley (1984).

## 5.5. POLARITY DETERMINATION IN ACENTRIC CRYSTALS

Chapter 7 describes methods for determining whether a given crystal contains a center of symmetry. Those without such a center are known as polar or acentric, and the need has frequently arisen for a simple technique for determining polarity. In the simple case of the sphalerite or cubic zinc sulfide structure (important for the study of III–V semiconductors), polarity determination corresponds to the ability to distinguish [111] from [$\bar{1}\bar{1}\bar{1}$], and hence to distinguish the "gallium side" of a GaAs wafer from the "arsenic side." Figure 5.4a shows the GaAs structure in projection and we see that, in the ideal case, for a crystal whose total charge is to be zero, the termination of a (111) slab at the shuffle (widely spaced) planes by Ga on one side must be accompanied by As termination on the other. (In practice, the high energy of the resulting dipole layer will be reduced by surface steps and other defects.)

In electron microscopy, one may wish to relate the orientation of a CBED pattern to the absolute orientation of the crystal, and hence, for example, to a lattice image of the same region. A method for achieving this was described by Taftø and Spence (1982), in which three- and four-beam patterns from Ge and GaAs were compared. Ge lies between Ga and As in the Periodic Table, and we may think of GaAs's sphalerite

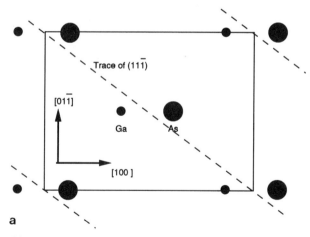

*Figure 5.4.* (a) The sphalerite structure projected along [011]. For GaAs, large disks are As and small disks are Ga. The diagram is in the correct absolute orientation with respect to Figs. 5.4b and 5.4c, allowing polarity to be determined. (b) The (−200) CBED disk for GaAs shown in the center is at the Bragg condition. The (000) disk is at the right, and the (−400) at the left. (c) The (200) CBED disk for GaAs shown in the center is at the Bragg condition. The (000) disk is at the left, and the (400) at the right.

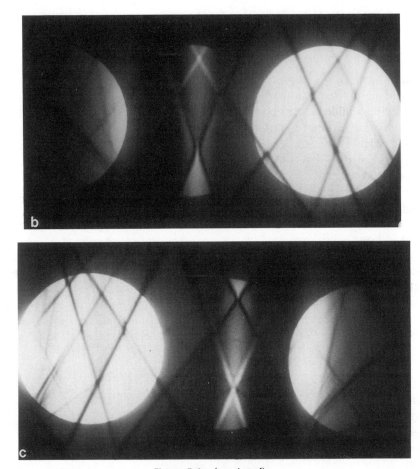

*Figure 5.4. (continued)*

structure as being derived from the diamond structure of Ge by small changes in the atomic number of half the atoms. This change results in a loss of the center of symmetry, and the loss of the ($h$00) mirror plane of symmetry present in Ge. For both Ge and GaAs, reflections for which $h + k + l = 4n$ are allowed. Additional weak "difference" reflections, such as (200), for which $h + k + l = 4n + 2$, are also permitted in GaAs. Since there is no mirror plane parallel to (200) in GaAs, we therefore expect the (200) and ($\bar{2}$00) CBED disks to differ for GaAs, but to be the same for Ge, in accordance with the test for a center of symmetry described in Section 7.1. (Here each disk must be recorded successively at the Bragg condition.)

A more useful quantitative rule can, however, be derived if the weak (200) reflection is coupled to certain HOLZ reflections. Since all these reflections are weak, with extinction distances normally much greater than typical sample thicknesses, quantitative predictions can be made concerning the intensities observed using, for example, the Cowley–Moodie polynomial series. In particular, if, in addition to the $(\bar{2}00)$, the $(\bar{1}\bar{1}, 1, \bar{1})$, $(9, 1, \bar{1})$ HOLZ reflections are simultaneously set to the Bragg condition, then a dark cross is seen in the $(\bar{2}00)$ CBED disk, as shown in Fig. 5.4b. If, however, a small tilt is used to bring the (200) and the $(11, 1, \bar{1})$, $(\bar{9}, 1, \bar{1})$ HOLZ reflection to the Bragg condition, the same cross appears bright, as shown in Fig. 5.4c. In Ge, identical bright crosses are seen in both cases. (We recall that this reflection is forbidden in Ge, and only occurs as a result of multiple scattering.)

As a result of the large extinction distances, the effect can be understood by considering only the structure-factor phases. We neglect coupling between the HOLZ reflections. The total phase change for $n$-fold multiple scattering is, from Eq. (7.3), for weak reflections at the Bragg condition,

$$\Psi = -\frac{n\pi}{2} + \sum_{i=1}^{n} \phi_i \qquad (5.14)$$

For the case of Ge, with an origin chosen at the center of symmetry, the two scattering paths into the (200) via the HOLZ reflections interfere constructively. These paths are (000) to $(11, 1, -1)$ to (200), and (000) to $(-9, 1, -1)$ to (200). There is no direct path, since the (200) is forbidden. For the $(-200)$ condition, the situation is the same since, by virtue of the center of symmetry, the structure factors are identical, and a bright cross also results. For GaAs, with the same origin, the direct path into the (200) scatters with the same phase (say 0) as the HOLZ reflections, and the pattern is bright at the Bragg condition. For the $(-200)$ geometry, however, the direct path is $\pi$ out of phase with the others and a dark cross is seen.

The outcome of this is that the absolute orientation of the crystal may be determined. In Fig. 5.4a the diagram has been correctly oriented with respect to the experimental patterns in Fig. 5.4b and 5.4c. The bright cross is seen on the side nearest the As termination for a (111) slab.

Similar effects are seen when the more sensitive, quantitative refinement methods of Section 4.2 are used. For example, reversing the polarity in the automated BeO refinement described in Section 5.1 was found to increase the value of the fitting index $\chi^2$ by a factor of three or four.

A quantitative analysis of these CBED geometries has also been given by Ishizuka and Taftø (1984), who thereby draw conclusions about the ionic state of GaAs. They find that there is weak charge transfer from Ga to As.

A second class of methods for determining polarity depends on asymmetries introduced into two-beam theory by absorption. We have seen in Section 3.9 that if absorption is included exactly (not using perturbation theory), then small differences appear between the **g** and −**g** rocking curves in two-beam theory. These are responsible for asymmetries in Kikuchi lines (Allen and Rossouw, 1989; Bird, 1990; Bird and Wright, 1989) and on inner-shell energy-loss spectra (Taftø, 1987), both of which may also be used to determine the polarity of acentric crystals. (Energy "absorbed" by inelastic processes from the elastic "channel" makes a positive contribution to peaks in the energy loss spectrum.)

## 5.6. THE INVERSION PROBLEM

The inversion problem refers to the problem of finding values of $U_g$ directly from measurements of the dynamical beam intensities, rather than by refinement based on forward computations. Two approaches have been made to this generally unsolved problem—the first based on closed-form inversion of Eq. (3.16) (for a finite number of beams), and the second on computational strategies in which algorithms are devised which might be shown to converge to a unique solution.

An analytical approach is given by Moodie (1979), for the three-beam case. Here closed-form expressions are given for the retrieval of the three structure-factor amplitudes and the three-phase invariant from the positions of certain lines in a three-beam CBED pattern along which the intensity has two-beam form.

Several computational schemes have been tested in unpublished work by Spence and Katz (1979). [See also Speer *et al.* (1990).] We write the scattering matrix defined following Eq. (3.26) in the form $\mathbf{S} = \exp(2\pi i \mathbf{A} t)$ for a centrosymmetric crystal without absorption. The problem then reduces to that of finding the entries $U_{\mathbf{g-h}}$ of the structure matrix **A** from a knowledge of the moduli of the entries in one column of **S**. We assume that the diagonal of **A** (the excitation errors), the accelerating voltage, and the thickness are known. Two cases might be considered—one in which the complex entries of one column of **S** have been determined [by the interferometric methods described in Section

8.4, for example, using a biprism in STEM (Cowley, 1991)], and one in which only the moduli of the elements are known. These data may be recorded as a function of thickness, orientation, and accelerating voltage. The following general comments may be made for centric crystals without absorption:

(1) The question of uniqueness must be investigated. Let $\mathbf{S} = \exp(2\pi i \mathbf{A} t) = \exp(2\pi i \mathbf{B} t)$. Then uniqueness is established by showing that $\mathbf{A} - \mathbf{B} = 0$. Since $\mathbf{A}$ and $\mathbf{B}$ commute (they have the same eigenvectors), $\exp[2\pi i (\mathbf{A} - \mathbf{B}) t] = \mathbf{S}\mathbf{S}^{-1} = \mathbf{I}$. This is only possible for all $t$ if $\mathbf{A} = \mathbf{B}$, and the inversion is therefore unique. (Otherwise, if it were not, in principle, two different structures could gave rise to the same diffraction pattern. In that case the inversion problem is not well posed.)

(2) The eigenvalues of $\mathbf{S}$ are $\lambda_i = \exp(2\pi i \gamma_i t) = r \exp(i \theta_i)$. Hence

$$2\pi i \gamma_i t = \ln \lambda_i = \ln r + i(\theta_i + 2n_i \pi)$$

gives $\gamma$ in terms of $\lambda$ if the $n_i$ are known. If complex $\mathbf{S}$ can be found at several thicknesses (or nonrelativistic wavelengths), the eigenvalues of $\mathbf{A}$ can therefore be found from the period in thickness $1/\gamma_i$ of the real part of $\lambda_i$. The eigenvectors of $\mathbf{S}$ are those of $\mathbf{A}$. Hence $\mathbf{A}$ can be found from $\mathbf{S}(t)$.

(3) Many other physical constraints may be placed on the retrieval process, and use can be made of other properties of the matrices (e.g., the trace of $\mathbf{A}$, related to the determinant of $\mathbf{S}$, is known, etc.). So far, however, all these computational approaches to the inversion problem have foundered on the difficulty of reconstructing all of $\mathbf{S}$ from a knowledge of some of its entries. Retrieving $\mathbf{A}$ from complex $\mathbf{S}(t)$, however, appears to be possible despite a problem in the assignment of particular eigenvalues to corresponding eigenvectors.

The use of the difference between data recorded at two slightly different accelerating voltages has also been considered (Spence and Katz, 1979). This is equivalent in the nonrelativistic regime to a small change in thickness $t$. A kinematic analysis may then be used if the data consist of complex amplitudes. These can be obtained from experiments using coherent overlapping CBED orders (Spence, 1978b).

# 6

# Large-Angle Methods

In this chapter we review the various instrumental techniques which have been developed to obtain an angular view of a diffracted order which is greater than the Bragg angle. These methods have been reviewed by Vincent (1989). Such an angular expansion is required for space-group determination of crystals with a large unit cell, in which overlap of low orders may occur at such a small illumination angle $\theta_c$ that little or no rocking-curve structure can be seen within the orders. It has also been discovered that many narrow high-order reflections may be observed simultaneously using large-angle techniques. The intensity of these reflections appears to agree with kinematic structure factors, and so can sometimes be used to assist in solving structures or to measure a static lattice modulation, for example, which modulates structure-factor intensities.

Closely related to the ZOLZ large-angle methods are the methods for recording many HOLZ reflections belonging to the same layer in reciprocal space on a single micrograph. All these techniques are related to the ronchigrams and shadow images described in Section 8.4, however they differ according to the angular range over which the illumination is coherent (in this chapter we deal only with "incoherent" conditions), and by the application of different techniques to prevent the overlap of orders. A fuller discussion of the distinction between the various cases is deferred until Section 8.5.

## 6.1. LARGE-ANGLE METHODS FOR THE ZOLZ

Following earlier work using bend contours in bright-field images of dome-shaped samples, and methods based on tilting of the sample, recent research has concentrated on two main techniques for recording a single CBED order over a very large angular range. These are (1) methods which use time-resolved electronic control of the beam-tilting coils, and (2) methods which depend on the use of a defocused probe.

Several groups have experimented with external control of the incident-beam tilt currents, which control the direction of the illumination onto the sample (Eades, 1980; Krakow and Howland, 1976; Tanaka et al., 1980). The beam-tilt coils may be driven from an existing STEM scan unit or from an external computer. The essential idea is to operate the microscope in the selected area mode, so that a "point" diffraction pattern is formed. The incident beam is then rocked over a certain angular range, while ensuring that the same selected area of the sample contributes to the diffraction pattern. For every incident-beam direction in this angular scan the intensity of the transmitted beam is displayed on a video monitor. This gives a wide-angle view of the (000) CBED disk. In practice, deflection coils both before and after the sample are required, since the beam must be both "rocked" and "unrocked," so that it remains fixed on the stationary detector. A single scan generator drives both the deflection coils and the display. Alignment of the system (so that the beam remains on the same specimen point during the scan) and compensation for image rotation are both important. The angular resolution of the system is about 0.1 mrad, and the area of the crystal which can be studied typically has a diameter of about 250 nm, limited by spherical aberration. The angular range of the scan is typically about 3°. We note that other diffracted beams in the "point" pattern can be excluded by using a small on-axis detector, so that an angular view which is larger than the Bragg angle can be obtained without any overlap of orders. It is also possible to display other diffracted beams ("dark field" disks) by deflecting the pattern at the detector to bring the spot of interest onto the axis. Using multiple detectors, several orders can also be displayed simultaneously over a large angular range.

Advantages of this "double-rocking" technique (Eades, 1980) include the great flexibility of the scan patterns which can be generated (if these are under software control) and the compatibility of the system with conventional energy filters. The axial spot may be simply passed through a serial ELS system set to the zero-loss (or other) peaks. One may then take advantage of the large dynamic range of a photomultiplier detector. The formation of a wide-angle CBED pattern from electrons which have excited inner-shell electrons, for example, using the vacuum generators HB5 can be found in Higgs and Krivanek (1981). The electronic control of signal levels and contrast also provides all the signal-processing advantages of scanning electron microscopy. "Hollow cone" illumination conditions (which result from the use of an annular illumination aperture) can be simulated readily (Tanaka and Terauchi,

1985b). In particular, if the beam is tilted until one of the HOLZ lines in the outer ring is centered on the axial detector, and the beam then scanned on the surface of a cone about the axis, a wide-angle view of HOLZ structure is obtained (Kondo et al., 1984).

Disadvantages of double-rocking methods include the need for new electronics, the inefficient serial detection, tedious alignment procedures, and the large specimen area from which the pattern is obtained.

The second method [known as the "Tanaka" or LACBED method (Tanaka et al., 1980)] allows parallel detection of the entire wide-angle pattern and requires no instrumental modifications. The pattern is again, however, obtained from a rather large area of sample. A clear description of the method is given in Eades (1984). Figure 6.1 shows the principle of the method, while Fig. 6.2 shows a pattern from (111) silicon taken at 100 kV by this method (Yamamoto, 1982). In Fig. 6.1, the CBED probe has been focused below the sample, forming an image of the electron source in the plane of the selected area aperture. A source image is formed in every diffracted order, as shown. The aperture can then be used to isolate one source image and so prevent other diffracted beams from contributing to the image. Because the source images are small at the crossover, the illumination cone can be opened up to a semiangle which is larger than the Bragg angle. The price to be paid for

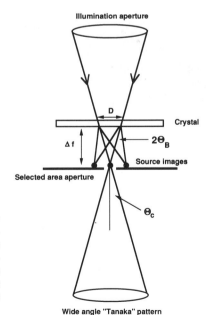

Figure 6.1. Ray diagram showing the formation of a "Tanaka" wide-angle CBED pattern. The probe images are assumed limited in size by the spherical aberration of the probe-forming lens. The viewing screen images the disk shown at the bottom of the figure. In the CBIM method, the viewing screen is conjugate to the downstream face of the sample.

*Figure 6.2.* Tanaka wide-angle pattern obtained from the silicon (111) zone axis at 100 kV (Yamamoto, 1982). The angular range covered exceeds the Bragg angle.

this is the large area of sample illuminated by the out-of-focus probe at the sample. In addition, different regions of the sample contribute to different parts of the diffraction pattern. Patterns may be obtained with the probe focused either above or below the sample—best results seem to be obtained with it below the sample (for TEM instruments).

The simplest procedure for obtaining these patterns (in principle) is to form an in-focus image of the electron probe in the plane of the selected area aperture, and then to adjust the sample height slightly to take the sample out of focus. Figure 6.2 was obtained using the specimen height controls on an EM400 in this way, but a better procedure is based on changing the objective lens focus, as follows. (1) Set up the microscope in the selected area mode using a medium-sized condenser aperture. Set the eucentric height and focus the image. (2) Overfocus the objective lens somewhat (excess lens current). The viewing screen will then be conjugate to a plane just below the sample and the image of the sample will appear out of focus. (3) Adjust the condenser lens until the probe comes into focus on this same plane, just below the sample. (4) Introduce the selected area aperture and center it about the central probe image (spot). (5) Switch to diffraction mode and replace the condenser aperture with the largest size available. This aperture controls the contribution of spherical aberration to the probe size. An important finding is that the use of the smallest selected area aperture together with the largest permissible defocus minimizes the contribution of inelastic scattering to the pattern. This effect has been studied in detail by Jordan *et al.* (1991) (see also Section 9.3). A similar technique is used to image other diffracted orders. Here the order of interest is brought onto the optic axis using the dark-field tilt controls.

If the geometric probe size and the effects of spherical aberration are both small [see Eq. (8.3)], the diameter $D$ of the region from which the pattern is obtained is given approximately by

$$D = 2\theta_c \Delta f \qquad (6.1)$$

so that the smallest defocus which allows separation of the orders should be used to minimize $D$. A small geometric source image (consistent with sufficient intensity on the viewing screen) also facilitates the separation of orders—this is controlled by the demagnification settings of the condenser lenses. Under the large-angle conditions at which these patterns are formed the effects of spherical aberration (proportional to $\theta_c^3$) dominate those due to diffraction [see Eq. (8.3)]. By assuming that the geometric source size is negligible, the size $D$ of the sample region contributing to the Tanaka pattern may be calculated for various illumination semiangles $\theta_c$ by setting the size of the axial aberrated probe image equal to the distance between the centers of these images. Thus

$$0.5 C_s \theta_c^3 = 2\theta_B \Delta f = \theta_B D / \theta_c \qquad (6.2)$$

Here it is assumed that the objective lens is focused on the plane of least confusion for the probe-forming lens, and that the probe images just touch in the plane of the selected area aperture; $\theta_B$ is the first-order

Bragg angle. We see that $D$ increases as the fourth power of the illumination semiangle $\theta_c$. Tables of values for $D$ and $\theta_c$ are given by Eades (1984). It is found that rather small selected area apertures are needed. Typically, for $\theta_c = 40$ mrad, a selected area aperture of about 10-micron diameter is needed. For a first-order $d_{hkl}$ spacing of 0.2 nm this gives a wide-angle pattern from an area of $D = 1$ micron at a defocus of 13 microns at 100 kV.

The Tanaka LACBED method thus has the great advantage of simplicity and parallel detection. Its disadvantages include the large area from which information is collected, and the fact that the pattern is superimposed on a kind of image if there are variations in thickness or orientation under the probe. (For the observation of dislocations, this "disadvantage" becomes an advantage, as in the CBIM technique.) Applications of the method can be found in Fung (1984). A variation of this method has also been demonstrated which makes it possible to record simultaneously on a single micrograph most of the CBED pattern, together with several diffracted orders at the Bragg condition (Terauchi and Tanaka, 1985).

A completely different large-angle method has been developed by Taftø and Metzger (1985). Here the aim is not symmetry determination, but rather the measurement of high-order structure factors. Consider a systematic row of reflections. The method depends on the fact that, because the extinction distances are large for high-order reflections, these reflections have very narrow rocking curves and, if the specimen thickness is much less than the extinction distance, may be described to a good approximation by two-beam or kinematic theory. The use of a systematic row also minimizes off-row dynamical interactions. In their study of a static displacement modulation in $V_2D$, they used a beam divergence many times greater than the first-order Bragg angle. Thus many beams are observed simultaneously at the Bragg condition. But the beam divergence used is less than that for the high-order beams (e.g. 11, 0, 0) of interest. Thus, while the direct beam does not overlap with these beams, they do overlap among themselves. However, the contribution of an adjacent beam to one at the Bragg condition is small, since its excitation error at that point is large. By comparing these high-order reflections with kinematic calculations, these workers were able to measure the static modulation of the lattice as 0.0070(5) nm. In a second application of the same method, Kolby and Taftø (1991) used high-order $(00l)$ reflections ($l = 8$ to 70) in the intermetallic $Al_{11}Ti_4Zn$ to refine the atomic displacements of the Al and Ti atoms. An accuracy in the fractional coordinate of $0.006 \pm 0.002$ was achieved. The method has also been applied similarly to the refinement of atom positions in $Al_3Zr$ (Ma et al., 1991). A third example of this approach is provided by the work of Boe and Gjønnes (1991), who compared the high-order $(00l)$ systematics in $YBa_2Cu_3O_{7-x}$ with calculations for three different models of oxygen ordering.

A useful method for providing continuous variation of the illumination semiangle by varying the strength of the "minilens" used in Philips microscopes is described by Green and Eades (1986).

## 6.2. LARGE-ANGLE METHODS FOR THE HOLZ

We have seen that, using electronic scanning of the incident-beam direction, it is possible to obtain an expanded angular view of the outer HOLZ lines (Kondo et al., 1984). It is also possible to obtain similar information without scanning from zone-axis patterns formed using an illumination semiangle equal to the angular radius $\alpha_1$ of the first-order Laue zone (Vincent and Bird, 1986). Consider the effect of increasing the illumination semiangle $\theta_c$ on the ZOLZ and FOLZ reflections. The effect is to open up an annular band of FOLZ lines, centered around those which appear at the zone axis. At the same time, the ZOLZ becomes filled with a continuous distribution of bright overlapping orders. The use of $\theta_c = \alpha_1$ is shown to be an optimum condition, which prevents the zero-layer reflections from overlapping with the HOLZ while also preventing overlap between reflections in different HOLZ. All electrons diffracted by planes in layer $n$ of the reciprocal lattice are confined to an annulus whose inner and outer radii are $\alpha_n$ and $\alpha_{n+1}$. Again, advantage is taken of the fact that the rocking curve for HOLZ reflections is very narrow due to the steep inclination of the Ewald sphere as it crosses HOLZ lattice points. Thus, although different reflections in the same HOLZ overlap geometrically, the intensity contribution of neighboring reflections is negligible. We recall the consequences of overlapping orders—if orders overlap, it means that there is an intensity contribution to one point in the diffraction pattern from two different points within the illumination aperture, via different Bragg scattering paths. In this case, one of the Bragg paths into the HOLZ from the source makes a negligible contribution because its excitation error is large, and the lines are narrow due to the curvature of the Ewald sphere. Independently of this overlap effect, dynamical scattering between all the reflections normally occurs. The intensities of these lines, however, in the case of the silicon [114] pattern at 300 kV are found to agree closely with electron structure factors, suggesting that a kinematic analysis of these intensities may be possible at some point along their length. (We have seen, in Sections 5.3 and 3.6, the special effects which can occur when HOLZ lines cross.) The good fit may also be related to the use of a sparse (low-symmetry) zone, in which dynamical interactions are minimized.

# 7

# SYMMETRY DETERMINATION

In this chapter we describe the use of CBED patterns to determine the point and space group of a small crystal. Since this determination can be made from nanometer-sized regions of crystal, the method has proven extremely useful, in combination with X-ray microanalysis, for identifying microphases in multiphase materials, and for the study of the symmetry changes which accompany phase transitions (Ecob *et al.*, 1981b). Review articles outlining the procedure can be found in Buxton *et al.* (1976), Eades *et al.* (1983), Goodman (1975), Tanaka (1989), and Steeds and Vincent (1983). The method requires one or more CBED patterns to be recorded at the highest-symmetry zone axes, including the outer HOLZ ring(s).

A summary of the general procedure is as follows: first, the point group and crystal system are determined. Then the Bravais lattice centering is obtained. Finally, the translational symmetry elements are identified. Roughly speaking, the ZOLZ detail in CBED zone-axis patterns indicates the mirror and rotational symmetry elements in the two-dimensional crystal structure when projected along particular zone axes, while the HOLZ detail reveals similar symmetry elements present in three dimensions. Additional tests can be made for the presence of a center of symmetry, since Friedel's law does not apply to dynamical electron diffraction (Goodman and Lehmpfuhl, 1968). The lattice centering is found by projecting the HOLZ spots onto the ZOLZ. In Section 7.3, we describe the remarkable dynamical extinctions which are used in the last step to reveal glide and screw symmetry elements, thereby determining the entire space group in most cases. Experimental considerations are outlined at the end of Section 7.1 and in Section 9.1. A simple example is provided.

We commence with a very brief reminder of some crystallographic facts. Appendix 3 contains more detail. Good introductory accounts of crystallography, including an explanation of how to read the *International Tables for Crystallography*, can be found in McKie and McKie (1986),

Sands (1969), Jackson (1991), and Stout and Jensen (1968). The abbreviated "teaching edition" of the *International Tables for Crystallography* and the book by Sands are particularly recommended. We shall describe crystals without a center of symmetry as "acentric" and those possessing a center of symmetry as "centric." The terms "vertical" and "horizontal" are used for convenience and refer to the orientation of a sample in the microscope, for which a preferred coordinate system is defined by the electron beam.

A crystal consists of a Bravais lattice plus a basis or "molecule," laid down at every lattice point. This molecule is known by crystallographers as the asymmetric unit.

In order to build a crystal which is periodic in three dimensions, it is necessary that the surroundings of every lattice point be identical. Only 14 distinct Bravais lattices may be constructed subject to this constraint, and these are given in Appendix 3. The Bravais lattice is a set of imaginary points in space, characterized by its arrangement in space and the disposition of its symmetry elements. It is always centric. It should be distinguished from the crystal structure, which is an arrangement of atoms in space. The lattice may be translated (but not rotated) at will through the structure. The Bravais lattices are differentiated by their symmetry, and may have only seven different (conventional) unit cell shapes (triclinic, monoclinic, orthorhombic, tetragonal, trigonal, hexagonal, and cubic). These define the seven *crystal systems*. The remaining seven lattices are generated by adding lattice points on the faces of the parallelepiped unit cell, giving different centerings, known as $P$ (primitive), $I$ (body-centered), $F$ (centered on every face), or $C$ (centered on two faces). The environment of every lattice point remains the same.

For a given Bravais lattice there are many choices of unit cell, including the "reduced," conventional, and primitive cells. A primitive cell is one which contains only one lattice point. A nonprimitive cell could be generated, for example, simply by doubling all the dimensions of a primitive cell. The conventional cells (some nonprimitive) given in Appendix 3 are chosen so as to express the full symmetry of the lattice, and Appendix 3 indicates how the symmetry elements impose relationships among the angles and dimensions of the conventional unit cell. An origin is usually taken at a center of symmetry, if it exists.

The Bravais lattice defines the geometry of the reciprocal lattice (by a Fourier transform), but the assignment of Miller indices will depend on the choice of unit cell used. The choice of a nonprimitive cell will introduce forbidden reflections. Additional forbidden reflections may also be introduced by the presence of the screw and glide translational symmetry elements. Geometric structure factors are those which depend only on the space group of the crystal, rather than the contents of the asymmetric unit. Systematic absences are forbidden reflections which arise from a particular choice of nonprimitive unit cell (such as the mixed indices in a cubic close-packed crystal), and from the presence of translational symmetry elements. (Non-translational symmetry elements do not give rise to systematic absences.) A study of these absent reflections therefore allows the Bravais lattice type and translational symmetry elements to be determined in X-ray diffraction. Additional "extinctions" known as "accidental" forbidden reflections [such as the (222) in Si] may be introduced if, for example, an assumption is made that the atoms are spherical, but these are not required by the space group. (These "extinctions" will therefore always be generated by computer programs based on tabulated atomic scattering factors.)

All crystals can be classified into one of 32 three-dimensional point groups (the *crystal classes*). These are defined using the point-group symmetry operations—mirrors (written $m$), $n$-fold rotations (written $n$, where $n$ is an integer), and inversions (written $\bar{n}$, and consisting of rotation by $360/n$ plus inversion through the origin). From these classes are derived the seven crystal systems.

All crystals can be classified into one of 230 space groups as listed in the *International Tables for Crystallography*. The space groups combine the symmetry elements of the point group with the translational symmetries (screw axes and glide planes). The (symmorphic) point group of a crystal is obtained by replacing each screw axis $n_\rho$ by a rotation axis $n$, and each glide element with the corresponding mirror plane $m$.

Application of a symmetry operation to an atom generates others at equivalent positions. Atoms are said to lie on "special positions" if they fall on a symmetry element or at their intersections. Then the number of new atom positions generated by a symmetry operation will be reduced. The number of such positions for an atom on a special position is known as its multiplicity, and given a Wyckoff symbol, starting with $a$ for the highest symmetry.

All two-dimensional patterns can be classified into 10 two-dimensional point groups.

There are ten projection diffraction groups. These are used to classify ZOLZ CBED patterns.

There are 31 diffraction groups. These are used to classify CBED patterns which include HOLZ interactions. A crystal of a given point group (crystal *class,* not *system*) will give rise to a different diffraction group at different zone-axis orientations.

A loose statement of Friedel's law is that, for X-rays, the intensity of the **g** reflection is equal to that of the $-$**g** reflection. Thus, if they were observable on a plane (and if we ignore anomalous dispersion), X-ray diffraction patterns would always show a center of symmetry (unlike CBED patterns). It is therefore not usually possible to determine point groups by X-ray diffraction, since acentric crystals produce centric patterns. These patterns belong to the eleven Laue classes. [For a complete discussion of Friedel's law in dynamical theory, see Miyake (1955) and Goodman and Lehmpfuhl (1968).] The law has also been formulated by stating that kinematic diffraction is invariant under an inversion of the crystal with respect to the incident beam. This law does not apply to dynamical electron diffraction, making it possible to determine polarity in acentric crystals, as discussed in Section 5.5 (Taftø and Spence, 1982).

It is important to distinguish between a crystal lattice (a mathematical abstraction) and a crystal structure (an arrangement of atoms in space). The space groups of some common simple crystal structures are as follows: the cubic close-packed metals, sodium chloride, calcium fluorite, $Fm3m$; hexagonal close-packed metals, $Im3m$; diamond, $Fd3m$; graphite, $P6_3/mmc$; cesium chloride, perovskite, $Pm3m$; sphalerite, $F\bar{4}3m$; wurtzite, $P6_3mc$; rutile, $P4_2/mmm$; spinel, $Fd3m$.

Most naturally occurring crystals are either monoclinic or orthorhombic.

## 7.1. POINT GROUPS

Three symmetry elements are used to define the point group of a crystal: the mirror planes of symmetry (with Hermann–Mauguin symbol $m$, denoting the direction normal to the mirror plane), the $n$-fold rotation axes ($n$), and roto-inversion axes ($\bar{n}$). (For the roto-inversion center the symmetry operation consists of rotation by $360/n$ about an axis, followed by inversion through a point.) A mirror is thus equivalent to $\bar{2}$. The symbol $\bar{1}$ denotes a center of symmetry. The point symmetry elements of a crystal form a group, with the associative, identity, inverse, and closure (but not commutative) properties. Thus, for example, the operation of the threefold axis in the cubic gold crystal structure acting on one of the mirror planes generates two additional mirror planes. The symbol $2/m$ denotes a twofold axis parallel to the normal to a mirror plane. (The / symbol indicates parallel symmetry elements.) The position of a symbol in the specification of a point group indicates the direction of the symmetry element. The different conventions for each crystal system are given in the *International Tables*. For example, a 3 (or $\bar{3}$) in the second place always indicates the cubic system, and refers to the body diagonals.

The first symbol then refers to the cube axes, while the third refers to the face diagonals. In the orthorhombic system, the three symbols refer to the three orthogonal axes. Thus *mm*2 denotes mirrors normal to $x$ and $y$ with a (redundant) twofold axis along $z$. The Laue groups are obtained by adding a center of symmetry to each of the point groups. This gives the symmetry of X-ray diffraction patterns, to which Friedel's law applies.

In many practical cases one has a crystal in which there is uncertainty about only a single symmetry element. Then it is simply a matter of recording a CBED pattern along a suitable zone axis and observing the resulting symmetry of the whole pattern, including HOLZ detail. As an example, it was determined at an early stage by X-ray diffraction that the space group of $YBa_2Cu_3O_{7-x}$ was either *Pmmm* (orthorhombic) or *P4mm* (tetragonal). Twinning and spatial variations in oxygen stoichiometry made further determination difficult. The matter was resolved using CBED patterns along the $c$ axis, with a probe size smaller than the twin spacing, for samples of differing stoichiometry. These clearly revealed the presence of a third mirror in the orthorhombic phase ($x = 0$), which is obtained by cooling from the high-temperature tetragonal phase ($x = 1$) (Eagelsham *et al.*, 1987; Graham *et al.*, 1987).

Usually, something will be known about the crystal, and experienced workers use a combination of shortcuts, simple reasoning, and experience to isolate the possible space groups after looking at "whole patterns" from as many high-symmetry axes as they can find. The *angle* between these axes, which can be read from the goniometer stage, provides additional important information. The internal symmetries of CBED disks at the Bragg condition may also be used.

Particular tests exist for individual symmetry elements. The following general rules apply and can be obtained by combining the reciprocity theorem with the symmetry elements of the crystal. Rotation axes are seen directly in CBED patterns when the beam is aligned with the rotation axis, as shown in Fig 3.11(a). Note, for example, in such figures how the ZOLZ detail shows sixfold symmetry, which the diamond structure possesses only if it is projected along (111). The whole pattern, however, shows the true threefold symmetry of the three-dimensional lattice when account is taken of the detail in the HOLZ lines, both those crossing the central disk and those in the outer ring. This HOLZ detail results from diffraction events with a component along the beam path and is therefore sensitive to the three-dimensional crystal symmetry.

Mirror planes of symmetry in the crystal are also seen directly as mirror lines in the CBED pattern, if the beam lies in the mirror plane of symmetry. A vertical glide plane produces a mirror line in the CBED pattern. A horizontal twofold axis or twofold screw axis in the ZOLZ along **g** imparts a mirror line of symmetry onto disk **g** if it is at the Bragg condition, and this line runs normal to **g**. Horizontal three; four-, and sixfold axes produce no useful symmetries. (The four- and sixfold axes, however, include a twofold axis.) A horizontal mirror plane or glide plane (strictly running through the midplane of the crystal slab) produces a centric distribution of intensity in every CBED disk at the Bragg condition. (This is given the diffraction group symbol $1_R$ in Table 7.1.) Such a horizontal mirror is always present in the projection approximation.

Unlike X-ray diffraction patterns, CBED patterns are very sensitive to the existence or absence of a center of symmetry (Goodman and Lehmpfuhl, 1968). An early test for a center of symmetry was based on a comparison of the **g** and $-$**g** disks, recorded successively at the Bragg condition. [These distributions are related by translation (not rotation).] A simpler method, not requiring tilting, is based on determination of the diffraction group as described below, since no diffraction group can come from both a centered and a noncentered point group (Eades, 1991). The central disk is a special case. Under general three-dimensional diffraction conditions, the presence of a center of symmetry in the crystal does not impose a centric distribution on the (000) disk. However, in the projection approximation, the (000) disk always has a center of symmetry, even in acentric crystals, due to the reciprocity theorem. The absence of a center in the (000) disk of a centric crystal may therefore be used as a test for three-dimensional scattering.

In those cases where nothing at all is known about a crystal, a systematic procedure based on group theory (Buxton et al., 1976; Eades, 1988b) may be followed. The tables which can be derived from this work assume a parallel-sided slab of crystal normal to the beam, no backscattering, and the validity of the reciprocity theorem. The analysis below will be based solely on patterns taken at zone axes, and a suitable tabulation is shown in Tables 7.1 and 7.2. It should be emphasized that the method is sensitive to the symmetry of the physical crystal as a whole, including its boundaries and defects. In some cases (as shown in Section 8.5), inclined boundaries or defects may eliminate symmetry elements present in the infinite crystal, and the presence of defects must always be checked for using TEM imaging. (The use of a smaller probe minimizes the contribution from defects.) Since every disk at the Bragg condition is centric in the projection approximation (ZOLZ detail only), the absence of a center in such a disk may be used as a test for defects (Steeds, 1979). (A twofold axis is equivalent to a center in two dimensions.) Under ultrahigh vacuum conditions, forbidden "termination" reflections will

# SYMMETRY DETERMINATION

*Table 7.1.* Table Showing the Relation Between the Observed Symmetries in Convergent-Beam Diffraction Patterns and the 31 Diffraction Groups which Correspond to the 32 Different Three-Dimensional Point Groups. (Loretto, 1984)[a]

| Observed symmetry in zero-order zone | Projection diffraction group | Possible diffraction groups | Symmetries of high-order information | |
|---|---|---|---|---|
| | | | whole pattern | zero-order disk |
| 1 | $1_R$ | 1 | 1 | 1 |
| | | $1_R$ | 1 | 2 |
| 2 | $21_R$ | 2 | 2 | 2 |
| | | $2_R$ | 1 | 1 |
| | | $21_R$ | 2 | 2 |
| m | $m1_R$ | $m_R$ | 1 | m |
| | | m | m | m |
| | | $m1_R$ | m | 2m |
| 2mm | $2mm1_R$ | $2m_Rm_R$ | 2 | 2mm |
| | | 2mm | 2mm | 2mm |
| | | $2_Rmm_R$ | m | m |
| | | $2mm1_R$ | 2mm | 2mm |
| 4 | $41_R$ | 4 | 4 | 4 |
| | | $4_R$ | 2 | 4 |
| | | $41_R$ | 4 | 4 |
| 4mm | $4mm1_R$ | $4m_Rm_R$ | 4 | 4mm |
| | | 4mm | 4mm | 4mm |
| | | $4_Rmm_R$ | 2mm | 4mm |
| | | $4mm1_R$ | 4mm | 4mm |
| 3 | $31_R$ | 3 | 3 | 3 |
| | | $31_R$ | 3 | 6 |
| 3m | $3m1_R$ | $3m_R$ | 3 | 3m |
| | | 3m | 3m | 3m |
| | | $3m1_R$ | 3m | 6mm |
| 6 | $61_R$ | 6 | 6 | 6 |
| | | $6_R$ | 3 | 3 |
| | | $61_R$ | 6 | 6 |
| 6mm | $6mm1_R$ | $6m_Rm_R$ | 6 | 6mm |
| | | 6mm | 6mm | 6mm |
| | | $6_Rmm_R$ | 3m | 3m |
| | | $6mm1_R$ | 6mm | 6mm |

[a] See text for full discussion.

Table 7.2. Relation Between the Diffraction Groups and Crystal Point Groups (Buxton et al., 1976)

| Diffraction group | 1 | 1̄ | 2 | m | 2/m | 222 | mm2 | mmm | 4 | 4̄ | 4/m | 422 | 4mm | 4̄2m | 4/mmm | 3 | 3̄ | 32 | 3m | 3̄m | 6 | 6̄ | 6/m | 622 | 6mm | 6̄m2 | 6/mmm | 23 | m3 | 432 | 4̄3m | m3m |
|---|---|---|---|---|---|---|---|---|---|---|---|---|---|---|---|---|---|---|---|---|---|---|---|---|---|---|---|---|---|---|---|---|
| | Tr | Tr | M | M | M | O | O | O | Te | Te | Te | Te | Te | Te | Te | Tg | Tg | Tg | Tg | Tg | H | H | H | H | H | H | H | C | C | C | C | C |
| 6mm1_R | | | | | | | | | | | | | | | | | | | | | | | | | | | × | | | | | |
| 3m1_R | | | | | | | | | | | | | | | | | | | | × | | | | | | | | | | | | |
| 6mm | | | | | | | | | | | | | | | | | | | | | | | | | × | | | | | | | |
| 6m_Rm_R | | | | | | | | | | | | | | | | | | | | | | | | × | | | | | | | | |
| 61_R | | | | | | | | | | | | | | | | | | | | | | × | | | | | | | | | | |
| 31_R | | | | | | | | | | | | | | | | | × | | | | | | | | | | | | | | | |
| 6 | | | | | | | | | | | | | | | | | | | | | × | | | | | | | | | | | |
| 6_Rmm_R | | | | | | | | | | | | | | | | | | | | | | | | | | × | | | | | | × |
| 3m | | | | | | | | | | | | | | | | | | | × | | | | | | | | | | | | × | |
| 3m_R | | | | | | | | | | | | | | | | | | × | | | | | | | | | | | | × | | |
| 6_R | | | | | | | | | | | | | | | | | | | | | | × | | | | | | | | | | |
| 3 | | | | | | | | | | | | | | | | × | | | | | | | | | | | | × | | | × | |
| 4mm1_R | | | | | | | | | | | | | | | × | | | | | | | | | | | | | | | | | |
| 4_Rmm_R | | | | | | | | | | × | | | | × | | | | | | | | | | | | | | | | | | |
| 4mm | | | | | | | | | | | | | × | | | | | | | | | | | | | | | | | | | |
| 4m_Rm_R | | | | | | | | | | | | × | | | | | | | | | | | | | | | | | | | | |
| 41_R | | | | | | | | | | | × | | | | | | | | | | | | | | | | | | | | | |
| 4_R | | | | | | | | | | × | | | | | | | | | | | | | | | | | | | | | | |
| 4 | | | | | | | | | × | | | | | | | | | | | | | | | | | | | | | | | |
| 2mm1_R | | | | | | | | × | | | | | | | × | | | | | | | | | | | | | | | × | | × |
| 2_Rmm_R | | | | | | | × | × | | | | | | | | | | | | | | | | | | | | | | × | | |
| 2mm | | | | | | | × | | | | | | × | × | × | | | | | | | | | | × | × | | | | | | |
| 2m_Rm_R | | | | | | × | | × | | | | × | | × | × | | | × | | × | | | | × | | × | | | | | | |
| m1_R | | | | × | × | | × | × | | | × | | | | × | | | | | | | × | × | | | | × | | × | | × | × |
| m | | | | × | | | | | | | | | | | | | | | | | | | | | | | | | | | | |
| m_R | | | × | | × | | | | | × | | | | × | | | | | × | | | × | | | | | | | | | | |
| 21_R | | | × | | × | | × | | | | × | | | × | | | × | | | × | | | × | | | | × | | × | | | × |
| 2_R | | | | | | × | | × | | | | × | | | | | | × | | × | | | | × | | | | | | | | |
| 2 | | | × | | | × | × | | × | | | × | × | | | | | × | | | × | | | × | × | | | × | | × | | |
| 1_R | × | × | × | × | × | × | × | × | × | × | × | × | × | × | × | × | × | × | × | × | × | × | × | × | × | × | × | × | × | × | × | × |
| 1̄ | | × | | | × | | | × | | | × | | | | × | | × | | | × | | | × | | | | × | | × | | | × |
| 1 | × | | | | | | | | | | | | | | | | | | | | | | | | | | | | | | | |

Crystal point groups

also produce misleading symmetries. Loosely speaking, these arise if the crystal contains incomplete unit cells at its surface. These expose the distinction between the symmetry of an infinite crystal and that of the finite slabs used in CBED (Ishizuka, 1982).

A formal procedure based solely on zone-axis patterns would consist of three steps:

1. Determination of the symmetry of the projection diffraction group, using ZOLZ detail.
2. Determination of the diffraction group, using HOLZ detail.
3. Determination of the point group from the above information, using tables.

We discuss these steps in turn, with respect to the example of BeO given in Figs. 7.1a–7.1e.

(1) A set of zone-axis patterns should be recorded at all the highest-symmetry zones which can be found. From Table 7.2, we see that the higher the symmetry of the pattern, the fewer will be the possible point groups. Patterns will be needed at both large and small camera length, with exposures which reveal the HOLZ and ZOLZ detail separately. We thus assume that it is possible to separate the HOLZ features in a CBED pattern from the ZOLZ features. All two-dimensional patterns may be classified into one of ten classes, the ten two-dimensional point groups. These are listed in the first column of Table 7.1. Here $n$ denotes an $n$-fold rotation axis normal to the pattern, and $m$ a mirror line in the plane of the pattern. The first step is to determine into which of these classes the experimental pattern falls, when only the ZOLZ detail is considered. Thus we ignore the fine HOLZ lines crossing the central disk, and the outer HOLZ rings. Figure 7.1a shows one of many similar high-symmetry patterns which could be obtained at the same axis from our BeO crystal, using a 100-nm-diameter electron probe. A survey of many of these suggests that they contain two orthogonal mirror lines of symmetry and a sixfold axis. No one pattern, however, showed precisely this symmetry, due perhaps to the presence of defects or inclined surfaces. Our conclusion that the ZOLZ symmetry is therefore 6 mm rests on a subjective judgement, based on experience. This emphasizes the important point that the CBED method, applied to real materials and using a "large" electron probe, is always likely to apparently underestimate the symmetry of the crystal, since it gives the true symmetry of the actual piece of material under the probe rather than that of the ideal crystal structure. Results closer to the ideal can be obtained using a smaller probe, which will minimize defect and boundary

*Figure 7.1.* (a) BeO CBED pattern recorded at 100 kV. The inner reflections only are shown, for a high-symmetry zone axis. Crushed samples were used. (b) Similar to Fig. 7.1a, but showing the HOLZ ring observable only with cooling to −183 °C. (c) BeO CBED pattern from a second high-symmetry zone axis (actually [1-100]). Outer HOLZ ring is seen. (d) Inner ZOLZ reflections from Fig. 7.1c. (e) BeO [10-10] zone axis, normal to $c$ axis, with first-order reflection (0001) at the Bragg condition, showing $G$–$M$ black cross AB, and dark radial lines A in every second order along the $c$ axis (see Fig. 7.2).

condition effects. The second column of Table 7.1 then indicates the name of the projection diffraction group. This label takes into account the additional effects of reciprocity (Pogany and Turner, 1968) and horizontal mirror planes present in all ZOLZ patterns, and thus renames the classes in a systematic way (Buxton *et al.*, 1976). There are more diffraction groups than two-dimensional point groups because the diffraction group takes into account the internal symmetry of a CBED disk at the Bragg condition. The projection diffraction group for our example is thus $6mm1_R$.

(2) The last four entries in column three of Table 7.1 now indicate the possible diffraction groups for our pattern. These may be distinguished using the HOLZ information, taken separately in the zero-order disk (or outer HOLZ ring) and for the pattern as a whole. From Fig. 7.1b

# SYMMETRY DETERMINATION

*Figure 7.1. (continued)*

we see that both the outer HOLZ ring (or the HOLZ lines in the central disk, not shown) and the whole pattern (including this HOLZ detail) have symmetry 6*mm*, so that the diffraction groups for this pattern may be either 6*mm* or 6*mm*$1_R$.

Our aim is to uniquely specify a point group on Table 7.2. The two diffraction groups we have identified now allow two possible point groups

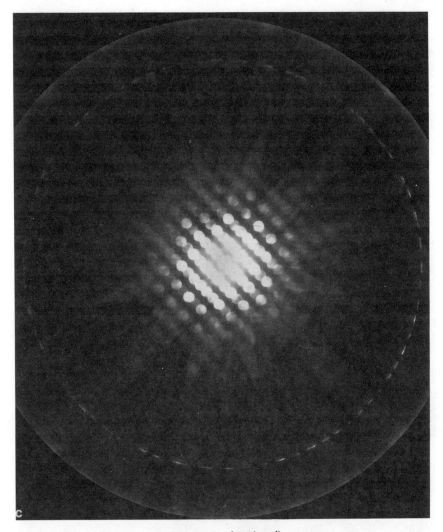

*Figure 7.1. (continued)*

on Table 7.2, $6mm$ and $6/mmm$. To distinguish these possibilities we need a second CBED pattern taken from the same crystal in a different orientation. Figures 7.1c and 7.1d show a high-symmetry CBED pattern which was obtained with the beam normal to the direction used in Figs. 7.1a and 7.1b. A similar analysis from Table 7.1 shows that the diffraction group for this pattern is $m1_R$, since the outer HOLZ ring in Fig. 7.1d shows $2mm$ rather than $m$ symmetry.

# SYMMETRY DETERMINATION

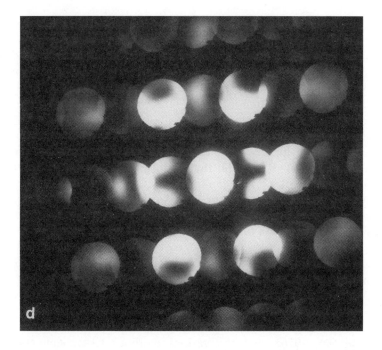

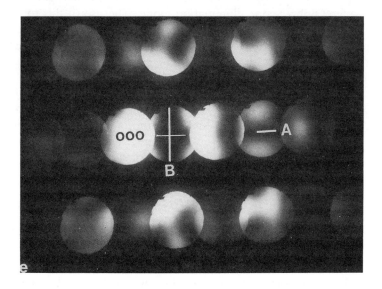

Figure 7.1. (*continued*)

In more complicated cases, it may be necessary to use knowledge of the unit cell (giving the crystal system) to obtain more zone-axis patterns, or to use the internal symmetries of CBED disks at the Bragg condition.

(3) Turning to Table 7.2, we may now determine the point group for the crystal by tracing across from the diffraction group symbols on the left. Only the $6mm$ column intersects both the $m1_R$ and $6mm$ rows. The point group is therefore $6mm$.

In practice, when confronted with an unknown crystal, one looks initially for mirror lines of symmetry, possibly in the Kikuchi pattern. A double-tilt stage with a large tilt range is essential. A quick look over the full allowable angular excursion of the stage should reveal the main high-symmetry axes. If no mirror lines are seen anywhere, the crystal is likely to be triclinic. Once a mirror line is found, one can tilt along it to find the intersection with other symmetry elements. Rotation axes normal to any mirror line may exist. A condenser aperture should be used which causes the disks to just touch for the largest lattice spacing. See Section 9.1 for more details.

Other systematic procedures have been described which exploit symmetries in certain CBED disks at the Bragg condition ("dark field disks") which may involve fewer steps and micrographs (Tanaka *et al.*, 1983; Buxton, 1976), and the various methods have been compared (Howe *et al.*, 1986). In particular, if the incident-beam tilt controls are put under electronic control and scanned, it is possible to obtain a time-resolved hollow-cone illumination mode. The use of this to provide "whole pattern" symmetry information is described by Tanaka and Terauchi (1985b). A wide-angle method has also been demonstrated which makes it possible to record simultaneously on a single micrograph the whole pattern, the central disk ("bright field"), and diffracted orders at the Bragg condition ("dark field") (Terauchi and Tanaka, 1985).

Similar principles may be applied to the study of quasi-crystals, provided a large probe is used. Methods for computing CBED patterns from quasi-crystals are discussed in Cheng and Wang (1989), and experimental patterns can be found in Tanaka *et al.* (1988).

## 7.2. BRAVAIS LATTICE DETERMINATION

The next step in a general space-group determination should be to index the pattern, determine the Bravais lattice, and identify a unit cell. Worked examples of this procedure can be found in Raghavan *et al.* (1983) and Ayer (1989). The preceding analysis will have determined the crystal system (see Table 7.2). Identifying a cell and its centering will greatly reduce the number of possible space groups. The 14 Bravais lattices may be primitive ($P$), body-centered ($I$), face-centered ($F$), centered on two sides ($C$), or trigonal. The corresponding real-space lattices are then defined by the vector relations given in Appendix 3, or by a Fourier transform. If there are no absent layers of reciprocal lattice points, it is usually possible to reconstruct the geometry of the three-dimensional reciprocal lattice by noting the relationship between the position of the HOLZ points and those in the ZOLZ "below." Recall Fig. 2.8. For this purpose, a CBED pattern recorded with a small condenser aperture along a high-symmetry axis should be used, using a small camera length. Lines may be drawn on a print or photocopy of the pattern, parallel to the ZOLZ rows of spots, across the CBED pattern, and connecting points in the HOLZ. The electron-optical distortions present in small-camera-length patterns can make this analysis difficult. These lines will pass either between the ZOLZ spots or through them. Two orthogonal sets need to be drawn, and their intersections will define the points in the FOLZ. In this way it may be possible to tell if the reciprocal lattice is $P$, $F$, or $I$. The real-space lattice is then $P$, $I$, or $F$, respectively. A TOLZ will assist this process. Indexed examples are given in Appendix 4, together with enough information to draw the HOLZ spots. Indices can then be assigned to the pattern, starting with the shortest three-dimensional reciprocal lattice vector. The diameter of the FOLZ gives the height of the first layer of spots [Eq. (2.13) with $K_t = 0$], from which the three-dimensional reciprocal lattice vectors can be deduced. A worked example of this procedure can be found in Ayer (1989). Questions of accuracy clearly arise in distinguishing possibilities. For example, it has been pointed out by Eades (1991) that the error in a measurement of the $c$ axis (taken parallel to the beam) will differ from that of the other axes, because of the square-root dependence in Eq. (2.13). There may also be difficulties due to absent planes of reflections arising from translational symmetry. For example, in the reciprocal lattice for diamond, all structure factors in the sixth (nonzero) reciprocal lattice plane along [111] are zero due to the diamond glide symmetry element.

It is possible to write a computer program which will determine the Bravais lattice type, unit-cell dimensions, and crystal system from a spot diffraction pattern rather than a CBED pattern (LePage, 1987; Johnson, 1976). Even with the most careful calibration to allow for lens distortions, this approach may lead to errors. However, it is important to establish that it is possible in principle to find a systematic procedure for this purpose. Assuming that the Bravais lattice has been found, many choices of unit cell are then possible. A Buerger primitive cell has been defined

(Buerger, 1956) by the shortest three noncoplanar translations. This cell may be identified by a computer, but this choice is not unique. The Niggli cell, however, is unique, and further allows the lattice type and crystal system to be immediately identified. A systematic procedure for determining the Niggli cell is given by Krivy and Gruber (1976).

From the pattern shown in Fig. 7.1a, the lattice was found to be primitive. The possible space groups are now restricted to those listed in the *International Tables for Crystallography* (Volume A) which belong to the point group 6*mm* with a primitive lattice. (The point group is given as the third entry on the top line of the left-hand page for each space group.)

## 7.3. SPACE GROUPS

A space group is constructed by placing a point group at each of the lattice points of the Bravais lattice and adding any additional translational symmetry elements (the screw and glide elements). Thus, for example, $P6_3/mmc$ describes a primitive hexagonal lattice obtained from the $P6/mmm$ point group by replacing the sixfold axis by a $6_3$ screw axis and one mirror plane by a $c$-axis glide plane. The symbol $n_p$ is here used to denote a screw axis with a symmetry operation consisting of a rotation of $2\pi/n$, followed by a translation by $p/n$ in the direction of the axis. A $c$ glide consists of translation by half the lattice constant along $c$, followed by reflection in a plane containing $c$. Diagonal glides are denoted by $n$, diamond glides by $d$.

The aim of the final step in the analysis is to determine if either of the translational symmetry elements—screw axes or glide planes—are present and, if so, to find their orientation. When combined with the point group and lattice determination described in previous sections, this will usually provide enough information to determine the space group. Clear accounts of the practical procedure used are given in Tanaka *et al.* (1983) and Eades (1988a). In X-ray diffraction these translational symmetry elements introduce certain systematic absences. (These are additional to those caused by centering of the Bravais lattice, which can never arise from multiple scattering.) The CBED method is based on the fact that, *for certain incident-beam directions,* some of these reflections which are absent due to the presence of translational symmetry remain absent, despite multiple scattering, for all sample thicknesses and accelerating voltages. This causes a dark band or cross to be seen in CBED patterns known as a Gjønnes–Moodie or G–M line, as shown in Fig. 7.1e. A full explanation for this effect, first observed by Goodman

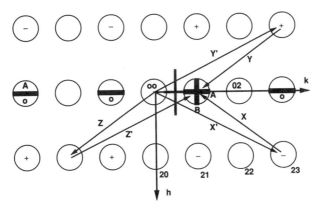

*Figure 7.2.* The principle of dynamically forbidden reflections. A projection normal to the c axis of the wurtzite structure is shown. The gray cross shows the locus of $K_t$ along which all scattering paths cancel into the black cross AB in the first-order reflection (01) at the Bragg condition.

and Lehmpfuhl (1964), was eventually given (following earlier papers by Fujimoto, Cowley, Miyake, Takagi, and others) by Gjønnes and Moodie (1965). An elegant derivation of the theory is also given using group theory in Portier and Gratias (1981).

First we will discuss the simplest case and give an example, then elaborate complications and exceptions. We take a beam direction approximately normal to the page in Fig. 7.2, and consider a screw axis running normal to the beam along the axis $(0, k)$, and/or a glide plane containing the beam direction which intersects the ZOLZ along $(0, k)$. The glide translation may be, for example, along $k$. Initially, we consider only ZOLZ interactions in the projection approximation. In projection, this screw axis becomes a glide line and the projected two-dimensional symmetry becomes the plane group *pmg* [see Kelly and Groves (1970) for diagrams of these symmetry elements]. Because of the screw axis (or glide element), the geometric structure factors have the signs shown in Fig. 7.2 with

$$U(h, k) = U(-h, k) \quad \text{for } k = 2n$$

and

$$U(h, k) = -U(-h, k) \quad \text{for } k = 2n + 1$$

while

$$U(0, k) = 0 \quad \text{for } k = 2n + 1 \quad (7.1)$$

in the ZOLZ (see Appendix 3.4). Hence alternate reflections along the screw axis are kinematically forbidden. To show this, these reflections contain a zero in Fig. 7.2. In addition, the signs of the structure factors alternate in the rows above and below the $(0, k)$ axis as indicated in the figure. For a glide plane containing the beam direction, every second row of reflections in a plane parallel to the glide plane is forbidden. These rows run normal to the glide translation. Then the situation after projection is the same as that just described for the screw. We will see that these relationships among the structure factors lead to certain extinctions, even in the presence of multiple scattering.

If HOLZ features are to be included in the analysis, the three-dimensional structure factors must be used. For a $2_1$ screw axis along $(00l)$ the relationship linking them is

$$U(h, k, l) = (-1)^k U(-h, k, -l) \qquad (7.2)$$

while for the same glide plane

$$U(h, k, l) = (-1)^k U(-h, k, l)$$

Here the unique $c$ axis is now taken as $(00l)$. The screw axis might be, for example, a $6_3$ screw axis. Repeated operation of this screw element and the $4_1$, $4_3$, $6_1$, $6_3$, $6_5$ screws shows that they all include a $2_1$ screw. (We note that $4_2$, $3_1$, $3_2$, $6_2$, and $6_4$ screws do not. Therefore they do not produce G–M lines, which is one reason why 49 of the 230 space groups cannot be completely identified by G–M lines.)

In their development of the multislice formulation of the dynamical theory, Cowley and Moodie had been able to show that the general expression for the intensity of a diffracted beam could be written (Moodie, 1972)

$$\Phi_\mathbf{g} = \sum_{n=1}^{\infty} E_n(\mathbf{g}) Z_n(\mathbf{g}) \qquad (7.3)$$

where the terms in $E_n$ depend only on structure factors $U$, while the functions $Z_n(\mathbf{g})$ involve polynomials which are functions of the geometric factors only (excitation errors, wavelength, lattice constants). When the conditions (7.1) are imposed on Eq. (7.3), it is found that the resulting series of terms can be arranged in pairs with products of $U(h, k)$ of equal magnitude and opposite sign. This occurs only if the excitation errors also preserve certain symmetry, in particular if $\mathbf{K}_t$ lies on the bold black cross in Fig. 7.2 (in the projection approximation). Here the $(0, 1)$ reflection is at the Bragg condition. Physically, the extinction along the radial line A occurs because of the exact cancellation by destructive interference of all pairs of multiple scattering paths, such as X and Y in Fig. 7.2, which are related by crystal symmetry. Along the line B [where the Bragg condition is maintained for the (0, 1) reflection] the cancellation occurs due to paths such as X and Z, since paths Z are equivalent to paths Y. Thus, within the projection approximation, such experimental CBED patterns should show black lines in alternate orders along $c$, and a black cross at the first-order reflection at the Bragg condition (or in any such diffracted beam that is tilted to the Bragg angle). Figure 7.1e shows our experimental pattern from BeO.

If three-dimensional scattering is included, theory predicts that the screw axis produces only the B line of extinction across the (0, 1) disk, while a glide plane produces only the radial line of extinction in the $(0, k)$ disks for $k$ odd. The two cases may thus be distinguished in principle. However, at the high-symmetry, low-index zone axes that are preferred for CBED study, it is usually the case that the three-dimensional effects are weak. Thus, in practice, the projection approximation applies and, for either a screw or a glide, there is extinction along A and B. In Tanaka's convention, we call the radial line of zero intensity $A_2$ if the scattering is two-dimensional, and $A_3$ under three-dimensional (HOLZ) conditions. A similar convention holds for the B lines.

For a glide plane normal to the beam, the period of the potential is halved by projection, leading to the extinction of alternate reflections in the ZOLZ. (The FOLZ point pattern will then appear twice as dense as the ZOLZ, allowing easy identification of this case.) In a CBED pattern, the theory thus predicts a point of extinction at the center of the black cross, due to cancellation between paths Y and Z. This dark spot has recently been observed for the first time by Tanaka's group (Tanaka *et al.*, 1987).

In summary, if ZOLZ interactions only are considered, either the twofold screw axis or the glide plane described above will produce both the A and B G–M lines of extinction shown in Fig. 7.2. Under three-dimensional scattering conditions, the screw produces only the B line while the glide produces only the A lines.

These are the theoretical results. In practice, however, it may be difficult to ensure three-dimensional scattering in order to distinguish these two cases. An increase in thickness or a change in accelerating voltage may achieve this. Several other methods have therefore been suggested to distinguish screws from glides. If a glide plane is present, the $A_2$, $B_2$, and $A_3$ type G–M lines will remain if the crystal is rotated about $h$, but are destroyed if it is rotated about $k$. For a screw axis $A_2$, $B_2$, and $B_3$ G–M lines are preserved if the crystal is rotated about $k$, but are destroyed by a rotation about $h$ (Steeds and Vincent, 1983). Second, it has been shown that the symmetry of the fine HOLZ lines which cross these disks can sometimes be used to distinguish the two cases (Tanaka, 1989). If the HOLZ lines are symmetric about $A_2$ type G-M lines, a glide plane is present. If the lines are symmetric about the $B_2$ line, a screw axis exists. Finally, knowledge of the diffraction groups can be used to limit possibilities, since only certain combinations of the point and space group elements are permitted (Eades, 1988a). These are indicated in Table 7.3. This table assumes that the diffraction group is known, and then allows the correct deductions to be made concerning any screw or glide planes. An alternative and useful tabulation has also been provided by Tanaka (Tanaka *et al.*, 1988), in which it is assumed that the point group is known. Then his table shows the dynamical extinctions which are expected for each space group.

*Table 7.3.* Origin of Dynamical Extinctions in Zero-Layer Reflections[a]

| Single row of zero-layer extinctions | Perpendicular rows of zero-layer extinctons | Deduction | Tanaka symbol |
|---|---|---|---|
| $m_R$ | $2m_R m_R$ | Screw axes parallel to each row of extinctions | $A_2 B_2$ |
| $2m_R m_R$ | $4m_R m_R$ | | $B_3$ |
| $m$ | $2mm$ | Glide planes parallel to the zone axis and each row of extinctions | $A_2 B_2$ |
| $2m$ | $4mm$ | | $A_3$ |
| $2_R m m_R$ | $4_R m m_R$ | Glide, if parallel to whole pattern mirror, *or* Screw, if perpendicular to whole pattern mirror | $A_2 B_2$ $A_3$ *or* $A_2 B_2$ $B_3$ |
| | $2_R m m_R$ | Glide parallel *and* Screw perpendicular to whole pattern mirror | $A_2 B_2$ $A_3$ *and* $A_2 B_2$ $B_3$ |
| $m 1_R$ | $2mm 1_R$ | Glide plane *and* Screw axis parallel to each line of extinctions[b] | $A_2 B_2$ |
| $2mm 1_R$ | $4mm 1_R$ | | $A_3 B_3$ |

[a] The table permits the proper interpretation of zero-layer extinctions when the diffraction group is known.
[b] This is the case for double-diffraction routes in the zero layer. If the extinction is produced by double diffraction via HOLZ reflections, there are space groups for which the extinction can be produced by a glide or a screw alone (e.g., horizontal glide).

Perpendicular rows of black lines may also be seen if there is more than one translational symmetry element. No forbidden reflections occur if a screw axis is parallel to a twofold axis, or if a glide is parallel to a mirror. The G–M lines are distinguishable from other zeros in intensity by the fact that they are present for all thicknesses and accelerating voltages. There are, in addition, some more subtle dynamical extinctions, such as those occurring in the outer HOLZ ring, or that due to a glide plane parallel to the surface, mentioned above. A vertical glide plane whose glide vector is not parallel to the crystal surface is strictly not a symmetry element of the layer groups for slab crystals—the implications of this are discussed by Ishizuka (1982). Here it is shown that a vertical glide with a vertical translation does cause G–M lines if the HOLZ spacing is large. A symmetry classification for parallel-sided slab crystals has been given (the layer groups) (Goodman, 1984), and may thus be used as the basis for space-group determination by CBED.

These procedures will identify 181 of the 230 space groups. For a discussion of methods for identifying the others, see Eades (1988a).

For the BeO pattern shown in Fig. 7.1e we see that radial lines of absence A are obtained in every second reflection along $c$, indicating a $c$ glide. In addition, the first-order (0001) reflection is at the Bragg condition, and this shows a band B of extinction across the $c$ axis, indicating a $2_1$ screw axis along $c$. This is contained in the $6_3$ screw element. From Fig. 7.1e alone we may thus conclude only that there exists either a screw axis along $c$, or a $c$ glide, or both. Since the diffraction group for this pattern was determined to be $m1_R$, we can check from Table 7.3 that the observation here of a single row of zero-layer black lines and a cross indicates a "glide plane and screw axis parallel to each line of black crosses." The actual space group for the hexagonal wurtzite structure of BeO is $P6_3mc$.

The general procedure for an unknown crystal is as follows: First find the point group, the crystal system, and index the pattern. Then proceed as follows. (1) List all possible space groups consistent with this information. (2) Use the table on page 162 of Tanaka (1985a) to determine the directions in which to look in order to observe G–M lines for each possible space group. Check for screws which do not produce G–M lines (such as $3_1$). Invaluable references include Eades (1988a), Steeds and Vincent (1983), Tanaka (1989), and Tanaka *et al.* (1988). Examples of solved structures can be found in Chapter 9—readers are particularly referred to the examples of $Ni_3Mo$ (Fraser, 1983) and aluminum oxynitride spinel (Dravid *et al.*, 1990).

The "chirality" of a crystal, or its "handedness," can also be determined by CBED. (This corresponds to the difference between a right-handed and a left-handed screw. Space groups which differ in this way are known as enantiomorphs and are mirror images of one another.) Applications to quartz are described in Goodman and Johnson (1977), and to MnSi in Tanaka and Terauchi (1985a). In both cases the absolute orientation (i.e., the "handedness") can be determined. The effect of enantiomorphism is to reverse the sign of the three-phase invariant, leading to observable effects, as discussed at the end of Section 5.1. The topic is also discussed by Vincent *et al.* (1986).

## 7.4. PHASE IDENTIFICATION

In many cases, CBED may not be the most efficient method of phase identification when using an analytical electron microscope. Frequently, one has a limited number of possible crystalline phases and conveniently large crystals, or is able to make an extraction replica. Then a combination of energy-dispersive X-ray microanalysis (EDX) or energy-loss spectroscopy and selected area point diffraction patterns taken at high-symmetry zone axes may quickly distinguish between these possibilities. Using the X-ray powder diffraction file, a list of possible reciprocal lattice spacings and their ratios can be drawn up for the most likely phases. The newest EDX systems, using diamond, ultrathin window, or windowless germanium detectors, can provide invaluable additional information on any light elements present.

For very small (subnanometer) microcrystallites, or those whose cell contains a large number of atoms (so that distinction between phases is difficult), or both, CBED (combined with EDX and ELS) may provide the only method of identification. This must be based on a determination of the crystal's space group, as described above. Experimentally, the use of extraction replicas has been found invaluable (Steeds and Mansfield, 1984) for small precipitates embedded in a matrix. To some extent, a fingerprinting technique may be used in which CBED patterns from an unknown phase are compared with standard patterns from an atlas (Mansfield, 1989). This procedure depends on the neglect of thickness-dependent variations in CBED patterns, and on finding a pattern for the phase of interest in the atlas or in a laboratory collection of patterns. Reference books are also available which list lattice spacings and space groups for many alloys (Pearson, 1967). Readers are referred to the paper by Mansfield (1989) for more practical details. An automated method for determination of the Bueger reduced primitive cell from CBED ZOLZ and FOLZ patterns has recently been published by Le

Page (1992), based on least squares analysis, and claiming an accuracy of 1% in lengths and 2° in angles. EDX data is also used and the results compared with a data base. It is planned in this way to avoid trial and error searches. These methods might be extended by combining the accurate lattice parameter measurements described in Section 5.4 (based on Holz lines) with the Niggli cell reduction (Section 7.2).

Recently, a new highly accurate method for identification has been described (Zuo, 1993). This allows the unique Niggli reduced cell to be determined from measurements of at least six HOLZ and ZOLZ lines in the central disk.

# 8

# Coherent Nanoprobes. STEM. Defects and Amorphous Materials

In this chapter we discuss the special effects and new kinds of information which can be obtained in the coherent nanoprobe mode. This is followed by a discussion of the interpretation of conventional CBED patterns from stacking faults, dislocations, and multilayers. The chapter therefore falls into two parts: In the first sections (8.1–8.4) we are concerned with the smallest possible probes and their use for extracting information on the atomic arrangement at the core of a defect. This subject has been reviewed by Brown *et al.* (1988), Cowley (1978, 1992), and Spence and Carpenter (1986). In Section 8.5 on conventional CBED the aim is to extract information on the long-range, slowly varying part of the strain field. We will frequently be discussing microdiffraction from defects, so it is important to bear in mind that an emission point within the effective source (corresponding to illumination of the sample by a plane wave) no longer gives rise to a point in the diffraction pattern (as for a perfect crystal), but rather will produce a diffuse streak of scattering. If the illumination is coherent, the streaks generated by adjacent source points must then be added coherently, and the resulting intensity then depends on the aberrations and focus setting of the probe-forming lens. If the sample is weakly scattering (or consists of a small particle on a very thin substrate), the microdiffraction pattern becomes an in-line Gabor electron hologram.

The real materials used for CBED analysis frequently contain a host of small microphases and defects, and are often very difficult to prepare in the form of parallel-sided slabs of fixed crystallographic orientation, so the history of CBED has seen the continuous development of methods for obtaining smaller and smaller electron probes. We shall see in this chapter that the illumination conditions for the smallest (subnanometer diameter) probes then necessarily become fully coherent. The nature of these coherent CBED patterns is discussed in this chapter, together with methods for their computer simulation. If coherent CBED disks are allowed to overlap, it then becomes possible to form a scanning transmission (STEM) lattice image. By observing this STEM lattice image, it thus becomes possible (in thin crystals) to stop the probe on the region at which a CBED pattern is required. We shall see that it is quite possible by this method to obtain CBED patterns from different regions within a single unit cell, and that these show different site symmetries. For very thin crystals, the resulting patterns may be interpreted as electron holograms. In order to obtain sufficient intensity from a probe of subnanometer dimensions, an instrument fitted with a field-emission gun is needed for this type of work. For the analysis of perfect crystals, the most important benefit of a field-emission gun is the reduced contribution from defects, thickness variation, and bending under the probe. The interpretation of coherent CBED patterns formed with a very large illumination aperture ("ronchigrams") are also discussed, since these provide the simplest and most accurate method of aligning the instrument, and of measuring the optical constants of the probe-forming lens.

In Section 8.5 the study of defects using larger probes is reviewed. Using conventional CBED, or out-of-focus CBED patterns which produce a shadow image, it is possible to determine fault vectors for line and planar defects. The methods for doing so are reviewed. An intensity analysis of the HOLZ lines in these patterns also makes it possible to map out strains quantitatively, as also discussed in Section 5.4.

In materials science, perhaps the most important applications of the methods described in the first part of this chapter have been to the study of catalyst particles, and to the identification of microphases at interfaces. These may be as small as a single unit cell in thickness, but they still exert a controlling influence on mechanical and electrical properties. A second productive area of research has been the study of catalyst particles, which may be as small as a few nanometers in diameter. Twinning, and the crystallographic relationship between the particles and their substrate may be analyzed in this way. The shadow-imaging CBED methods described in Section 8.5 have made their greatest impact in the study of strains at semiconductor interfaces, where much work remains to be done. (The discussion of dynamical shifts on HOLZ lines in Sections 5.4 and 3.6 is relevant here.)

## 8.1. SUBNANOMETER PROBES, COHERENCE, AND THE BEST FOCUS FOR CBED

In order to study fine-grained materials, to avoid contributions from defects, and to obtain strain information by the methods of Section 5.4

from the smallest possible regions, it is often desirable to use the smallest possible probe. In this section we review the basic physics of the factors which ultimately limit probe size, and give some experimental examples of CBED patterns obtained with subnanometer probes. With conventional electron sources, there is usually insufficient intensity from such a probe, so that the use of a field-emission gun is assumed throughout this section.

We begin by introducing the concept of transverse coherence width $X_a$. For a review of coherence theory in electron microscopy, see Spence (1988a). Figure 8.1 shows a simplified ray diagram for CBED under the "critical illumination" conditions usually used. Note that the source is imaged onto the sample by the probe-forming lens, while the illumination aperture C2 (called the "objective aperture" on STEM instruments) is imaged onto the detector by the objective lens. C2 subtends a semiangle $\theta_c$ at the sample, while the geometrical electron source image (of

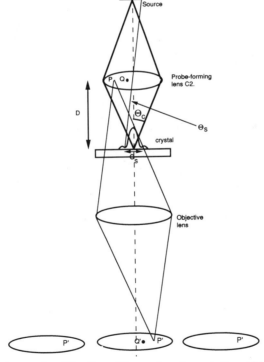

*Figure 8.1.* Electron microdiffraction. The angles $\theta_c$ and $\theta_s$ which determine the coherence conditions are shown. P' is conjugate to P. The source is imaged onto the sample and the illumination aperture is imaged onto the detector.

diameter $d_s$) subtends an angle $\theta_s$ at C2. It has been shown that the Zernike–Van Cittert theorem used in optics may also be used in electron optical systems, and this gives the transverse coherence width of the electron beam *in the plane of C2* as

$$X_a = \lambda/2\pi\theta_s = D\lambda/\pi d_s \qquad (8.1)$$

This provides an estimate of the distance between points in the plane of C2 at which the wavefield is capable of producing strong interference fringes in a Young's slit experiment with slit spacing $X_a$. We let the diameter of the C2 aperture be $2R_a$.

If $X_a \ll 2R_a$, the illumination aperture can be considered to be incoherently filled and so treated as an ideally incoherent effective source. The probe will then be partially coherent, since coherence always *increases* as one progresses along an optical system. (The imposition of a diffraction limit, for example, increases the correlation between the wavefunction at adjacent points.) This is the situation for conventional TEM systems using a tungsten or $LaB_6$ source under most operating contions. Then C2 can be treated as an ideally incoherent source, within which each point acts as a statistically independent emitter of electrons. (A useful exercise is to calculate $X_a$ for an $LaB_6$ source operating at the smallest probe size.) The coherence angle is $\alpha_c = \lambda/d_s$.

For this case, the coherence width at the plane of the sample is

$$X_s = \lambda/2\pi\theta_c \qquad (8.2)$$

Points closer together than $X_s$ at the sample are coherently iluminated. Points further apart than $\lambda/\theta_c$ may be considered incoherently illuminated. At intermediate spacings, the illumination is partially coherent. It is readily shown [by combining Eq. (8.2) with the Bragg law] that the condition that the coherence width $X_s$ at the sample be smaller than a lattice spacing $d_{hkl}$ is equivalent to the requirement that the CBED orders for these planes just overlap. Thus Bragg diffraction becomes possible (and the diffraction spots can be distinguished) when the coherence width at the sample becomes equal ot the $d$ spacing of interest.

If $X_a > 2R_a$, the illumination aperture is coherently filled and the irradiation can be considered to originate from a point source. Then the entire optical system beyond C2 is filled with perfectly coherent radiation and the probe may be treated as perfectly coherent. This is often a good approximation for field-emission-gun (FEG) instruments. The sample is then illuminated by an aberrated, converging spherical wave.

In this book, we shall refer loosely to "coherent CBED" as the case

$X_a > 2R_a$, and "incoherent CBED" as the case $X_a \ll 2R_a$, noting that these labels refer to the coherence conditions in the illumination aperture, and that in the second case the probe itself is partially coherent.

We see that coherence is increased by decreasing $d_s$, that is, by increased demagnification of the source, by using lower accelerating voltage and by decreasing the size of C2. In practice, an estimate of the coherence width at the sample can be obtained from the width of the band of Fresnel fringes seen in images of the edge of a sample, and from experiments using an electron biprism.

It follows immediately that, if the diffraction limit imposed by the objective lens can be ignored, then the CBED patterns from perfect crystals produced under coherent conditions will be identical to those produced under incoherent conditions if the orders do not overlap and if inelastic scattering is ignored. This follows from Fig. 8.1, since adjacent points such as P and Q are imaged independently and the intensity at the detector is then independent of any phase relationship between them. (For a perfect crystal, only scattering through multiples of the Bragg angle is permitted. Hence source point P cannot contribute at Q'.) For computational purposes, the two cases can therefore be treated in the same way provided there are no defects in the crystal. The focus condition does not enter the calculation. Experimental comparisons of CBED patterns obtained on a Phillips FEG EM400ST and those obtained on a similar machine fitted with an $LaB_6$ source confirm this finding, despite earlier suggestions of differences (Dowell and Goodman, 1976). An experimental comparison is complicated, however, by the fact that the use of a larger electron source both decreases coherence and also increases the size of the region of crystal from which the pattern is obtained. This may introduce additional effects due to thickness variation, bending, and defects.

For crystals containing defects, the elastic diffuse scattering around P' generated by source point P on Fig. 8.1 will overlap with that from Q at the detector. Different patterns will be observed depending on whether the complex amplitudes of this diffuse scattering from different source points are added (coherent case), or whether their intensities are added. This topic is discussed further in Section 8.3.

Calculations and experiments comparing CBED patterns from cold (V. G. HB5) and heated (EM400 FEG) field-emission guns are described by Zhu *et al.* (1987). Small differences were observed in patterns obtained from antiphase boundaries in $Cu_3Au$. It is also shown in this paper that the condition for coherent illumination ($X_a > 2R$) is equivalent to the requirement that the coherence width $X_s$ at the sample should be greater than the probe diameter.

Under "incoherent" conditions (i.e., $X_a \ll 2R_a$) the total probe diameter $d_0$ is given approximately at Gaussian focus by adding in quadrature the various contributions to $d_0$. Thus,

$$d_0^2 = d_s^2 + d_d^2 + d_{sa}^2 + d_c^2 + d_f^2 \tag{8.3}$$

where $d_s$ is the geometrical source image diameter, $d_d$ is the diffraction broadening equal to $0.6\lambda/\theta_c$, $d_{sa}$ is the contribution from spherical aberration (equal to $0.5 C_s \theta_c^3$ in the plane of least confusion, not the Gaussian image plane), and $d_c$ is the contribution from chromatic aberration, given by $(\Delta E_0/E_0) C_c \theta_c$, with $\Delta E_0$ the energy spread in the electron beam. This last term can be neglected as a first approximation. In Eq. (8.3) $d_f$ is the contribution $2\theta_c \Delta f$ from a small focusing error $\Delta f$. For a typical modern TEM instruments with $C_s = 2$ mm at 100 kV, the contributions of diffraction $d_d$ and spherical aberration $d_{sa}$ are equal at an angle of about 7 mrad. Equation (8.3) assumes that all these contributions are Gaussian and therefore cannot strictly be used for coherent conditions.

It seems natural to require a focus setting and illumination aperture size which will produce the most compact probe for CBED work. However, provided that the sample is a parallel-sided slab of defect-free crystal (and if inelastic scattering is ignored) it is easily shown that, for small focus changes, both coherent and incoherent *CBED patterns are independent of focus setting* $\Delta f$. (We assume that the diffraction disks do not overlap, i.e., $\theta_c < \theta_B$.) This result follows from the preceding discussion, which showed that the intensity distribution at P' and Q' is independent of the phase relationship [given by Eq. (8.4) below] between P and Q. (For very large focus changes, the disks eventually become small images of the sample—we are concerned here only with near-field effects.) We also show that the *CBED intensity distribution is independent of the position of the probe* under these conditions. (The fact that CBED patterns do change in practice when the probe is moved is mainly due to variations in thickness and crystal orientation under the probe.)

Despite the fact that CBED patterns are independent of small focus changes, in practice it is often desirable to use the smallest probe in order to reduce the chance of including a defect under the probe, and to minimize the variation in sample thickness and orientation which contributes to the pattern. We then require the focus setting which produces the most compact probe, and this is now discussed for the coherent case.

The smallest probe is obtained by minimizing all the quantities in Eq. (8.3) which are under experimental control. Thus we may make $d_s$

smaller than $d_d$ and $d_{sa}$ by combining a small physical source with large demagnification. Then the probe formation becomes "diffraction limited" and the illumination [from Eq. (8.1)] necessarily coherent ($X_a > 2R_a$). Then the quadrature addition of Eq. (8.3) does not apply, and detailed computations are required for the probe shape for particular values of the spherical aberration constant $C_s$, the defocus $\Delta f$, $\lambda$, and $\Theta_c$.

These calculations have been published by several authors (Crewe, 1985; Mory et al., 1987; Spence, 1978b). The intensity distribution of the probe at the sample is given by

$$I(\mathbf{r}) = \Psi(\mathbf{r})\Psi(\mathbf{r})^*$$

with (Spence 1988a)

$$\Psi_p(\mathbf{r}) = F\{A(\mathbf{K}_t)\exp[i\chi(\mathbf{K}_t)]\} \tag{8.4}$$

where the symbol $F$ indicates a Fourier transform, $A(\mathbf{K}_t)$ is a "top-hat" pupil function of width $\theta_c$ describing the illumination aperture, and the wave-front aberration function $\chi$ is given by

$$\chi(\mathbf{K}_t) = \pi(\Delta f \lambda K_t^2 + \tfrac{1}{2}C_s\lambda^3 K_t^4 + 2K_t c) \tag{8.5}$$

Here $c$ is the probe coordinate. The required focus setting $\Delta f$ can be defined as that which minimizes the radius of the probe area which contains, say, 70% of the beam intensity (Mory et al., 1987). Calculations based on Eq. (8.4) then show that the following values must be used to obtain this "most compact" probe:

$$\theta_c = 1.27 C_s^{-1/4} \lambda^{1/4} \tag{8.6}$$

and

$$\Delta f = -0.75 C_s^{1/2} \lambda^{1/2} \tag{8.7}$$

This gives the minimum probe diameter (containing 70% of the intensity) as

$$d(70\%) = 0.66 \lambda^{3/4} C_s^{1/4} \tag{8.8}$$

For experimental reasons, it may be easier to measure the probe displacement for which the intensity at a sharp edge in a STEM image falls from 80% to 20%. As judged by this criterion, the smallest probe is obtained with the constant 0.66 above replaced by 0.4. Experimental measurements of coherent probe widths, in rough agreement with the above theoretical estimates, can be found in Berger et al. (1987).

The total current in such a probe is discussed in Section 9.4. Thus

$$I_p = 0.25\beta\pi^2\theta_c^2 d_0^2$$

where $\beta$ is the source brightness. The coherent current is obtained by replacing $\theta_c$ by $\alpha_c$.

Each plane-wave component of the coherent, subnanometer probe illuminates the entire sample. The Fourier synthesis of all components concentrates the energy into a localized region. The amount of energy remaining in the tails of the distribution depends on the focus setting. As an example, for the Vacuum Generators HB501 machine with analytical pole piece ($C_s = 3.1$ mm), we find $d$ (70%) = 0.415 nm at 100 kV. In practice, extremely stable conditions are required to achieve this performance, and the effects of tip vibration, sample movement, stray fields, and electronic instabilities in the lens and accelerating voltage must also be considered. Section 8.3 shows some experimental patterns obtained from subnanometer regions of crystal.

Equations (8.6) and (8.7) can therefore be used to give the experimental conditions which should be used to obtain CBED patterns from the smallest possible regions of defect-free crystalline sample. But the resulting CBED pattern will be difficult to index and interpret (for reasons to be discussed below) if such a large illumination aperture is required that the CBED orders overlap. This will occur if, from Eqs. (2.4) and (8.6),

$$d_{hkl} = 0.7874 C_s^{1/4} \lambda^{3/4} \qquad (8.9)$$

This equation gives the largest $d$ spacing that can be studied without overlap of orders by a CBED instrument of given $C_s$ and $\lambda$, if the (diffraction-limited) probe is to be as small as possible.

Different operating modes of STEM instruments require different focusing conditions. For analytical microscopy, the scattering of the probe distribution $I(\mathbf{r})$ is treated by solving a transport equation using, for example, Monte Carlo methods. Thus, although the probe-formation process is treated coherently, the multiple scattering within the sample is treated incoherently, using the Boltzman transport equation rather than the Schrödinger equation. This incoherent scattering approach is used, for example, in the X-ray microanalysis of non-crystalline samples, and for these purposes Eqs. (8.6)–(8.8) also give the optimum probe-formation conditions. For crystalline samples, the possibility of Bragg scattering, coherent Bremsstrahlung, and channeling effects must be considered, all of which can produce "artifacts" in X-ray and energy-loss spectra (Buseck *et al.*, 1989). Provided the diffraction orders do not overlap, these effects (which can only be predicted using a coherent scattering model) are independent of focus setting, but again, use of the above conditions will collect spectra from the smallest region.

In quantum mechanics it is the observable quantities at the detector which are important. In X-ray microanalysis, for example, the probe is not detected. Unless it can be shown that the probe intensity distribution enters the quantum mechanical description of the complete scattering problem in a simple way, we should conclude that there may be a different resolution for each detector (each signal) and that, since this depends on the sample, it is not a property of the instrument alone. For analytical signals, it would seem that the high-angle dark-field STEM image resolution may provide the best experimental estimate of probe size, since the scattering theory used is similar in the two cases. (Inelastic delocalization effects must, however, also be considered.)

Since the STEM instruments under consideration are also capable of forming a scanning transmission lattice image, we consider in the next section a different focusing condition which produces the most faithful image of a crystal in STEM. This lattice image may be used to position the probe for CBED work.

## 8.2. LATTICE IMAGING IN STEM USING OVERLAPPING CBED ORDERS

If CBED orders overlap, and if the angular range over which the illumination is coherent exceeds the Bragg angle, it becomes possible to form a phase-contrast STEM lattice image. The relevant theory is given in Spence and Cowley (1978) and a review in Spence (1988b). The image is formed by detecting part of the CBED pattern (the overlap region) and by displaying this intensity as a function of probe position, as the probe is scanned over the sample. Figure 8.2 shows the geometry. We assume a point source, and therefore complete coherence. The focused probe is a diffraction-limited image of this point source, as formed by an imperfect lens. The probe wavefunction is given by Eq. (8.4), and we may think of the phase factor in Eq. (8.5) as filling the exit pupil of the probe-forming (objective) lens. We see that the orders just touch if $\theta_c = \theta_B$. If the orders do overlap slightly, as shown, and if the illumination aperture is coherently filled across A–B, it is possible for radiation from two different points A and B to reach the same detector point D. The interference at D will therefore depend on the phase factor in Eq. (8.5), and hence on the probe position, on $C_s$, and on the focus setting $\Delta f$. In

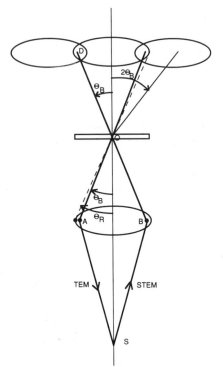

*Figure 8.2.* Lattice imaging in STEM. Two illumination angles $\theta_B$ and $\theta_R$ are shown. The second allows a small overlap of orders, so that a detector at D records a two-beam lattice image as the probe is scanned. Interchanging S and D shows that this is equivalent to inclined illumination TEM lattice imaging.

fact, the intensity at D varies sinusoidally in this case with the period $c = d_{hkl} = \lambda/2\theta_B$ of the lattice. It also depends on the phases of the crystal structure factors, so that experiments with overlapping CBED orders may be used to measure these quantities (Spence, 1978c).

However, we can see immediately that a lattice image will be formed, by using the theorem of reciprocity. On interchanging the point source S and a point detector at D, we have exactly the arrangement used to form two-beam lattice images in a TEM instrument, using inclined illumination from D. (It may be useful to imagine the crystal being scanned under a stationary probe to see this more clearly.) We note in passing that if $C_s = 0$ and $\Delta f = 0$, then the full width at half maximum height (FWHM) of the Bessel function probe would be

$$d_{FW} = 0.61\lambda/\theta_c \qquad (8.10)$$

Using the Bragg law [Eq. (2.4)] and setting $d_{FW} = d_{hkl}$, we find that the FWHM for the probe just equals the lattice spacing $d_{hkl}$ if

$$\theta_c = 1.2\theta_B \qquad (8.11)$$

COHERENT NANOPROBES                                                              179

In this sense, it becomes possible to resolve the lattice in a STEM instrument when the probe size becomes comparable to the *d* spacing of interest, and this occurs with a 20% overlap of the orders. In practice, the effects of spherical aberration, tip vibration, and electrical instabilities cannot be neglected. It is also more useful to think of lattice resolution in STEM as limited by the angular range over which radiation incident on the sample is coherent.

To form a bright-field axial three-beam image, it is necessary to use the geometry shown in Fig. 8.3, with an axial detector, since, by the reciprocity theorem, such an image is identical to that which would be formed in a high-resolution TEM instrument if reciprocal aperture, source, and detector sizes are used. It is for this reason that the illumination aperture in STEM is known as the objective aperture, since it may be thought of (by reciprocity) as limiting the number of "beams" which contribute to a STEM lattice image. A Scherzer focus condition

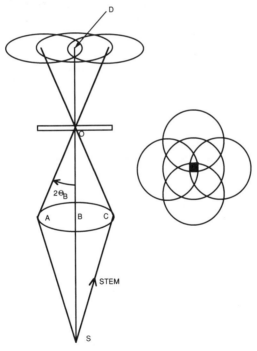

*Figure 8.3.* Axial three-beam lattice imaging in STEM. By opening up the illumination angle to twice the Bragg angle, three orders overlap at the axial detector D. The appearance of a two-dimensional coherent CBED pattern used for axial five-beam lattice imaging is shown at the right.

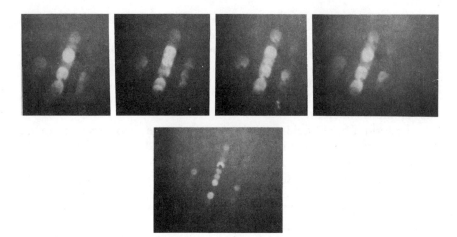

*Figure 8.4.* Coherent CBED patterns taken with a 0.5–1 nm probe as it is scanned across a single unit cell of $Ti_2Nb_{10}O_{29}$ (Cowley, 1981a). The period of the cell is 2.8 nm. Close examination of the original shows that only the overlapping portions of the patterns change as the probe is moved, and that the first pattern is identical to the last.

therefore gives the most faithful representation of a weakly scattering sample in STEM. The focus and aperture condition to be used is then

$$\Delta f = -1.15 C_s^{1/4} \lambda^{-3/4} \quad \text{and} \quad \theta_c = 1.51 C_s^{-1/4} \lambda^{1/4} \quad (8.12)$$

The lattice images for STEM and TEM may therefore be computed in the same way, with the integration over the source required in TEM becoming the integration over the detector in STEM. Figure 8.4 shows a series of experimental coherent microdiffraction patterns obtained as a subnanometer probe is moved across a single unit cell of $Ti_2Nb_{10}O_{29}$ (Cowley, 1981a). The period of the cell is 2.8 nm. Close examination of the figures shows that only the overlapping portions of the patterns change as the probe is moved, and that the first pattern is identical to the last. (Figure 8.7, to be discussed later in another connection, shows a STEM lattice image of $YBa_2Cu_3O_7$, together with CBED patterns taken from different parts of the cell.) The probe may be stopped at any point in the image and used to form a CBED pattern. Instrumentation requirements for this are described in Section 9.3.

A dark-field or minimum phase-contrast focus condition also can be defined (Cowley, 1981b), for which the numerical factor in Eq. (8.7) becomes −0.44 instead of −0.75. The optimum focus for annular dark-field STEM imaging is found to be midway between this condition and the Scherzer condition (Mory *et al.*, 1987).

## 8.3. CBED FROM DEFECTS—COHERENT NANOPROBES. SITE SYMMETRY, ATOM POSITIONS

Tables 8.1 and 8.2 summarize the various ways in which CBED may be used to study defects. Since the measurement of strains falls under this heading, the material in Sections 2.4 (on the geometry of HOLZ lines), 3.6 (on theory of dynamical HOLZ line shifts), and 5.4 has also been included. There are four cases.

1. A coherent nanoprobe may be used, with overlapping orders. Then information on local symmetry and atomic positions may be extracted, in a similar way to lattice imaging.
2. The probe may be much larger than the defect. Then one finds HOLZ and Kikuchi line splittings, which may be used to determine the fault vector of a defect as described in Section 8.5.
3. Mixed-mode operation of the microscope is possible, as in the CBIM and LACBED methods, in which a shadow image of the sample is seen superimposed on the CBED pattern. Section 8.5 has more information on this topic.
4. The strain field may be slowly varying on a scale which is large compared to the probe. Then the column approximation may be used, and HOLZ-line shifts (without splitting) result, so that strains may be measured, as described in Section 5.4.

In this section we consider case (1). [Cases (2) and (3) are considered in Section 8.5.] It should be borne in mind, however, that for the study of the statistical properties of defects, the use of point diffraction patterns and diffuse elastic scattering may provide far more information than incoherent CBED patterns, in which this diffuse scattering is convoluted with the incident cone of illumination angles. For a review of the study of diffuse scattering from defects, see Cowley (1981b). For a recent application to oxygen ordering in oxide superconductors, see Zhu et al. (1990).

Only in the coherent case, using subnanometer probes and overlapping orders, does it become possible to determine the detailed atomic arrangement at the core of a defect by CBED. Here the aim becomes the determination of individual atomic coordinates, rather than the statistically averaged properties of an ensemble of defects or the determination of a fault vector. All the scattering originates from the local region of the defect and, as shown in Section 8.1, the intensity distribution depends on the absolute probe position and the spherical aberration constant and

Table 8.1. A Summary of Defect Analysis Techniques by CBED[a]

| Case / Section in this book | Probe size | Coherence at illumination aperture | Information extracted |
|---|---|---|---|
| 1 | | | |
| 8.3 | Nanoprobe | Coherent orders overlap | Atomic coordinates and site symmetry about probe center |
| 2 | | | |
| 8.4 | Larger than defect | Incoherent | Fault vectors |
| 3 | | | |
| 8.4 | Mixed-mode CBIM/LACBED | Incoherent | Fault vectors Image information |
| 4 | | | |
| 2.4, 3.6, 5.4 | Strain varies slowly on scale of probe size | Incoherent or coherent | Quantitative local strain measurement |

[a] Coherence at the illumination aperture (rather than at the probe) is used because an incoherent probe is not achievable in practice (see Section 8.1). The possibility of determining translations at vertical interfaces is also discussed in the text.

Table 8.2. Hybrid Imaging and Diffraction Modes in CBED[a]

| Method | Detector conjugate to ··· | Spatial resolution | Angular resolution | Illumination |
|---|---|---|---|---|
| LACBED | Back-focal plane of post-specimen lens | Limited by probe size | Diffraction limited by post-specimen lens | Incoherent $\Theta_c > \Theta_B$ |
| CBIM | Sample | Diffraction limited by post-specimen lens | Limited by probe size. May be diffraction limited | Incoherent $\Theta_c < \Theta_B$ |
| Shadow image, large source | Back-focal plane of post-specimen lens | Limited by probe size | Diffraction limited by post-specimen lens | Incoherent |
| Shadow image, point source | Back-focal plane of post-specimen lens | Use holographic reconstruction if sample very thin | Diffraction limited by post-specimen lens | Coherent |

[a] A periodic object is the only object whose shadow image, if projected from a point, forms a perfect unaberrated image. In every case above the source is conjugate to an out-of-focus plane near the sample.

defocus setting of the objective lens. It may be useful here to summarize our findings so far on *coherent* (nanoprobe) CBED.

1. Patterns from perfect crystals in which the orders do not overlap are identical to incoherent patterns and are not sensitive to changes in focus or probe position.
2. Any defect or discontinuity in a crystalline sample renders these patterns sensitive both to the probe position and to the focus and aberration constants of the probe-forming lens. This applies both to the case of overlapping and nonoverlapping orders. "Defects" here includes bending of the crystal and thickness variations under the probe.
3. Patterns from perfect crystals with overlapping orders are sensitive to probe position, focus, and aberrations only in the regions where the orders overlap.

A "perfect crystal" here means specifically a parallel-sided slab of defect-free crystalline material which produces an ideal "point" diffraction pattern when illuminated by a plane wave.

Large dynamical computations are required for the interpretation of these patterns, and the instrumental parameters and absolute probe position must be known. The problem of interpreting these patterns therefore becomes rather similar to that of interpreting high-resolution lattice images. Figure 8.5 indicates the principle of the method (Spence, 1978a). The probe wavefunction given by Eq. (8.4) is used in the first slice of a multislice calculation. The "unit cell" ABC (containing many cells of the perfect crystal) is made large enough so the probe wavefunction falls to a small value at the boundary of the cell. Thus the size of the cell used depends on the focus setting and on the aberrations of the lens. The defect strain field should also fall to negligible amplitude

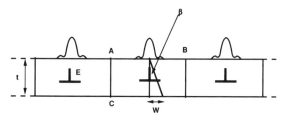

*Figure 8.5.* The principle of periodic continuation used to calculate coherent microdiffraction patterns from crystals containing defects. The supercell ABC and the incident probe, containing a defect E, are continued periodically. Probe broadening W (discussed in the text) is indicated.

at the boundaries A and B if calculations are to be made with the probe near the boundary. The additional broadening W of the probe due to elastic scattering is given by the Takagi triangle construction, and is approximately $W = 2t\theta_B$, where $\theta_B$ is the largest Bragg angle for which appreciable scattering occurs (Humphreys and Spence, 1979). The asymptotic behavior of the probe may be improved for computational purposes by multiplying Eq. (8.4) by a Gaussian function, to simulate the effects of using a detector of finite angular resolution.

Figure 8.6 shows computations by this method for a 0.3-nm probe situated over dislocation cores in iron (Spence, 1977). A dislocation dipole has been used, so that the strain field falls off rapidly toward the boundary of the cell. Patterns are shown for two probe positions and two focus settings. We note the reduced diffuse scattering when the probe is situated over the more perfectly crystalline material between the cores. The scattering is also seen to depend on the focus setting, due to the overlap of diffuse scattering originating from different points within the coherently filled effective source. Also note that the mirror plane of symmetry running across Fig. 8.7a is preserved in the resulting microdiffraction patterns. A STEM detector filling the region between these disks would produce a high contrast image of the dislocation cores.

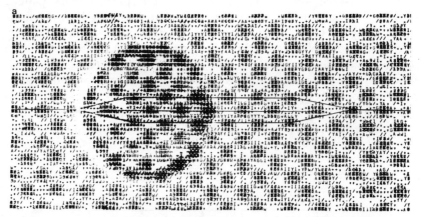

Figure 8.6. (a) The atomic potential for two [100] unit edge dislocations in iron, projected along the dislocation line. The supercell dimension is $17 \times 37$ angstroms. Atomic coordinates from Sinclair (1971). The incident electron probe is also shown, computed using Eq. (8.4), with $C_s = 2$ mm at 100 kV and optimum aperture. (b) Four simulated coherent CBED patterns which result for two different probe positions and two different focus settings. Thickness-8 nm. Note that the more perfect crystal at probe position B generates less diffuse scattering. At large defocus, these discs would sharpen into (110) spots (Spence, 1978c).

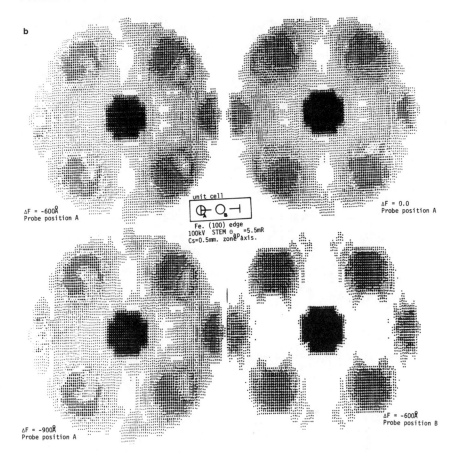

Figure 8.6. (continued)

The four outer disks shown would sharpen up into (110) Bragg spots under plane-wave illumination. Larger computations of this type, including the contribution of phonon scattering, have been used for simulating high-angle dark-field STEM lattice images, as described by Loane et al. (1991).

We now consider how the resolution limit to information contained in these patterns compares with that in a STEM or HREM lattice image. Because of the sensitivity of these patterns to any relative displacement between the field-emission tip and the sample, detail in the high angle scattering will be washed out by vibration. Electronic instabilities in lens current and accelerating voltage (which appear as fluctuations in focus setting) have a similar effect. This can be seen by treating the case of an

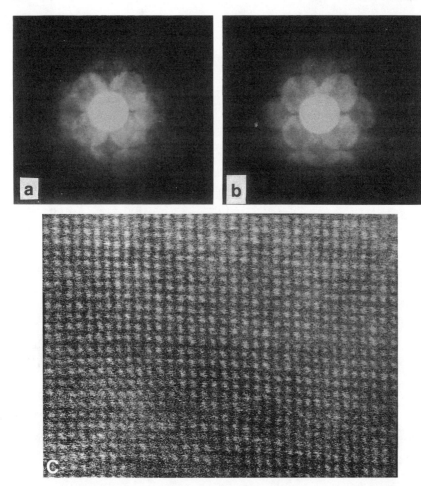

*Figure 8.7.* (a) The CBED pattern obtained with a coherent 0.3-nm probe situated over the Cu(1) site within the unit cell of YBa2Cu3O7. (b) The probe has been moved to the O(4) site. (c) The bright-field STEM lattice image used to position the probe. The detector used approximately fills the central disk (Ou and Cowley, 1988).

amorphous sample as a weak phase object so that, if a plane wave is incident on the sample, the scattering becomes

$$\Psi(\mathbf{u}) = \delta(\mathbf{u}) + i\sigma\bar{V}_p(\mathbf{u}) \tag{8.13}$$

where $\bar{V}_p(\mathbf{u})$ is the Fourier transform of the projected sample potential and $\mathbf{u}$ is a scattering vector. The complex amplitude distribution for the coherent CBED pattern is obtained by convoluting this expression with the inverse Fourier transform of Eq. (8.4). At the dark-field focus setting the important term in the intensity distribution becomes

$$\mathbf{I}(\mathbf{U}) = \int \left| \int \sigma \bar{V}_p(\mathbf{u}) A(\mathbf{U} - \mathbf{u}) \cos[\chi(\mathbf{U} - \mathbf{u}, c)] \, d\mathbf{u} \right|^2 G(c) \, dc \quad (8.14)$$

Here $G(c)$ is a Gaussian function describing the statistical distribution of probe positions due to sample or tip vibration. At large angles (large $U$), small changes in $c$ (the probe coordinate) or focus (due to electronic instabilities) cause large changes in the convolution—the intensity is dominated at high angles by the rapidly oscillating portion of the transfer function. This effect can be avoided by inserting a smaller objective aperture [narrower $A(u)$] to exclude the rapid oscillations in $\chi$, but only with a consequent increase in the (diffraction-limited) probe size. In general, the highest resolution information which can be extracted from a coherent microdiffraction pattern from noncrystalline material is approximately equal to the finest detail contained in a STEM image of the same sample, and is limited to the "information limit" (Spence, 1988b) of the objective lens transfer function. But the instabilities contribute to Eq. (8.14) in a different way than those in high-resolution transmission electron microscopy, allowing for different possibilities for the extraction of information (Bates and Rodenburg, 1989; Cowley, 1976; Cowley and Spence, 1979). Since the coherent CBED pattern with overlapping orders is an in-line Gabor electron hologram, holographic techniques may also be used in the single scattering regime (Lin and Cowley, 1986), discussed further in Section 8.4 below.

We have seen in the previous chapter that conventional CBED patterns contain information on the translational symmetry of crystals. Now many minerals and ceramics have projected unit cells larger than, say, 2 nm on a side. The smallest probe size on the Vacuum Generators HB501 is about 0.25 nm. We may therefore ask how this translational symmetry information can be present if the probe size becomes smaller than the unit cell. As we have seen, this can only happen if the orders overlap, whereupon (if the overlap is large) it no longer becomes possible to index the pattern and so to determine the space lattice. In addition, the intensity distribution becomes sensitive to the probe position within the cell and to other instrumental parameters. In fact, the symmetry information then contained in such a pattern is site symmetry information, as reckoned about the center of the probe (Cowley and Spence, 1979). (We here assume a perfectly aligned microscope without astigmatism.) Thus, for example, a center of symmetry would be seen in the overlapped CBED pattern only if the probe were accurately centered over such a symmetry element in the crystal. A similar result applies to $n$-fold rotation axes and mirror planes, as seen in Figs. 8.6 and 8.7.

Experimentally, by searching for such (projected) symmetry elements in the coherent CBED pattern, it has proved possible to position the probe over a particular site within the unit cell (Spence and Lynch, 1982). The result was confirmed by electron energy loss spectroscopy. Figure 8.7 shows two CBED patterns obtained from a thin crystal of $YBa_2Cu_3O_7$. In Fig. 8.7a the 0.3-nm probe has been situated over the Cu(1) site, while in Fig. 8.7b the probe falls over the O(4) site. Figure 8.7c shows the STEM lattice image used to position the probe. Differences in the site symmetry can be directly read out from the patterns (Ou and Cowley, 1988). Translational symmetry elements are also expressed in nanodiffraction patterns if the samples are thick enough. Figure 8.8. shows coherent CBED patterns obtained using a 1-nm-diameter probe from a thin crystal of rutile for three thicknesses. The pattern from the thickest region clearly shows the characteristic dark bands of a dynamical extinction in the forbidden (100) reflections (Cowley, 1985). It is the coherence width $X_s$ given in Eq. (8.2) which determines the spacings in the sample which contribute to the CBED pattern. The probe size determines the intensity of their contribution.

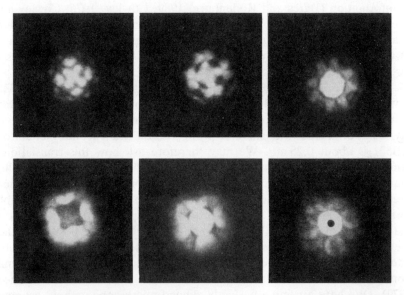

*Figure 8.8.* Coherent CBED patterns obtained using a 1-nm probe from three thicknesses of rutile. Dynamically forbidden reflections are seen in the thickest (right-hand) pattern (Cowley, 1985).

Thus, for coherent patterns with overlapping orders, we might expect that these dynamical extinctions would disappear as the probe was moved away from the glide plane or made smaller than the glide translation. (A "disappearing extinction" here refers to the appearance of intensity in an otherwise forbidden reflection.) As described in Section 9.3, efficient, bakeable two-dimensional microdiffraction recording systems are essential for this work.

Further examples of the study of defects by coherent CBED may be cited as follows:

(1) The loss of symmetry which would result from the threefold dissociation of a screw dislocation in iron has been investigated using computed dynamical coherent microdiffraction patterns (Spence, 1977; 1978c).

(2) A useful effect was discovered from computations and observations of coherent CBED patterns from very small crystals (Cowley ad Spence, 1981; Pan *et al.*, 1989). This allows them to be used to find the fault vector which characterizes planar faults in crystals, in the spirit of the "$\mathbf{g} \cdot \mathbf{b}$" analysis used in TEM imaging. Specifically, it was found that CBED disks which would normally be uniformly filled with intensity show annular rings of intensity instead, if the probe is placed near the edge of a crystal. Figure 8.9 shows the effect. We will refer to this effect loosely as "spot splitting." In subsequent work on planar faults, it was found that, if the probe was situated over a fault (with the beam in the plane of the fault), then only those reflections for which $\mathbf{g} \cdot \mathbf{R} \neq 0$ were split. Here $\mathbf{R}$ is the fault vector describing the translation needed to bring the crystal on one side of the fault into coincidence with that on the other. (Such a vector may not always be defined.) Thus by noting which disks are split, $\mathbf{R}$ may be deduced. For planar faults, the splitting is normal to the plane of the fault. Examples of this general approach can be found as follows: For a study of antiphase boundaries in $Cu_3Au$, see Zhu and Cowley (1982). Here it was found possible to deduce the nearest-neighbor atomic coordination across a boundary from a study of coherent CBED patterns. Figure 8.10 shows a typical pattern containing split spots from the copper–gold superlattice reflections. For similar work on stacking faults and twins in steel, see Zhu and Cowley (1983).

By comparison with conventional dark-field TEM imaging, this approach has two important advantages—first the $\mathbf{g} \cdot \mathbf{R}$ conditions for many reflections may be obtained from a single diffraction pattern, and second, the method may readily be combined with STEM imaging. GP zones (thin precipitates) in Al–4% Cu have also been studied by this method, providing support for a particular atomic model of the precipitate (Zhu and Cowley, 1985).

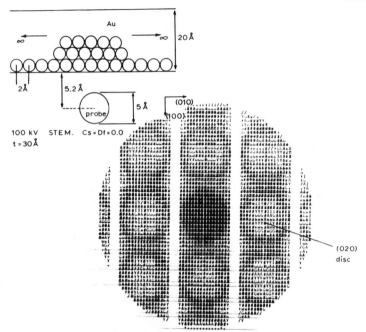

*Figure 8.9.* Two-dimensional computed coherent CBED pattern for a probe of diameter 0.5 nm placed 0.52 nm outside of a gold bar 2 nm wide and 3 nm thick, as shown inset.

(3) Platelets in diamond have been studied by this type of nanodiffraction (Cowley *et al.*, 1984). The patterns were compared with computed patterns for various models proposed for the atomic structure.

(4) Oxide formation on chromium has been studied by coherent nanodiffraction (Watari and Cowley, 1981). The epitaxial relationship between the oxide and the metal was deduced, together with the structure of the oxide in its early stages. Conclusions can be drawn about the mechanism of oxide growth. This type of study of the microcrystallography at interfaces demonstrates the unique power of nanodiffraction. A second example concerns the Ag/MgO interface, in which a new spinel phase was identified just a few atomic layers thick. This has important consequences for our understanding of the mechanical properties of this metal–ceramic interface, since the interface energy of the intergranular phase must be considered.

(5) The structure of a particular type of very small oxygen precipitate (keatite) in silicon has been determined by coherent CBED (Kim *et al.*, 1987).

(6) Microdiffraction studies on catalyst particles have proved informative. For work on Rh particles (about 2–3 nm in diameter) on Ce

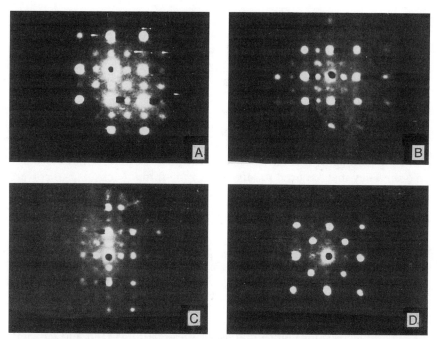

*Figure 8.10.* Experimental coherent CBED pattern obtained by placing a 0.5-nm probe over four different types of antiphase boundaries in a thin crystal of copper–gold. The inner superlattice spots are "split" (Zhu and Cowley, 1982). The direction of the boundary is parallel to the gap in the annulus of intensity (see also Fig. 8.9).

oxide substrates, see Pan *et al.* (1987). Epitaxial relationships between particle and substrate, twinning and oxidation may all be investigated. Work on Pt and Pd particles on various substrates is described by Lynch *et al.* (1981) and Pike *et al.* (1987).

Finally, we note the development of a new technique for the determination of atomic positions using a series of coherent CBED patterns from overlapping regions (Konnert *et al.*, 1989). Autocorrelation functions are computed for each probe position, resulting in atomic coordinate determinations with an accuracy of 0.02 nm. Applications to noncrystalline materials are discussed. A recent review of computing methods for coherent microdiffraction patterns is given by Krakow and O'Keefe (1989). The interpretation of these patterns from amorphous materials is discussed further in Section 8.6.

## 8.4. RONCHIGRAMS, POINT PROJECTION SHADOW LATTICE IMAGES, AND HOLOGRAMS

A ronchigram is a coherent CBED pattern recorded with a very large objective aperture, or with the aperture removed entirely. If the sample is crystalline, gross overlap of the CBED orders results. These patterns are called ronchigrams because the geometry used to obtain them is identical to that used to test optical lenses and mirrors (Malacara, 1978; Ronchi, 1964). They have a number of interesting properties and uses for coherent CBED, as follows:

1. They may be used to measure the spherical aberration constant and defocus setting of the probe-forming lens. These quantities are required for the interpretation of the patterns and any resulting STEM lattice images (Lin and Cowley, 1986).
2. They may be used to align the probe-forming lens and to correct astigmatism. This is often the most tedious and time-consuming aspect of CBED work using subnanometer probes.
3. If the sample is crystalline, the central portion of this pattern will be shown to consist of a high magnification lattice image of the sample (Cowley, 1979). Thus, a lattice image can be formed without scanning.
4. For very thin sample, the ronchigram is an in-line Gabor electron hologram. All the image-processing techniques previously developed for holography in optics may therefore be applied to the interpretation of these patterns (Lin and Cowley, 1986).
5. An understanding of these patterns is basic to the interpretation of bright and dark-field STEM lattice images, and to the design of special detectors for this purpose.

We commence with a simple geometric explanation of the formation of projection lattice images and Fourier imaging, then summarize the relevant theory and give experimental examples. The interpretation of these patterns as holograms is further discussed at the end of this section. Figure 8.11 shows a ray diagram for coherent CBED with an out-of focus probe focused in front of a thin crystal. Bragg diffraction of the incident cone generates additional cones deflected by multiples of twice the Bragg angle. If $\theta_c > 2\theta_B$, these will overlap and interfere at the center of the detector. By tracing these deflected cones back toward the source, a set of virtual sources may be defined, which are necessarily coherent with the physical source. As shown in the lower figure, these virtual sources lie on the reciprocal lattice, and the situation is thus identical to that found in the back-focal plane of a TEM when used for lattice imaging. If an ideal point source were available, the arrangement would therefore produce an unaberrated lattice image of the crystal, without using either lenses or scanning. This point-projection method for electron lattice imaging was

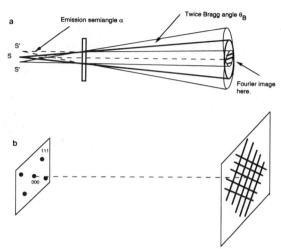

*Figure 8.11.* The upper figure (a) shows the coherent overlapping orders produced by a point electron source focused in front of a thin crystal and the virtual sources which result from Bragg diffraction if $\theta_c > 2\theta_B$. The lower figure (b) shows how these virtual sources lie on the points of the reciprocal lattice, and therefore may produce an ideal lattice image without using lenses or scanning (if a true point source were available). The (110) ZOLZ of a face-centered cubic lattice is shown. (From Spence and Qian, 1991.) Figure 6.1 shows the underfocused case.

first proposed by Cowley and Moodie (1957), who provided the theory and experimental results using coherent light. They named the resulting images "Fourier images." It is important to note that only strictly periodic detail is faithfully imaged by this technique, since it relies on Bragg diffraction. Defects are not seen in Fourier images [see Spence (1988a) for a review], however, if the object is weakly scattering (or compact), the patterns may be interpreted as holograms of defects. Figure 8.12 shows ronchigrams obtained from a thin crystal of beryl, using an out-of-focus 0.3-nm-diameter probe. According to the preceding discussion, the central region shows an aberrated lattice image of the 0.8-nm crystal planes. The magnification of the image is seen to depend on the focus. No scanning was used to obtain these images.

Similar images have been obtained at very low accelerating voltages (about 300 volts) using a sputtered tungsten field-emission tip instead of a focused probe as a "point" electron emitter, again without using any lenses (Fink *et al.*, 1991). The (small) aberrations of the tip are, however, those of the corresponding virtual source. Because of their similarity to ronchigrams in STEM, these images may be interpreted using much of the following theory (Spence and Qian, 1991). However, the multiple scattering inside the sample must be treated by the theory of transmission

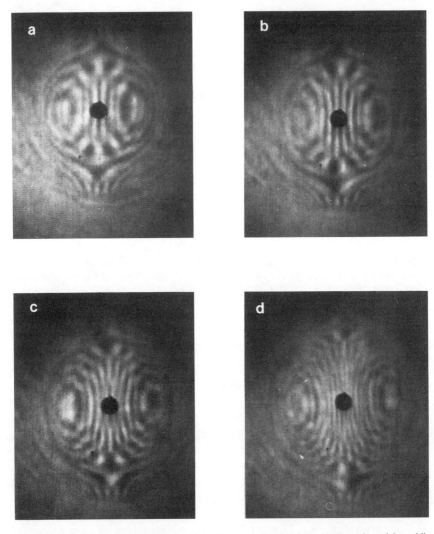

*Figure 8.12.* Ronchigrams obtained with a 0.3-nm probe for four focus settings (a) to (d) from a thin crystal of the mineral beryl. The fringes in the center are an aberrated point projection lattice image of the crystal. The lattice spacing is 0.8 nm. Outer ellipses allow measurement of the spherical aberration constant and defocus for the probe-forming lens (Cowley, 1986).

low-energy electron diffraction (TLEED) (Qian *et al.*, 1992). (For thick samples, a "transmission function" may not be definable in a useful way, and this means that the usual weak-phase object treatments cannot be used.) The TLEED theory includes backscattering (inside and outside the crystal) and exchange corrections to the crystal potential. The previous description must now be modified to take into account the aberrations of the lens and the finite source size. First, we treat the general case of a nonperiod object, which provides a useful practical method of electron-optical alignment. According to Eq. (8.4), the diffraction-limited probe wavefunction incident on the sample is

$$\Psi_p(\mathbf{r}) = F\{A(\mathbf{K}_t) \exp[i\chi(\mathbf{K}_t)]\} \qquad (8.14)$$

with $A(\mathbf{K}_t)$ the illumination aperture pupil function, and we will take $c = 0$, since the probe is not scanned. Here $|\mathbf{K}_t| = \theta/\lambda$, and it is assumed that the illumination aperture is coherently filled. The transmission function of the sample is taken to be

$$q(\mathbf{r}) = \exp[+i\sigma V_p(\mathbf{r})] \qquad (8.15)$$

where $V_p(\mathbf{r})$ is the projected sample potential, and the crystalline sample is assumed sufficiently thin so that there is no variation in the intensity of the rocking curves within each CBED order. Equation (8.15) includes multiple scattering effects within the approximation of a "flat" Ewald sphere. Then, if $Q(\mathbf{K}_t)$ is the Fourier transform $F\{q(\mathbf{r})\}$ of $q(\mathbf{r})$, the intensity distribution at the detector is

$$I(\mathbf{K}'_t) = |Q(\mathbf{K}_t) * A(\mathbf{K}_t) \exp[i\chi(\mathbf{K}_t)]|^2 \qquad (8.16)$$

Here $\mathbf{K}_t$ is a two-dimensional coordinate in the illumination aperture plane while $\mathbf{K}'_t$ is the coordinate at the detector screen. The asterisk denotes convolution. Numerical computations based on this expression are in excellent agreement with experimental ronchigrams recorded on the Vacuum Generators HB5 machine. The reader is referred to Cowley and Disko (1980) for more details. These characteristic patterns are obtained by placing the probe on the edge of a noncrystalline sample and removing the objective aperture. They may be used for the final stages of lens alignment and a stigmatism correction for coherent CBED, much as Fresnel fringes at a sample edge are used for alignment prior to TEM lattice imaging.

It is a remarkable fact that Eq. (8.16) can be reduced to exactly the form of the expression used to describe an out-of-focus HREM image in the phase-grating approximation. This follows by evaluating the convolution in (8.16) for the case where $A(\mathbf{K}_t)$ is replaced by a constant (Cowley, 1979). Thus, we may conclude that *a shadow image formed by a point source, distance Z from a transmission object, produces an identical image to that produced by plane-wave illumination of the same object on a plane distance Z beyond the object.* This result, illustrated in Fig. 8.12A below, applies only within the phase grating approximation, since, under general multiple scattering conditions, the transmission function used with spherical wave illumination is not equal to that which can be used for plane-wave illumination (Spence and Cowley, 1978). For thin samples, the effect of limiting the range of coherent illumination angles $\alpha_c$ in the point-projection geometry is equivalent to limiting the objective aperture in HREM. Thus the point-projection geometry provides magnification with negligible aberrations if a field-emission tip of the type used by Fink *et al.* (1991) is used. The aberrations of nanotip emitters are computed in Scheinfein *et al.* (1992). Implementation of this scheme requires the use of sufficiently high accelerating voltage to allow exchange and virtual inelastic scattering corrections to the scattering potential to be neglected. At low voltages (as in LEED) these corrections can be significant, so that Poisson's equation cannot be used to relate the measured optical potential to the charge density. Image interpretation for this "transmission LEED" case is discussed in Qian, Spence, and Zuo (1992).

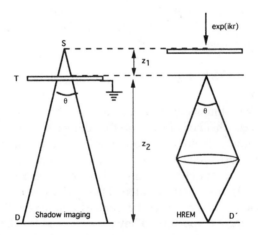

*Figure 8.12A.* Equivalence of shadow image $D$ formed with spherical wave $S$, and high resolution electron microscope image, out-of-focus by $z_1$, at $D'$ formed with plane wave illumination. For equal magnifications, the images at $D$ and $D'$ are identical in the absence of aberrations for thin samples. As $z_1$ goes to zero, the magnification becomes infinite, and the pattern at $D$ becomes a CBED diffraction pattern. For finite $z_1$ the resolution limit of the shadow image can be expressed by a *convolution* with the source function. For $z_1 = 0$ it is expressed by *multiplication* with a similar function. The focusing error $z_1$ cannot be eliminated using additional lenses, but can be by treating the shadow image as a hologram if a strong reference beam exists. The field emission point-projection microscope was invented by Morton and Ramberg (1939). See also Spence (1992d) for more details.

It is important to emphasize that electron holography does not allow one to look inside a sample. For extended thin crystals, the beam energy required for in-line holographic conditions is sufficiently high to ensure the validity of the projection approximation, in which case the scattered wavefield is independent of the $z$ coordinate of atoms in the sample. The failure of this approximation and the holographic condition $\Psi = 1 + \varepsilon$, where $\varepsilon \ll 1$, can both be achieved at very low energies (less than 1 kV) for isolated molecules, or at the edges of thin crystals. At these energies multiple scattering is dominant within the sample. However, all reconstruction schemes correct for focus errors by application of the free-space Fresnel propagator, and this cannot be used to propagate back inside the sample. Within the sample, propagation occurs according to the Schrödinger equation, involving dynamical dispersion and potential effects. Nevertheless, at these low energies, $\Psi$ depends strongly on the $z$ coordinates of the atoms in the sample, so that image matching experiments based on trial structures could, in principle, be used to determine the three-dimensional structure.

The preceding comments apply to general nonperiodic thin transmission samples. We now consider in more detail the special case of a thin extended crystal.

From the interior of a thin crystal, patterns such as that shown in Fig. 8.12 are obtained. These may also be simulated with the aid of Eq. (8.16), using a periodic potential. We now show that the resulting pattern is a true image and discuss its aberrations (Cowley, 1979). It is convenient to transform to new spatial coordinates $x$ and $y$ in the detector plane. Let $q(\mathbf{r})$ have Fourier coefficients $F_{h,k}$. Then $Q(\mathbf{K}_t)$ consists of a set of delta functions on reciprocal lattice sites. If the objective aperture has been removed, we may take $A(\mathbf{K}_t) = 1$. The convolution in Eq. (8.16) then becomes

$$\Psi(x, y) = \sum_{h,k} F_{h,k} \exp\left[i\chi\left(\frac{x}{R\lambda} - \frac{h}{a}, \frac{y}{R\lambda} - \frac{k}{b}\right)\right] \quad (8.17)$$

where $R$ is the distance from sample to detector, $a$ and $b$ are the cell constants for the projected unit cell, while $h$ and $k$ are integers.

To show that Eq. (8.17) produces a faithful, magnified image of the sample, we consider first points near the optic axis such that the effects of spherical aberration are negligible. Then, with $C_s = 0$, and using Eq. (8.5) with $c = 0$, Eq. (8.17) becomes, in one dimension,

$$\Psi(x) = \exp(i\phi) \sum_h F_h \exp\left(\frac{2\pi hx}{Ma}\right) \exp\left(\frac{\pi\lambda\Delta f h^2}{a^2}\right) \quad (8.18)$$

Where $\phi$ is a constant phase factor and $M = R/\Delta f$. For the special infinite set of focus settings

$$\Delta f_n = 2na^2/\lambda \qquad (8.19)$$

the last term in Eq. (8.18) becomes unity since $n$ and $h$ are integers. The intensity distribution at the detector therefore becomes

$$I(x) = \left|\sum_h F_h \exp\left(\frac{2\pi hx}{Ma}\right)\right|^2 \qquad (8.20)$$

which is a magnified copy of the modulus squared of the transmission function of the object $q(\mathbf{r})$. The magnification $M = R/\Delta f$ is that of a geometric point-projection image, projected from the defocused probe position above (or below) the sample. These Fourier images also occur in two dimensions if certain restrictions are placed on the ratio of the projected cell dimensions and angle. In general, the images are periodic in probe defocus $\Delta f$, and it may be shown that additional half-period and reversed contrast images also occur at intermediate focus settings. The reader is referred to the original papers for more detials (Cowley, 1979; Cowley and Moodie, 1957). In fact, for the particular transmission function used in this example (a phase object), the "in-focus" Fourier image at $\Delta f_n$ will show no contrast, since the modulus of Eq. (8.15) is unity. It has been shown, however, that contrast is obtained near the in-focus setting (Cowley and Moodie, 1960).

For experimental ronchigrams, the effects of spherical aberration cannot be neglected. The preceding discussion explains the appearance of the inner lattice fringes in Fig. 8.12. Moving away from the center horizontally, we see two "eyes," within which the intensity is relatively constant. This may be understood with reference to Fig. 8.13, which provides a geometrical interpretation of Eq. (8.17) for the axial case of the three Bragg beams. Here, the lens aberration function $\chi$ has been sketched across the illumination aperture. The on-axis image intensity at O arises from interference along paths AO and A'O (which are Bragg scattered in passing through the sample), together with the axial ray. An off-axis point such as P involves paths CP and C'P, and therefore samples the aberration function at different points. We have seen that for points near the axis, lattice fringes are produced. For points further from the axis the effects of spherical aberration become dominant. The "eyes" in the experimental pattern (Fig. 8.12) result from rays leaving near the stationary phase turning points S and S'. Then there is relatively little phase change with change in detector position. For a given lattice spacing

# COHERENT NANOPROBES

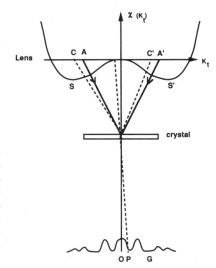

Figure 8.13. Ray diagram showing formation of ronchigram and point-projection lattice images. The on-axis image point results from interference between paths AO and A'O. An off-axis image point P selects rays originating at C and C' if only Bragg scattering is allowed. Stationary-phase points are shown at S and S'. The wave-front aberration function for the probe-forming lens is $\chi(K_t)$.

$d_{hkl}$, this defocus occurs for

$$-\Delta f = C_s \lambda^2 / d_{hkl}^2 \qquad (8.21)$$

The experimental pattern is also seen to consist of a set of outer ellipses. It has been shown (Lin and Cowley, 1986) that values of the defocus and spherical aberration constant can be deduced from measurements of these ellipses.

For samples sufficiently thin that a first-order expansion of Eq. (8.15) can be employed, Eq. (8.16) describes just the conditions required for Gabor in-line holography, since Gabor's original proposal described a small source in front of a transmission object, as in Fig. 8.11. Early experimenters, using an 80-kV electron beam, used a lens to form this probe, as in the CBED geometry. Experimental results and an analysis of the effects of aberrations for this type of "projection" holography can be found in Haine and Mulvey (1952). Using holographic reconstruction techniques, it is therefore possible to reconstruct the complex wavefunction at the exit face of the sample. This process is known as "image reconstruction." The more complex problem of object reconstruction is concerned with reconstructing the crystal potential from this image wavefunction, and this depends on the scattering approximations used. We briefly review the recent literature on this topic, much of which is only indirectly related to the subject of this book. The requirement for very thin samples may mean that thin-film elastic relaxation effects render any defects seen in holography experiments unrepresentative of bulk material, and surface structure can be expected to contribute significantly. The aim of this work is the direct determination of atomic positions by digital processing of coherent nanodiffraction patterns from very thin samples.

It will be recalled that the "twin image" problem in the in-line hologram geometry led subsequently to the development of off-axis methods. [This problem is circumvented in the Fraunhofer holography method for small objects (Munch, 1974).] Considerable literature exists on the subject of electron holography [for a review, see Hanszen (1982)]. A recent review distinguishes twenty possible holographic modes, by considering all possible STEM and HREM modes and their reciprocally related equivalents (Cowley, 1992b). The bulk of the holography work previously described in the literature is concerned with imaging magnetic-field distributions in magnetic materials and superconductors, or with the correction of the effects of lens aberrations on high-resolution images. (These also form in-line holograms if the scattering is weak.) Work on Fraunhofer, near-field Fresnel and off-axis holography using biprisms also exists, but the geometry in these experiments does not usually correspond to that used in electron microdiffraction. The formulation of the microdiffraction problem [Eq. (8.16)] presents new opportunities for holographic reconstruction, and this has been explored by several authors. The digital reconstruction of electron holograms in this geometry is discussed by Cowley and Walker (1981), and a reconstruction scheme using coherent microdiffraction patterns is described by Lin and Cowley (1986). The measurement of structure-factor phases under dynamical conditions using coherent overlapping orders is described by Spence (1978b). A procedure for determining atom positions using autocorrelation functions derived from nondiffraction patterns obtained from overlapping regions is described by Konnert *et al.* (1989). This method is found to be capable of determining atomic positions to within an accuracy of 0.2 Å. An algorithm which is capable of retrieving the phase of a weakly scattering object from coherent nanodiffraction patterns by successive Fourier analysis is described by Bates and Rodenburg (1989b). This algorithm has subsequently been used in trial computations to retrieve information beyond the point resolution of the probe-forming lens (Friedman and Rodenburg, 1992). [See also the discussion of Eq. (8.14) and the following, where this resolution is related to that of a STEM lattice image.] The use of a biprism in the STEM geometry has been described by Cowley (1992) and Lichte (1991). In this way, by passing one probe through the sample and another outside the sample, a reference wave is made to interfere with the coherent CBED pattern, allowing reconstruction of the object wavefunction.

## 8.5. CBED FROM DEFECTS—INCOHERENT. DISLOCATIONS AND PLANAR FAULTS. MULTILAYERS. CBIM

The first CBED observations of faulted crystals appear to be those of Johnson (1972), who found that patterns taken along the $c$ axis of graphite showed a threefold axis instead of the sixfold axis of the perfect crystal if the probe was situated over a horizontal stacking fault. This finding is consistent with a lateral shift of $(1/3, -1/3)$ at the fault. More specifically, HOLZ interactions in the faulted crystal destroy three of the six vertical mirror planes and the center of symmetry present in the perfect crystal. The result confirms that CBED is sensitive to the symmetry of the sample as a whole (including its defects), rather than that of the ideal infinite bulk crystal. Consistent with the reciprocity theorem, it was also found that the (000) distribution for a fault at depth $h$ in a crystal of thickness $t$ was equal to that from a fault at depth $(h - t)$. The threefold symmetry was computed to be strongest for a fault in the middle of the foil. Similar methods have since been used to study twinning in $ZrS_3$ and to identify its two forms (Gjønnes, 1985).

Subsequent work in this field may be divided between studies on line defects and on planar defects or interfaces. The earliest work was based on conventional CBED patterns with the probe focused onto the sample—more recent work has been based instead on either the LACBED or CBIM techniques, which combine CBED information with images of the defect. We shall review the literature in roughly historical sequence, with emphasis on the underlying concepts.

Work on dislocations commenced with the experimental finding that HOLZ lines become split into many fine parallel lines if an in-focus probe was situated over a dislocation (Carpenter and Spence, 1982). This result applies if the probe encompasses all of the dislocation core and much of the surrounding strain field (case 2 in Table 8.1), in contrast to the more slowly varying strain fields and smaller probes used to measure strains from HOLZ line shifts (case 4). We may assume a unit dislocation running normal to the beam at an arbitrary depth in the sample. By indexing at least three HOLZ (or Kikuchi) lines which are not split, it is therefore possible to find the three components of the Burgers vector $\mathbf{b}$ of a dislocation (Carpenter and Spence, 1982). Unsplit lines are those diffracted from planes $\mathbf{g}$ which are unaffected by the dislocation, and these must satisfy the condition $\mathbf{g} \cdot \mathbf{b} \times \mathbf{l}$, where $\mathbf{l}$ is the dislocation line direction. (In practice the $\mathbf{g} \cdot \mathbf{b} = 0$ condition is usually sufficient.) These authors point out that, since three-dimensional diffraction is involved, planar fault vectors with a component in the beam direction can be found, making the technique a useful complement to HREM imaging on microscopes with limited tilting stages. They discuss the effects of dissociation and probe size, and point out that, unlike imaging experiments, the effects of the dislocation on many beams are revealed simultaneously in CBED patterns. They also describe a method for finding the sign of **b**, and provide two-beam calculations for the HOLZ splitting. These calculations are based on the reciprocity theorem, applied between the source and a typical point within a HOLZ line crossing the (000) disk. The theorem then shows that, for a point source, the perturbation to the HOLZ line intensity will be equal to the intensity at the center of a bright-field diffraction contrast image of a dislocation, as the direction of the illumination is varied.

More detailed computations and experiments have since been undertaken by Cherns and Preston (1986), Fung (1985), Preston and Cherns (1986), Tanaka et al. (1988), Wen et al. (1989), and Zou et al. (1991). In particular, it was shown by Cherns and Preston (1986) that there are advantages in using LACBED patterns (see Section 6.1) since, by separating the spatial and orientation information, these also reveal the dislocation line direction, and allow the probe to be positioned more accurately. It is also then possible to determine **b** and its sign. (The spatial information is obtained by defocusing the image of an extended source, to form a shadow image in every beam. Since information on detailed atomic positions is not sought, the penalty in loss of spatial resolution is unimportant.)

The interpretation of these LACBED patterns from defects is based on the analogy with images of dislocations crossing a bend contour (Hirsch et al., 1977). From Fig. 6.1 we see that each local region of sample is illuminated from a different direction as a result of the probe focus defect. The simplest case to consider is a systematics row of reflections with a screw dislocation running parallel to this row. Then **b** and **l** are parallel to **g**. We take an image coordinate **x** running normal to **g** and the beam direction. Each point in the sample then defines an incident-beam direction and a local value of the (depth-dependent) strain. For a high-order reflection we may use the two-beam or kinematic theory, together with the column approximation (Hirsch et al., 1977). The kinematic theory gives the intensity of a "beam" diffracted from a crystal containing a strain field $R_x(z)$, which varies only with $z$, as proportional to

$$I_\mathbf{g}(x) = \left| \int_0^t V_\mathbf{g} \exp\{-2\pi i [\mathbf{g} \cdot R_x(z) + S_g \cdot z]\} \, dz \right|^2 \qquad (8.22)$$

This expression is based on the column approximation. This argues that, because electrons are strongly forward-scattered, the image at one point below a column at $x$ in the true crystal is equal to the intensity of a Bragg beam diffracted by an equivalent crystal in which the strain is given by $R(z)$ for all $x$. Only horizontal shearing is permitted in the equivalent crystal, and there is one such equivalent crystal for each image point. A single rocking curve calculated for one value of $x$ therefore corresponds to a microdiffraction pattern computed for a probe size about equal to the column width of the column approximation (typically a few angstroms). The effects of larger probes can be incorporated using a local average over $x$. The failure conditions for the column approximation in microdiffraction for very small probes are discussed by Carpenter and Spence (1982).

An expression is obtainable from the theory of linear elasticity for the strain field $R_x(z)$ around the dislocation, and this can then be used in Eq. (8.22). Figure 8.14 shows the results of calculations based on two-beam theory. Since many orders are observed simultaneously in CBED patterns, a high-order reflection may be used for which the two-beam theory is a good approximation. In Fig. 8.14, the dislocation core runs up the page in the center of each pattern. Thus, for a suitably oriented defect, the LACBED method can provide a one-dimensional real-space shadow image along $x$, together with an orthogonal one-dimensional rocking curve at each point $x$. Note that the "splitting" of the intensity distribution seen here is not the same effect as that which occurs on HOLZ lines in perfect crystals due to the degeneracy of the dispersion surfaces (Section 5.4), which occurs on a much coarser scale. In Fig. 8.14, the effects of using a larger, in-focus probe can be understood by integrating over $x$.

We see from Fig. 8.14 that the value of $\mathbf{g} \cdot \mathbf{b} = n$ can be determined directly from the patterns. It is equal to one more than the number of subsidiary maxima between the dark lines entering from each side. If the sign of $\mathbf{g} \cdot \mathbf{b}$ is reversed, the effect is to reflect the patterns about a horizontal central mirror line at $w = 0$ on the diagrams. Again, contrast is best for a dislocation in the middle of the sample, and the two-beam results given above fit best for thicknesses of less than half an extinction distance—thus first-order reflections at the Bragg condition are avoided. The $\mathbf{g} \cdot \mathbf{b}$ rule is found to be insensitive to the depth of the dislocation. If, in Fig. 8.14, we consider the intensity along a vertical line at a constant value of $x$ (corresponding to a fixed position of a very small probe), we see that the position of the Bragg maximum shifts in angle ($w$) as the probe moves closer to the dislocation core. This is similar to the effect discussed for HOLZ line shifts in Section 5.4, although here the effect is solely due to strain and no composition change is involved.

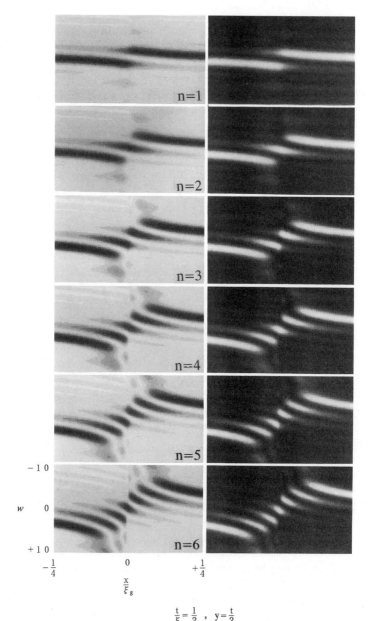

$$\frac{t}{\xi_g} = \frac{1}{2} \ , \ y = \frac{t}{2}$$

*Figure 8.14.* Bright and dark field two-beam calculations for CBED disks from a crystal containing an undissociated screw dislocation. The vertical axis corresponds to deviation parameter $w = s_g \xi_g$, while the horizontal axis corresponds to distance $x$ from the dislocation, which runs up the page at $x/\xi_g = 0$ (Tanaka *et al.*, 1988).

# Si  200kV

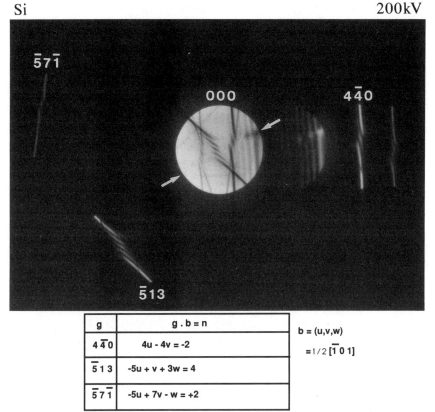

| g | g · b = n |
|---|---|
| 4 4̄ 0 | 4u - 4v = -2 |
| 5̄ 1 3 | -5u + v + 3w = 4 |
| 5̄ 7 1̄ | -5u + 7v - w = +2 |

b = (u,v,w)
= 1/2 [1̄ 0 1]

*Figure 8.15.* Experimental CBED pattern from silicon containing a dislocation. Probe size about 200 nm. The table below shows the values of **g · b** deduced from (one plus) the number of subsidiary fringes (Tanaka *et al.*, 1988).

Figure 8.15 shows an experimental pattern from silicon containing a dislocation running in the direction of the arrows. Three non-collinear weak reflections have been excited, and the (000) disk shows the clearest image of the dislocation. (A weaker "reflex" image showing the same dislocation line direction is formed in every order.) Three-beam conditions (where Bragg lines cross) are avoided. Thus three values of **g · b** can be found, which is sufficient to solve for **b**, as shown below.

Calculations based on Eq. (8.22) depend on five parameters (thickness, wavelength, structure factor, orientation, and image position) which define the local strain field under the column approximation. The reciprocity calculations of Carpenter and Spence (1982) are equivalent to those described above taken along the vertical line at $x = 0$ in Fig. 8.14.

In both cases the effects of probe broadening can be incorporated by introducing a local average over the $x$ coordinate.

More extensive calculations and experimental images can be found for partial dislocations and dislocation loops in Tanaka *et al.* (1988). The determination of the direction and sense of **b** from the geometric distortions in CBED patterns are further discussed by Wen *et al.* (1989).

Similar principles apply to the study of planar faults and interfaces, as first described by Rez (1979). Stacking faults are characterized by a phase angle $\alpha = 2\pi \mathbf{g} \cdot \mathbf{R}$, where **R** is the fault vector. For fcc crystals $\alpha = 2\pi/3$, $-2\pi/3$, or 0. Faults may be inclined to the surface and may be either extrinsic or intrinsic. Two-beam and kinematic expressions (based on the column approximation) for the diffracted intensity $I_g(S_g, x)$ may be found in texts (Hirsch *et al.*, 1977) and may be used directly to interpret LACBED images (Tanaka, 1986). A complete two-beam analysis can be found in Wang and Wen (1989), together with experimental LACBED patterns from stacking faults in austenitic stainless steel. For example, conventional single-beam diffraction contrast images show fringes running parallel to the fault trace for an inclined fault, as the intensity varies with fault depth. A horizontal stacking fault therefore produces no intensity variation in an image. In CBED, however, we observe the variation of $I_g(S_g, x)$ with $S_g$ and therefore sees fringes from a horizontal fault. For an inclined fault observed by LACBED we again have an image appearing in every diffracted order. Consider the two-beam case where the trace of the fault (its intersection with the surface) is parallel to **g**. Then one may examine the intensity variation $I_g(S_g, x)$ along **g**, giving the rocking curve for constant fault depth. The intensity variation normal to **g** gives the intensity for constant excitation error as a function of fault depth. A study of these rocking curves shows several methods for deducing the fault type from the patterns (Cherns and Preston, 1989; Tanaka, 1986). The use of HOLZ lines appears to be the best method. A HOLZ line in the (000) disk usually shows a single sharp minimum in the perfect crystal. The introduction of a stacking fault causes two minima to appear, one deeper than the other. The deepest minima occurs to either side of that which occurs in the perfect crystal, depending on the sign of $\alpha$. For $\alpha = 2\pi/3$ ($\alpha = -2\pi/3$) the deepest minima lies on the side for which $w < 0$ ($w > 0$). This rule is independent of the depth of the fault, however faults at the mid-plane of the sample again give the most pronounced splitting. Since HOLZ reflections have large extinction distances, this rule is also usefully independent of sample thickness, since the required condition $t/\xi_g < 3/5$ will always be satisfied in practice. (We have seen several techniques in this book which depend on this fact, such as the polarity determination rule described in Section 5.5.) By noting which lines are split and which are not in the central disk, it is therefore possible to determine **R**, as for the dislocation case. Worked examples can be found in Tanaka *et al.* (1988). By comparison with imaging, the CBED method has the advantage of allowing faults lying parallel to the surface to be solved. A clearer recording of the HOLZ lines is obtained using the hollow cone method, since this removes all ZOLZ detail from the central disk.

More detailed dynamical calculations have been performed for particular systems. For stacking faults in austenitic stainless steel, the HOLZ line splittings have been studied as a function of thickness, fault depth, displacement vector, and absorption coefficients by Chou *et al.* (1989). The results may be interpreted, as in the theory of diffraction contrast imaging, in terms of a redistribution of excitation between dispersion surfaces. For transverse faults in the layer compound 2Hb–$MoS_2$, a detailed analysis has been given by Jesson and Steeds (1990), who use the dynamical method of Vincent *et al.* (1984) described in Sections 3.10 and 5.1. Here the ZOLZ interactions are treated exactly by the Bloch-wave method, and a single HOLZ interaction treated as a perturbation using plane waves. The ZOLZ states are interpreted using hydrogenic orbitals of appropriate symmetry. From both these papers we may conclude that there is no fundamental difficulty in performing detailed dynamical intensity computations for CBED patterns from faulted crystals, but that considerable computational effort (as described in Section 4.2) is required to go beyond the determination of fault vectors.

A shadow image is formed in LACBED, so patterns recorded at interfaces may be used to measure the spatial variation of strain across an interface. In a recent example, the misfit strains at $NiSi_2$/Si interfaces have been studied in this way (Cherns *et al.*, 1988). Similar results for LACBED shadow images of Si/SiGe interfaces are described by Cherns and Preston (1989), and for silver halide particles by Vincent *et al.* (1988). All of the precautions concerning dynamical shifts due to composition variation and thin-film elastic relaxation discussed in Sections 3.8 and 5.4 apply to studies of this type. Since, however, as emphasized by Carpenter and Spence (1982), three-dimensional strain information is obtainable using HOLZ lines, it is possible to work with sample geometries for which these thin-film relaxation effects are minimized. An excellent example of this approach is provided by recent work on multilayers, which we now discuss.

Semiconductor multilayers may be studied by CBED with the plane of the interfaces normal to the beam (Cherns and Preston, 1989; Vincent *et al.*, 1987). These layers produce "sideband" reflections within a particular diffracted order arising from the multilayer periodicity. Thus, in a kinematic analysis, the rocking curve is the Fourier transform of the depth dependence of the multilayer potential, and so contains information on the multilayer spacings and composition profile. By choosing a sufficiently high-order reflection, kinematic conditions can be approximated and computations based on Eq. (8.22), with the integral broken into segments for each layer. Compositional sensitivity is further enhanced through the use of reflections which depend on the difference between scattering factors for the two elements present in a binary semiconductor alloy. The method has the important advantage that elastic relaxation effects, normally dominant in cross-section samples, are minimized. In addition, by dark-field imaging using the sideband reflections, the spatial variations in layer thickness can be observed. (These steps and ledges are particularly difficult to interpret in HREM images taken in the cross-section geometry.) Under conditions of strong multiple scattering, these "sidebands" will themselves split into many fine lines due to the multiplicity of ZOLZ dispersion surfaces as shown in Fig. 3.11, and this effect has been observed and analyzed by Gong and Schapink (1991) for GaAs/AlAs multilayers.

The case of a bicrystal in which the grain boundary contains the beam direction is treated by Schapink *et al.* (1986). The symmetries of the resulting patterns can be related to the classification of bicrystals given by Pond and Vlachavas (1983). Because each crystal may not intercept all rays in the incident cone, it was found that the double-rocking method was required to obtain useful results.

A second CBED shadow image method known as convergent beam imaging, or CBIM, has been developed by Humphreys *et al.* (1988a,b). We may distinguish this method from the LACBED method in an oversimplified way, as follows. In the CBIM method the microscope is focused onto the specimen plane—that is, the detector is conjugate to the sample (see Table 8.2). In the LACBED method, the detector screen is conjugate to the diffraction pattern in the back-focal plane of the post-specimen lens. Thus, from Fig. 6.1, we see that in CBIM the angular width of HOLZ lines depends on the probe size, while (since the sample is in focus) the spatial resolution is given by the normal performance of the electron microscope. The angular resolution is estimated to be typically 0.0001 radian. Conversely, in LACBED, the angular width of HOLZ lines is limited only by any diffraction limits imposed by the post-specimen lens (normally negligible) since the diffraction pattern is in focus, while the spatial resolution is limited by the probe size. Thus there are contributions to a particular HOLZ line from an area at least as large as the defocused probe and extended throughout the thickness of the sample [see the application of the uncertainty principle to changes in wavevector components in the beam direction—discussed following Eq. (2.12)]. Experimentally, however, LACBED patterns are usually obtained at some intermediate focus condition, giving compromise performance.

Figure 8.16 shows an experimental CBIM pattern from a series of interfaces between $Ge_xSi_{1-x}$ and Si (Eagelsham, 1989). The HOLZ lines are seen to curve as the lattice rotates at the interface, perhaps as a result of strain relaxation. Applications to the Si–SiO$_2$ interface are also described in this paper. Similar thin-film relaxation of a tetragonally distorted SiGe alloy has been observed by Humphreys *et al.* (1988a) to cause a slowly varying rotation of the [013] zone axis, estimated to be about 0.001 rad. The authors discuss the advantages of cooling samples to improve HOLZ-line visibility.

All these shadow-imaging methods have the disadvantage that strains due to composition variations cannot be distinguished from those due to sample bending as a result of elastic relaxation. From a CBIM

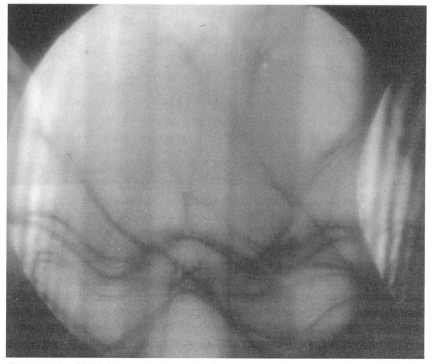

*Figure 8.16.* CBIM image of alternating $Ge_xSi_{1-x}$ and Si layers, showing HOLZ line oscillations resulting from strain relaxation (Eagelsham, 1989).

image alone, it is not possible to distinguish HOLZ-line displacements due to lattice parameter variations from those due to bending. (These bending effects are, however, much less than those present in the much thinner samples used for lattice imaging.) It has been suggested that best results are obtained by using a shadow image to locate the probe, followed by in-focus CBED patterns for the strain analysis.

## 8.6. MICRODIFFRACTION FROM AMORPHOUS MATERIALS

Many structural models have been proposed for glassy or amorphous materials, and the problem of finding a characterization signal which distinguishes among these models has occupied many researchers for many years. The most important models include the continuous random network (CRN) model due to Zachariasen (1932) [see also Polk (1971)], the microcrystalline model [see Rudee and Howie (1972) for a review], and the amorphous model [see, for example, Freeman *et al.* (1976)] in which a mixture of the diamond and wurtzite structures is used. There has been considerable effort aimed at distinguishing these models using high-resolution electron microscopy, and the field has been reviewed recently by Howie (1988). For many years, the lower limit which could be put on microcrystallite size was equal to that of the resolution of the electron microscopes used. The projection problem presents formidable obstacles to image interpretation. Experimental evidence makes it clear that the type of order observed depends strongly on the conditions under which the material was synthesized and on the possible presence of hydrogen.

Electron diffraction patterns obtained from a relatively large region of amorphous material illuminated by a plane wave show broad rings or halos, which may be analyzed to determine the number of nearest neighbors and their interatomic distances using the radial distribution function (RDF). This function, which may be extracted from electron diffraction patterns, is defined by

$$J(r) = 4\pi r^2 n(r) \tag{8.23}$$

where $n(r)$ is the number of atoms per unit volume falling within the range between $r$ and $r + dr$, from a given atom as origin. For the microcrystalline model in the limit of large microcrystals, the diffraction pattern can be understood as an orientational average of the pattern for a perfect crystal (as for a Debye–Scherrer ring pattern) with additional broadening due to strain. The electron diffraction RDF technique has been extensively developed by Graczyk et al. (1982) and by Cockayne and co-workers (Cockayne and McKenzie, 1988), who have used zero-loss energy filtered data. In this book, however, we are chiefly concerned with the microdiffraction mode in which a small focused probe is used so that the diffraction spots become disks. The question then arises as to whether coherent nanodiffraction patterns can provide any new information on the structure of glassy materials.

Radial distribution functions provide valuable characterization data which impose severe constraints on proposed models for amorphous material, however the RDF does not provide a unique specification of the structure and it provides only one-dimensional information. Different structures can give rise to very similar RDFs. For a small coherent focused probe, however, we have seen that the diffraction pattern from disordered material is highly sensitive to the detailed arrangement of atoms (see Fig. 8.6) and to the probe position. In particular, for a probe of interatomic dimensions, we might expect that the patterns would be sensitive to the angles between atomic bonds, in addition to the bond lengths. This information is not obtainable from either RDF or EXAFS studies (but may be from near-edge inner-shell spectroscopies). Experimental nanodiffraction patterns from amorphous materials thus show that the broad halos of intensity are broken up into irregular spots as the probe is made smaller by focusing (Brown et al., 1976). Experimental patterns from nanometer-sized regions of amorphous germanium have also been obtained by Cowley (1981c), where a detailed analysis is given for the dependencies of these spot widths on bond-angle variations. The small probe results in an orientation average over a much smaller number of atoms. The variation in intensity around the diffraction rings provides two-dimensional information and may be related to an "angular" Patterson function, giving the probability of finding a bond in a particular orientation, given a certain bond direction at the origin.

The use of pattern recognition techniques has also been described as a means of identifying particular kinds of order in near-amorphous materials (Monosmith and Cowley, 1983). Here a mask is used to collect all microdiffraction patterns which express a particular structural feature as the probe is scanned over the sample. In this way a table showing the frequency of occurrence of each type of microcrystal or atomic configuration can be built up. The entire process is executed under computer control on a Vacuum Generators HBS STEM instrument.

A related approach has been adopted by Rodenburg and co-workers (Rodenburg and Rauf, 1989; Rodenburg, 1988). Here the nanodiffraction patterns have been analyzed by analogy with laser speckle patterns. The correlation function

$$C(\phi) = \sum_n \int_0^{2\pi} I_n(\Phi)I_n(\Phi + \phi)\, d\Phi \qquad (8.24)$$

is formed, where $I_n(\phi)$ is the intensity measured at azimuthal angle $\phi$ (for a given scattering angle) from the $n$th probe position. Many thousands of patterns are added together. Peaks in $C(\phi)$ are interpreted as a measure of angular order. In the absence of lens astigmatism, a strong peak is expected at $\phi = 180°$ in very thin samples where Friedel's law holds as a consequence of single scattering conditions. For a microcrystal smaller than the probe in two-beam conditions, a strong peak is seen at $\phi = 360°$, corresponding to Bragg diffraction. The signal between these extremes is characteristic of the degree of angular order of the sample.

As with all coherent microdiffraction patterns from imperfectly crystalline material, in which diffuse elastic scattering can contribute to one point in the detector from several points within the illumination aperture, the principle difficulty is that the patterns depend on instrumental parameters and on the probe coordinates. The sensitivity to these parameters increases with scattering angle. The spherical aberration coefficient and focus setting must therefore be known accurately in order to interpret these patterns. In this sense, the interpretation of nanodiffraction patterns from amorphous materials is akin to that of high-resolution electron microscope lattice images. If new methods for measuring instrumental parameters (such as the ronchigrams described in Section 8.4) can be developed, there would appear to be many possibilities for the development of this application of microdiffraction.

*Note added in proof.* The uncertainty principle relates the width of a HOLZ line to the width of crystal which contributes to it. Thus, for a HOLZ line, one tenth of the width of the central disc, the contributing width of crystal is ten times the width of the probe, if this is diffraction limited and $C_s = 0$. The width of crystal must be measured normal to the HOLZ planes [see Zuo and Spence (to be published) for details].

# 9

# Instrumentation and Experimental Technique

Electron microscope instrumentation is changing very rapidly, so this chapter outlines only the general features which are desirable in an electron microscope intended for microdiffraction work. Many of the instrumental requirements (particularly in the pole-piece region) will be found to be similar to those for energy-dispersive X-ray microanalysis. With the field-emission instruments used for the smallest probes, the need for a large tilting range may conflict with the electron-optical requirements for subnanometer probe formation, which requires a very small pole-piece gap. Several manufacturers are now prepared to offer the customer a series of compromise designs, from which a tilt range and minimum probe size combination can be selected. Certainly, for most of the applications described in this book which use the "incoherent" mode, a double tilt holder offering at least $\pm 40°$ in one direction and $\pm 60°$ in the other is desirable.

Specimen preparation for electron microscopy has been treated in a number of texts and review articles (Chew and Cullis, 1987; Goodhew, 1975; Hirsch *et al.*, 1977). For CBED work on small precipitates, carbon extraction replicas are often used to avoid matrix effects. In thinned samples containing particles of interest, one otherwise seeks particles which extend over a hole and thereby retain their epitaxial relationship. Crushed samples are often useful since, initially at least, they offer the cleanest surfaces. Ion-beam thinning has become extremely popular in combination with dimpler grinding devices, particularly for thin films on a thick substrate viewed in plan form. Surface layers may be damaged, but for the relatively thick samples used for CBED this is less of a problem than for lattice imaging.

In practice there are many experimental difficulties. The determination of the space group of a small precipitate embedded in a dense matrix, for example, can be extremely difficult (or impossible) due to multiple scattering in the matrix. And beginning researchers are often dismayed to find how quickly their patterns fade as contamination builds up, particularly on older microscopes with inadequate vacuum systems or when using dirty samples. Movement of the probe during an exposure may also degrade a pattern by introducing an integration over a range of thicknesses, defects, or strained regions. In this chapter we also therefore discuss the experimental procedures which minimize these and other practical problems. Review articles which discuss experimental technique include Loretto (1984), Spence (1988a), Vincent (1989), and Williams (1987).

## 9.1. INSTRUMENTATION

Most modern analytical transmission electron microscopes provide a CBED or nanoprobe mode. This development has resulted partly from the fact that the requirements for EDX (large tilt, small probe, low contamination) exactly match those for CBED. The best initial advice a student can be given is to read the manual for the microscope carefully! The new instruments are controlled by computers, and a video monitor provides most of the human interface with a small number of multifunction panel knobs. Then a certain amount of time is required to familiarize onself with all the menus and "pages" displayed. The new digital goniometers have the great advantage that a particular set of sample coordinates (stage $x, y, z$) and two tilts may be stored. When later recalled, the sample holder moves immediately to the required sample position (or particle). Sample height and orientation are also reset.

An instrument designed exclusively for CBED would attempt to achieve the best performance compromise among the following considerations:

The information required from a CBED pattern is elastic scattering, and the useful information in this increases with thickness (giving more oscillations in ZOLZ disks, for example). However, inelastic scattering depletes the elastic wavefield as thickness increases, degrading CBED contrast, until no elastic Bragg scattering remains. A "best thickness" therefore exists, which increases with accelerating voltage. Radiation damage, however, sets an upper limit to the voltage which can be used. One solution lies in the use of an imaging energy filter, which dramatically improves the quality of CBED patterns by removing most of the inelastic scattering (see Section 9.3).

The detail in ZOLZ patterns will also be washed out if the probe is so large as to cover a range of sample thicknesses or orientations, or moves during the exposure. While this is not so important for symmetry determinations (if the changes are small), it matters crucially for quantitative comparisons with computer calculations. The width of the elastic probe distribution also broadens as it travels through the sample [see Humphreys and Spence (1979) for calculations], placing limits on the smallest probe which can be used with the rather thick samples needed for CBED. Probe and sample drift (due to thermal instabilities) must therefore be minimized.

Contamination increases rapidly as the probe becomes smaller, due both to hydrocarbons brought in on the sample and to contaminants in the vacuum system [see Rackham and Eades (1977) for a discussion of contamination in CBED]. Unless a field-emission gun is used, the intensity of the patterns for a given thickness also falls off rapidly as the probe size decreases below about 20 nm. Below this size contamination problems are dealt with on dedicated STEM instruments by maintaining a vacuum of better than $10^{-9}$ torr in the sample regions. Specimen preheating facilities may also be needed to prepare sufficiently clean sample surfaces. The entire machine may then need to be baked to maintain this vacuum level. This in turn complicates detector design, and will preclude the use of film in a fully bakeable instrument (see Section 8.3). The use of differentially pumped machines offers one solution to this problem. Here a field-emission gun is used, but a vacuum of about $10^{-8}$ torr is maintained at and below the sample, allowing the use of film recording. The zirconiated field-emission sources discussed below offer another possibility. Contamination may then be reduced through the use of a cooling holder.

The stability of the emission intensity for field-emission guns is usually inferior to that of $LaB_6$ sources, unless uncompromising ultrahigh vacuum conditions are maintained in the field-emission gun in order to prevent the backstreaming of ions. If a serial readout system is used (such as scanning the microdiffraction pattern over the entrance aperture of a serial energy-loss spectrometer), this source instability will limit the accuracy of the results. A glance at a STEM image recorded over a minute or so using a field-emission gun will indicate the problem. A comparison of the intensity at the beginning and end of the scan (at the same point on the sample) must be used to monitor source intensity variations. An imaging energy filter combined with an area detector, such as film or a YAG/CCD system, largely solves this problem of electron source instability. A new type of heated field-emission source is currently under development for TEM instruments, based on the successful application of zirconiated tungsten tips in lower-voltage SEM instruments (Swanson, 1991). This source is claimed to be more tolerant to vacuum conditions, and to provide higher total currents than a cold field-emission gun at the larger probe sizes sometimes needed for CBED work. When used with an imaging filter, so that small source intensity fluctuations can be tolerated, this new source may be ideal for CBED work. Section 9.4 contains more details.

A double-tilt cooling holder is desirable for CBED work. While one group has consistently advocated the use of tilt-rotate holders (because of their larger tilt range), most researchers use double tilt sample holders with orthogonal tilt axes. (These simplify the mental spherical trigonometry calculations otherwise required when changing orientations.) A cooling holder also increases the visibility of the outer HOLZ ring by reducing thermal diffuse scattering in most samples. The magnitude of the improvement depends on the Debye temperature of the sample. Thus the outer HOLZ reflections in Fig. 7.1b could not be seen at all at room temperature.

A flexible condenser-objective lens design is required, allowing a wide range of probe sizes and convergence angles. A method of controlling the size of the illumination angle electronically (by varying the strengths of the lenses) is highly desirable and has been described (Green and Eades, 1986). An angular field of view of at least 12°, and preferably 20°, is desirable in order to include HOLZ rings. These may be required to pass through an imaging filter device (see Section 9.3). It is important to be able to translate the pattern electronically on the screen, so that different parts can be recorded at large camera length.

Owing to the distortions inherent in electron lenses, it will be necessary to calibrate the CBED patterns by comparing a pattern from a known structure with calculated diffraction angles. Differences of 10% or more are common at large angles, so that angular measurements taken from CBED prints may contain this error. A correction chart can be drawn up for particular lens currents.

A continuously variable accelerating voltage supply is highly desirable, allowing one to seek particular HOLZ-line intersections and symmetry conditions. Relative strain variations can then be measured at the microscope (see Section 5.4).

There is a trade-off between the probe size and the tilt range that depends on the aberrations of the objective lens. In dedicated field-emission STEM instruments the very small pole-piece gaps required to form a subnanometer probe will limit the tilt range. Thus, one requires the smallest probe compatible with an adequate tilt range of, say, 45° and a 1-nm probe. This is readily obtainable at present. It has been pointed out that by using smaller samples and grids (say 2 mm), a more favorable trade-off can be obtained.

In summary, and taking into account the above considerations (and the discussion of the following sections), an ideal instrument for microdiffraction would appear to be one which provides probe sizes in the range of 1–50 nm, using a field-emission gun (possibly in a differentially pumped column) and an imaging energy filter, and fitted with film, image plate, and CCD detector systems (see Sections 9.3 and 9.4). A double tilt liquid-helium holder should be available. This can also be operated with inexpensive liquid nitrogen and will reduce contamination rates. A digitally controlled goniometer is desirable, fitted with a clutch to allow rapid manual scans about one axis.

A similar instrument, fitted with an $LaB_6$ source and using slightly larger probe sizes, comes close to this ideal at much lower cost. However, it should be emphasized that for some difficult materials CBED patterns of sufficient quality to allow comparisons with calculations can only be obtained on field-emission machines. For example, in $YBa_2Cu_3O_7$, the high density of twins together with the spatial variation in oxygen stoichiometry meant that meaningful agreement with dynamical calculations could only be obtained on a field-emission instrument (Zuo, 1991a). The new zirconiated tungsten field-emission tips may turn out to be ideal for CBED, since they offer greater total current and the "crossover," below which it becomes beneficial to use a field-emission tip, shifts to a larger probe size (see Section 9.4).

A useful method for providing continuous variation of the illumination semiangle by varying the strength of the "minilens" used in Philips microscopes is described by Green and Eades (1986).

## 9.2. EXPERIMENTAL TECHNIQUE

In this section we describe the experimental procedures used to obtain sharp CBED patterns on modern analytical electron microscopes. The methods used for obtaining wide-angle CBED patterns are described in Chapter 6.

The procedure for setting up an electron microscope for CBED consists essentially of focusing the beam onto the specimen in the image mode, then switching to diffraction mode. In more detail we proceed as follows. Having aligned the microscope carefully and introduced the sample, the first step, after setting the thermostat on the cooling holder, is to set the holder to the eucentric height. The duty cycle of the cooling holder thermostat should be checked once the required temperature has been obtained, to ensure that thermal drift (resulting in relative probe movement) is minimized. The astigmatism of the probe-forming lens must be carefully adjusted and attention paid to optimizing the filament conditions. The brightest source is critically important for work with the smallest probes, so that the gun alignment and condenser astigmatism settings are particularly important.

The probe must now be focused accurately onto the sample, while at the same time both are arranged to be conjugate to the viewing screen. This is done by first focusing the sample (at the eucentric height) using the objective lens in the image mode. Then, still in the image mode, the probe is focused using the condenser lens. With a typical modern symmetrical Reicke–Ruska lens, there may be some linkage between the condenser- and objective-lens settings. On switching to the diffraction mode, a CBED pattern should be seen. The diffraction focus control should now be used to focus the edge of the illumination aperture. The probe size will be fixed by the first condenser lens, and on many modern digital instruments this size is indicated directly on the panel. For an $LaB_6$ source, a probe diameter of about 30 nm provides a good compromise between the loss of intensity which occurs as the probe size is reduced, and the need for a small probe in order to avoid defects, thickness variation, and bending.

With the microscope in the diffraction mode and the probe fully focused, it will be found that shadow images of the sample can be formed in each diffracted order by defocusing the probe. These images are useful for locating the probe on the sample, but must disappear at exact focus, even if the probe is passed over an edge. The final focusing for CBED is carried out using the diffraction lens to bring the edge of the illumination aperture into sharp focus, and simultaneously to make the Kikuchi lines sharp (Christenson and Eades, 1987). Some small additional condenser-lens focusing adjustments may be needed to align the HOLZ lines inside the CBED disk with the Kikuchi lines outside the CBED disk.

If the illumination angle can be controlled electronically, it will usually be chosen so that disks just touch in their closest direction. [We note that overlap of orders does not alter the symmetry of an "incoherent" CBED pattern, and such patterns can readily be computed. If an $LaB_6$ source is used, the intensities with the region of overlap can simply be added together (see Chapter 8).] Unlike lattice imaging, the precise location of the optic axis is not critical if a large probe is used, so that a different angular view of the CBED pattern may be obtained simply by translating the illumination aperture, which tilts the incident beam. In this way it is possible, for example, to go from the zone axis orientation, at which the "dark line" dynamically forbidden reflections are seen, to the first-order Bragg condition, which may show a G–M cross. With smaller probes, the resulting misalignment will translate the probe on the sample. A different angular view can also be obtained by adjusting the dark-field beam tilt controls.

It may often happen that no HOLZ lines are seen, even at small camera length, from a crystal with a small primitive unit cell (so that the FOLZ occurs at a high angle) and a low Debye temperature (so that the background of thermal diffuse scattering is large). Assuming that a range of thicknesses has been tried, the following steps can be taken to improve HOLZ-line visibility: (1) Cool the sample. (2) Adjust variable accelerating voltage (if available), or change to another voltage for the best HOLZ Bragg condition. (3) Reduce contamination. (4) Reduce sample bending, thickness variation, and defect concentration, or use a smaller probe. Any stacking disorder in the beam direction will strongly affect HOLZ intensities, as we saw in Section 8.5. (5) Use the LACBED method with a very small selected area aperture to minimize inelastic scattering. (6) Use an imaging energy filter.

For field-emission instruments fitted with a subnanometer probe, the position of the optic axis is critically important, and the alignment, focusing, and stigmatism of the probe-forming "objective" lens in a dedicated STEM follows similar principles to those used in lattice imaging (and is equally difficult). The coherent Kossel pattern (or "ronchigram") formed when the illumination aperture is removed then becomes a vital aid to assist in aligning the objective lens (Cowley, 1986).

The most common problems in CBED work arise from specimen bending and defects, which result in CBED patterns whose symmetry is too low. It is sometimes useful to examine the outer HOLZ ring while moving the probe over a range of thicknesses and imperfections until the highest symmetry is found. The visibility of these rings improves if a smaller illumination aperture is used.

Experimental methods for LACBED work are also discussed in Section 6.1.

## 9.3. DETECTORS AND ENERGY FILTERS

Reviews of detector systems for electron microscopy can be found in Cowley and Ou (1989), Egerton (1986), Herrman (1983), and Spence (1988a). We first discuss detector systems and energy filters for conventional TEM–STEM instruments, since these frequently form an integrated system. For such instruments, the detector region is not baked or maintained in very high vacuum conditions ($>10^{-8}$ torr). Using differential pumping, these machines may also be fitted with a field-emission gun (Tonomura, 1987). We then discuss the more complex bakeable systems needed for dedicated STEM instruments (such as the Vacuum Generators HB601 machines) in which film recording cannot easily be used.

For symmetry determination, electron micrograph film proves an excellant parallel detection system because of the very large number of pixels it offers per exposure. A review of the use of film in electron microscopy can be found in Valentine (1966) and Spence (1988a), where the more recent fast X-ray emulsions used for radiation-sensitive materials are also discussed. The detective quantum efficiency (DQE) of film is limited to about 0.7 by background fogging at low exposures, by shot-noise statistics at medium exposures, and by nonlinearities and saturation at higher exposures. The number of gray levels obtainable with film is less than 256 (8 bits). For symmetry determination, the use of film recording together with an imaging energy filter (see below) and a cooled double-tilt specimen holder will provide excellent performance. (The use of CCD cameras for symmetry determination is briefly discussed below.)

For the quantitative analysis of the intensity distribution in microdiffraction patterns from thick samples, an energy-filtering device combined with a detector system with large dynamic range is required. Egerton (1986) provides a clear review of the theory and application of energy filters. The thickness of the sample is important in determining the need for filtering. In very thin samples there is little ineastic scattering, and hence no requirement for elastic filtering. But these samples are known to be strongly influenced by thin-film relaxation effects (Treacy et al., 1985). Thus, for the study of thicker material which is more representative of the bulk, elastic energy filtering is required. The purpose of the filter is to remove from the microdiffraction pattern all those electrons which, on traversing the sample, lose more than a few electron volts in energy. The advantages of zero-loss energy filtering are outlined in more detail below in the discussion of imaging filters, however a glance at Figs. 9.2 and 9.3 will indicate the improvement in the quality of data to be expected. This improvement affects all the techniques discussed in this book, including strain measurement and defect studies. In order to obtain accuracy in structure-factor measurement comparable to that achieved in X-ray crystallography, such an energy filter, tuned to the elastic peak, is essential. However, the optimum choice of detector geometry for CBED depends on the type of analysis to be performed. In particular, we must distinguish between line and area detectors. Each of these may or may not be fitted with an energy filter.

For the collection of systematics data, line scans are needed. Line scanning readily accommodates an energy-filter device, such as the popular quadrant magnetic sector filter (Egerton, 1986). The CBED pattern is scanned over the entrance aperture of the energy-loss spectrometer, which is tuned to the elastic peak. The scanning can be performed by placing the image deflection coils of the electron microscope under the control of an external computer. Alternatively, in order not to disrupt the microscope electronics system, additional deflection coils can be built and installed, mounted on the 35-mm-film camera ports just above the viewing chamber.

If a serial rather than PEELS (parallel detection electron energy-loss spectrometer) is used, this system has the great advantage that the unsurpassed dynamic range of a photomultiplier–scintillator combination can be used, because this is fitted to most serial energy-loss spectrometers. (Since a photodiode array is used in commercial PEELS systems, they are less suitable for line scanning due to their more limited dynamic range.) Line scanning has the disadvantage that readout times can be long, placing severe demands on the stability of the electron source, the thermal stability of the stage and sample, and the anticontaminator. We have seen that the line-scan data produced are most useful for structure-factor refinement in centrosymmetric crystals (Section 4.2), for atomic-position parameter refinement (Sections 6.1 and 5.1), and can also be used for the analysis of HOLZ-line positions and profiles. The collection of line scan data should not be taken to imply that the matching calculations require only a single line of structure factors as input. Three-dimensional dynamical computations will be required to correctly position HOLZ lines crossing the systematics disks obliquely, for example, and to incorporate other dynamical interactions. For the collection of the three-beam intensity distributions which are needed to measure structure-factor phase triplets in acentric crystals (and for other applications), it is possible to devise a system in which the scan path may be curved. The CBED pattern is first imaged at video speed and the resulting (noisy) pattern displayed on a monitor. A "mouse" is then used to trace along the required hyperbolic Bragg lines needed for the analysis, and the final data collected at lower speed along this path. A film record is also taken, and this is used as described in Chapter 2 to assign a beam direction to each point of the scan. One recent such system was based on a PDP-11 computer controlling the post-specimen deflection coils of a Philips EM400T electron microscope, and a GATAN serial ELS unit. For a study of the MgO systematics reflections (Zuo et al., 1989a), typical experimental parameters were as follows: Two hundred points were collected, with a dwell time of 0.3 s at each point; camera length 6500 mm; entrance aperture to spectrometer 1 mm, giving an angular resolution of 0.15 mrad; probe size about 200 nm; accelerating voltage 120 kV; the energy window around the elastic peak was set to about 5 V. This system was used to produce the experimental data in Section 4.2.

Electronic area detectors suitable for microdiffraction have only recently become available for the efficient collection of two-dimensional data. These offer the important advantage over scanned readout systems of conferring immunity to small fluctuations in the electron source intensity during an exposure (such as those likely to occur with field-emission guns), and of reducing recording times by orders of magnitude. The most suitable of these for microdiffraction are based on either cooled, slow-scan charge-coupled device (CCD) cameras coupled to ytterium–aluminum–garnet (YAG) single-crystal scintillator screens (Autrata *et al.*, 1978), or the JEOL image-plate device. Until recently, however, there has been no simple way to combine these devices with energy filtering. At least two different commercial imaging filters are now available and can be used with CCDs; these are discussed below. We first examine the CCD detector and image plate.

*Room-temperature CCDs operating at video rates are not suitable for microdiffraction recording*, due to their very limited dynamic range. Microchannel plates become less suitable for direct electron detection than CCD–YAG systems at accelerating voltages above about 15 kV. Similarly, *video cameras directed at a fluorescent screen are not suitable detectors for microdiffraction patterns* in view also of their very limited dynamic range. This range cannot be increased significantly by image integration at video rates, since each readout from the photocathode introduces new noise, and a law of diminishing returns results. The recent trend, then, has been toward the use of a single-crystal YAG (or YAP, phosphate) screen, coupled by fiber optics to a CCD element. Although the luminous efficiency of YAG (about 5%) is less than that of P20 powder phosphor (about 20%), the YAG systems offer much more spatially uniform emission (and greater speed for scanning systems). Important advantages of YAG–CCD systems for quantitative CBED are their strict linearity (Herrman, 1990), absence of geometric distortions of the field (found in video systems), and relatively large dynamic range. The normalization of intensities in CCD systems is important—the intensity distribution collected must be divided by that which results from uniform illumination of the screen, in order to allow for small pixel-to-pixel gain variations. The maximum speed of CCD systems is currently (1992) limited to a few frames per second. The long exposures possible with them make them ideal for the study of weak outer HOLZ lines. For weak signals, they have the advantage over film of not being susceptible to fogging. The background (which may be subtracted) arises mainly from stray X-rays in the electron microscope and some cosmic rays. The stray X-ray problem is aggravated at higher accelerating voltages. (It may therefore be necessary to mount the CCD system off the optic axis of the microscope and to deflect the image for recording.)

An experimental CCD camera with a dynamic range of 16,000 (14 bits), based on a liquid-nitrogen-cooled CCD camera (designed for use in astronomy by the Photometrics Corp., Tucson, AZ) and a YAG screen, is described by Spence and Zuo (1988). This was found to be well suited to electron diffraction work. A $576 \times 382 \times 14$ bit CCD was coupled by fiber optics to a thin single crystal of YAG. The CCD pixel size was $23 \times 23$ microns, giving an active area of $13.2 \times 8.8$ mm. The camera is operated at $-130\,°C$. The fiber-optic fiber size was 6 microns, and this provides the required thermal insulation between the cooled CCD and the room-temperature YAG, which should not act as a cold trap. The YAG thickness was about 20 microns. The green emission of the YAG is reasonably well matched to the maximum in spectral sensitivity of the CCD. The maximum readout speed was 2.5 frames per second. Thus, in normal use, the microscope shutter is used in the same way as for a film recording, and very long exposures are possible because of the low background. With this system, the background was negligible at 100 kV, while at 400 kV it was found necessary to mount the camera off-axis. The saturation limit of a CCD is typically $10^5$ CCD electrons (per pixel) and the background about 10 electrons, giving a dynamic range of about $10^4$, requiring 14 bits. This is inferior to the performance of a photomultiplier but superior to the performance of film and video systems. $1024 \times 1024 \times 12$ bit (scientific grade, square pixel) CCDs are available at greater cost.

A thorough analysis of the DQE of a cooled, slow-scan CCD–YAG detector system, including Monte Carlo multiple-scattering calculations for the beam electron in the YAG, has now been completed by Daberkow et al. (1991). Their system consists of a $Ce^{++}$-doped $Y_3Al_5O_{12}$ scintillator, coupled by fiber optics to a $1024 \times 1024$ CCD array (THX 31156) operated at $-20\,°C$. These workers find a DQE of about 0.9 at 120 kV if the YAG thickness is about 50 microns. The DQE falls to 0.6 at 300 kV. Figure 9.1 shows the variation of overall DQE and resolution as a function of YAG thickness for various accelerating voltages, according to their calculations and experiments. The resolution is defined as the full width at half maximum height of the distribution of scattered electrons in the YAG. Their CCD pixel size was $19 \times 19$ microns, and fiber optics size 6–9 microns, with a numerical aperture of unity. They recommend 16-bit data collection and lower operating temperatures. This dynamic range (greater than 10,000) should be compared with the 8 bits (256) available from film. Possible imperfections in these systems include Newton interference fringes at fiber-optics interfaces, "chicken wire" patterns in the fiber optics due to subbundle formation, and Moire patterns. For an optimized system operating at 100 kV, these workers recommend a YAG

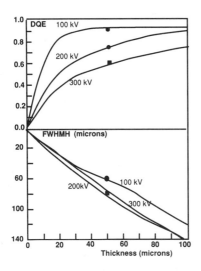

*Figure 9.1.* Detective quantum efficiency (DQE) and resolution (FWHM) of YAG–CCD detector systems as a function of YAG thickness for several accelerating voltages (Daberkow *et al.*, 1991). Continuous lines refer to theory, dots to experimental results.

thickness of 30 microns, a numerical aperture of 0.6, and a CCD array with 24-micron square elements. Fiber-optic cones have also been used to match detectors with different pixel sizes and may be needed at larger YAG thicknesses (higher voltages). Mirrors and lens-coupled systems may be preferred at high voltage to minimize radiation damage to the CCD array. Cross talk may be a serious problem.

An evaluation of CCD performance for electron microscopy can also be found in Roberts *et al.* (1982). Similar systems, some thermoelectrically cooled, are now available commercially (Mooney *et al.*, 1990; Tietz, 1991). One such system (Mooney *et al.*, 1990) under the control of a Macintosh computer also allows for direct control of the microscope from this computer. Most modern TEM machines are now fitted with an RS232 port, which may be connected directly to the Macintosh. The GATAN software, for example, allows the user to write new source codes in C, FORTRAN, or PASCAL which can be compiled and linked (under the Macintosh Programmers Workshop software) together with the GATAN software which controls the CCD camera. In this way, customized data-collection systems may be developed by the user. Image-processing software is also included. It would thus be possible, for example, to create zoom magnification (camera length) and deflection controls under the control of the mouse, so that selected regions could be sought, magnified, and recorded at will. Difference images between symmetry-related regions could quickly be formed. In summary, one now has all the flexibility and power of modern image-processing techniques running on work stations available at the microscope. This results in a change in working habits—one continues to collect data until the desired results are obtained instead of recording many micrographs for later analysis. Advances in the capacity of bulk storage media needed for storing images (such as magneto-optical media) have generally kept pace with our requirements, but data transfer rates onto these media may limit ultimate performance if very large CCD arrays are used.

A second promising area-detector system for microdiffraction is the image plate (Mori *et al.*, 1988). The plates are similar to film, but contain a photostimulable phosphor and must be read out using a special digital film reader. (The JEOL company may offer such a film-reading service.) A grain size of 100 microns is quoted. The medium offers similar dynamic range to CCDs (about $10^5$), good linearity, and many more pixels per image than CCDs. (The plate size is 102 mm × 77 mm.) It is also extremely sensitive ($10^{-14}$ C per cm$^2$) and so is ideal for radiation-sensitive materials. Its performance appears to be similar to that of medical X-ray film, but with the added advantages of smaller grain size, greater dynamic range, and better linearity. Recording times are typically about one-thousandth of that used for conventional micrograph film. The DQE appears to be close to unity (Ishikawa *et al.*, 1990). The sensitivity and resolution of the plates is discussed as a function of accelerating voltage by Isoda *et al.* (1991). Apart from the high cost of the film-reading device and the lack of on-line image-processing facilities at the microscope offered by CCD systems, this detector appears to be well suited for microdiffraction. Since the image plates replace the film in a (modified) microscope film camera, the image plate is compatible with CCD and ELS systems, and might be best used for the study of weak diffuse scattering and HOLZ features in conjunction with an imaging filter. An example of the use of the imaging plate in electron diffraction is given by Shindo *et al.* (1990), who studied the dynamical diffuse electron scattering from $Cu_3Pd$ alloys as a function of temperature.

We now discuss the use of imaging energy filters for microdiffraction. With such a filter, the entire two-dimensional diffraction pattern is elastically filtered simultaneously and no scanning is used. (A one-sided imaging "band-pass" filter would therefore be adequate, and may be simpler to construct than the differential imaging filters currently in use.) In one commercially available design (the Zeiss 912 Omega TEM–STEM) the electron beam follows an omega-shaped path through four quadrant filters placed between the fourth and fifth post-specimen lenses. Two lenses follow the filter. A full discussion of the design can be found in Lanio (1986). Electrons losing more than a preset amount of energy strike an aperture and cannot contribute to the image or diffraction pattern. Imaging energy-loss filters, which use a single-quadrant magnetic sector analyzer together with quadrupole lenses, are also available (Krivanek and Ahn, 1986; Krivanek *et al.*, 1991). These devices may be retrofitted to existing machines.

Figures 9.2 and 9.3 show the striking improvement which results in the quality of CBED patterns from the use of a filter, even when film recording is used (Mayer *et al.*, 1991). These patterns were recorded on the Zeiss Omega model 912 TEM–STEM. The exposure time used in Fig. 9.2 was 1 s for the unfiltered recording and 3 s for the filtered pattern. These times were arranged to produce an approximately equal optical density on the film, so that a valid comparison could be made. The image recording time for a $10^3 \times 10^3$ pixel image using this system is 1000 times less than that required by the scanned readout system described previously, for the same dose. (This assumes that a field-emission gun was used in the scanning system which was 1000 times

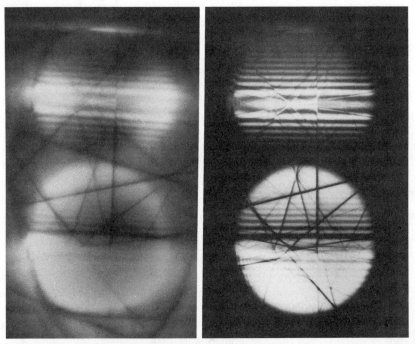

*Figure 9.2.* Silicon (000) and (220) CBED disks recorded (left) without and (right) with the Zeiss Omega filter. The filter has been set to an 8-eV window around the elastic peak, probe size is 100 nm, and sample thickness about 270 nm. Apart from the introduction of the filter, no other changes were made to the experimental conditions (Mayer *et al.*, 1991).

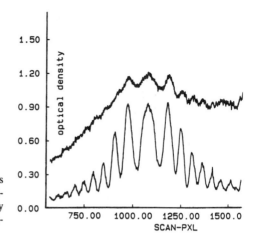

Figure 9.3. Densitometer traces taken from Fig. 9.2 showing the logarithm of the transmitted light intensity on the ordinate. Upper curve is without a filter.

brighter than the source used with the Omega filter.) The advantages of the Omega filter increase rapidly with the number of pixels. Figure 9.3 shows densitometer traces taken from the filtered and unfiltered micrographs. The quality of this data recorded on a CCD camera can be expected to be even better, while the serial filtered line-scan data shown in Section 4.2 (discussed above) has greater dynamic range.

The important inelastic processes contributing to the background in CBED patterns are phonon scattering, plasmon scattering, and single-electron excitation. [For a review of energy-loss processes in electron microscopy, see Spence (1988b) and Egerton (1986).] Phonon scattering involves relatively large inelastic scattering angles, but very small energy losses (perhaps 30 meV). These are not excluded by elastic filtering. Plasmon losses involve larger energies (about 15 eV) and small scattering angles. Since the experimental observation in Fig. 9.2 is that the background between the Bragg reflections is almost entirely removed by elastic filtering, we must conclude that this background is due to multiple, coupled phonon and plasmon scattering. The phonon scattering events provide the large angular change, and the associated plasmon losses then allow these electrons to be removed by elastic filtering. This interpretation is consistent with the relatively large thickness used ($t = 270$ nm).

For microdiffraction, the important advantages of elastic energy filtering are as follows:

1. The dramatic reduction in background allows much thicker crystals to be examined without incurring the penalty of radia-

tion damage, which would result if higher accelerating voltages were used.
2. The use of greater thickness (without background) means that HOLZ lines are sharper and low-order disks contain more oscillations, thus more information. This is new information, which was not previously extractable due to the presence of the background.
3. For systematics data, the imaging filter allows the possibility of integrating the rocking-curve intensities along the direction normal to the systematics line. If HOLZ lines can be avoided, a large increase in signal-to-noise ratio results.

It should be noted that an optimum sample thickness exists with energy-filtered data. For very thin crystals, there is little inelastic scattering and so no requirement for filtering, however the CBED disks show no useful contrast variation, and elastic relaxation effects may be severe [but see Goodman (1976) for the use of this data in layer compounds]. At very large thickness, all scattering is inelastic and no elastic signal can be recorded. In the simplest model (Hirsch *et al.*, 1977), the thickness dependence of, for example, the plasmon-loss electrons is given by the product of the multiply-scattered Bragg-beam intensity with an appropriate term of the Poisson distribution. The zero-loss intensity will be given, as a function of thickness, by the expressions in Section 3.2. and shown in Fig. 3.1. For beams other than (000) this rises from zero intensity at zero thickness, after which the thickness oscillations (with two-beam period $\xi_g = K/|U_g|$) are damped exponentially due to the term $U_0'$. Recording times for the elastically filtered "dark-field" CBED disk therefore increase with increasing thickness, and increase with decreasing thickness below $t = \xi_g/2$. The number of turning points in the CBED rocking curve per unit angle increases with thickness. Thus the optimum thickness to be used will depend on the ratio $U_g/U_0'$ and on the stability limits of the instrument. Practical experience shows that for inner reflections in light materials, best results are obtained at about $t = 5\xi_g/2$ (see Section 4.2), where the exposure time for the elastically filtered pattern is 3 to 4 times that of the unfiltered pattern. The minimization of contamination is therefore important when using energy filtering. The plasmon-loss contribution to the total scattering "preserves contrast" but is excluded from the elastically filtered data. However, as we have seen, it is heavily convoluted with phonon scattering at the large thicknesses used. More discussion of these points (with emphasis on the use of the energy-loss patterns) is presented by Reimer *et al.* (1990), who refer to the use of area-detected energy-filtered patterns as "electron spectroscopic diffraction."

For the large-angle shadow-imaging techniques described in Chapter 6 and Section 8.5, we have pointed out that the selected area aperture used in a plane conjugate to the electron source (see Fig. 6.1) performs an energy-filtering function. If such an image is formed on an instrument fitted with an imaging energy filter (which normally has insufficient energy resolution to exclude the phonon-scattered background), this aperture may be used to exclude the phonon scattering, while the energy filter excludes larger energy-loss processes which scatter through smaller angles. By varying the height of the aperture, the cutoff angle may be varied. This technique has been analyzed in detail by Jordan *et al.* (1991). In this way a very great improvement in the quality of LACBED patterns should be possible.

In many general-purpose electron microscopy laboratories, conflict arises between the need to use, on the same microscope, a video system for fast recording of lattice images (for focusing, for dynamic studies, and for radiation-sensitive materials), a cooled slow-scan CCD camera for quantitative image and diffraction measurements, and a parallel-detection energy-loss spectrometer. Interchanging these attachments is highly inefficient. The long-term solution would appear to lie in the combination of an imaging filter with a CCD camera capable of operating at several speeds, together with imaging plates also available for the film camera. The imaging filter, when operated in the dispersive mode, provides a parallel-detection spectrometer function in conjunction with the CCD camera. Zero-loss filtering improves the quality of all the results in every mode of the electron microscope [weak-beam imaging, reflection electron microscopy (Spence and Mayer, 1991), bright-field imaging, point diffraction patterns, etc.] and appears to have no disadvantages. For the future it would therefore appear that the combination of larger cooled CCD detectors with slightly greater dynamic range (say 18 bits), imaging energy-loss filters, and image plates may eventually replace both the scanned readout systems and the current PEELS systems. The electron-optical design problems involved are, however, considerable.

At the present state of development, for reasons of cost, laboratories may be forced to choose between, for example, film recording, a CCD system *without* energy filtering, or a serial ELS–PMT scanning readout system *with* filtering. In this regard, our experience has been as follows:

(1) For symmetry determination, the very large number of pixels which can be recorded on film has obvious advantages. When using a CCD camera, a very small camera length must be used if HOLZ detail is to be included and the number of pixels is limited. Methods are available,

however, for patching together CCD images, side by side, and the use of a CCD camera is compatible with film. The image processing and contrast enhancement possibilities of the CCD systems may then have advantages, and remain to be fully evaluated.

(2) For quantitative analysis, if sufficiently stable conditions can be obtained and if a particular problem can be solved using line scans, then the improved quality of data which can be collected using a photomultiplier will justify its use and is to be preferred. A CCD–imaging filter system is ideal, however, for the analysis of HOLZ lines crossing the central disk (where the position of the line intersections rather than the line intensities is required), for the study of weak diffuse scattering in point diffraction patterns, and for any problem requiring the analysis of complex two-dimensional intensity distributions, such as three-beam phase determination in acentric crystals.

For dedicated STEM instruments, a few groups have devised detector systems which are optimized for microdiffraction recording on the Vacuum Generators HB501 instrument. (The facilities provided with the machine are unsuitable for the quantitative analysis of nanodiffraction patterns.) The combination of microdiffraction recording capabilities with energy-loss spectroscopy and bright-field STEM imaging has challenged the ingenuity of many researchers [see Cowley and Ou (1989) for a review]. Any detector design optimized for microdiffraction must meet the requirements of bakeability, linearity, high dynamic range, DQE as close to unity as possible, and, preferably, elastic energy filtering of the microdiffraction pattern. ELS spectra may also be required. In addition, an axial fast detector system is also usually required for bright-field STEM imaging at video rates. This introduces the problem that modern PEELS systems use either CCD arrays or photodiode arrays, both of which are too slow to be used for the bright-field STEM detector [see McMullan and Berger (1988) for a discussion of these problems]. Finally, for STEM imaging and "Z-contrast" experiments, the ability to vary the shape of the detector to match a particular diffraction pattern may be required (see Fig. 8.3). For a given shape of detector, the scale of the pattern is varied by controlling the strength of the post-specimen lenses, or the specimen height. To avoid distortion in the patterns a strong post-specimen lens is required, preferably a symmetrical condenser–objective design. The number of pixels needed for the microdiffraction detector can be determined by dividing the angular resolution required (approximately equal to the HOLZ-line width) by the angular field of view. Energy filtered CBED in STEM is reported by Xu et al. (1991).

At least five designs have appeared in the literature which attempt to meet these criteria. Two approaches have generally been used—those based on clever fiber optics and those based on lens coupling. In lens-coupled systems a transfer lens is placed outside a viewing port. This images the CBED pattern, formed on an inclined reflection phosphor, onto a CCD camera placed at right angles to be beam. The phosphor screen is inclined at 45° to the beam and may contain a hole to allow passage of the central disk through to the ELS spectrometer. If this is a serial spectrometer which uses a photomultiplier, it may be used as the bright-field STEM detector. Both lenses and fiber-optics systems with a numerical aperture of about unity are available. Each beam electron generates several thousand photons in the fluorescent screen. Since the readout noise of cooled CCD detectors is much lower (a few electrons) than that of diode arrays (a few thousand electrons), a higher DQE should be obtained using a CCD camera.

A system which emphasizes the need for flexible configuration of the detector shape was described by Cowley and Spence (1979). Here the primary detection element is a phosphor screen laid down onto a fiber-optics plate, which conveys the image out of the vacuum system. (The UHV frit seal of the fiber-optics plate to the stainless steel vacuum chamber may present problems.) The detector is mounted on the optic axis. Exciting the ELS spectrometer bending magnet below the detector screen deflects the beam to one side, into the spectrometer. (The spectrometer and microdiffraction facilities cannot therefore be used simultaneously.) An electrostatic image intensifier is coupled directly to the fiber-optics plate and could be removed at this joint if high-temperature baking were required. The image formed on the output of the image intensifier is then dissected by a system of optical lenses, which form an image of the coherent CBED pattern on a glass plate in a light-tight enclosure. By placing small movable, inclined, circular mirrors or prisms on this plate, various portions of the CBED pattern can be directed to photomultipliers disposed about the enclosure. In this way, STEM images can be formed with any part of the CBED pattern (such as those where orders overlap). Alternatively, a Polaroid film recording of the pattern on the plate can be made, developed, and then cut up and reinserted as a mask. In this way a high-angle annular detector may by matched to the HOLZ structure of a particular crystal structure, for example, or images formed using diffuse scattering between the Bragg directions.

Such a system is extremely flexible for STEM imaging, but lacks energy filtering of the microdiffraction pattern and may suffer from excessive stray scattered light if the optical system becomes too complex. A second-generation device was therefore developed, as described by Cowley and Ou (1989). This used two fiber-optic rods with obliquely cut

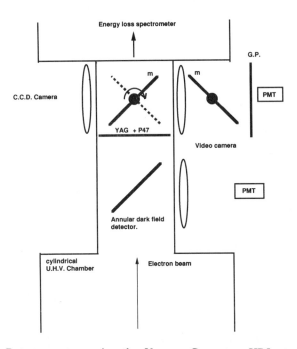

*Figure 9.4.* Detector system using the Vacuum Generators HB5 at Arizona State University. The mirrors m slide along an axis normal to the page. Some contain an axial hole, giving access to the spectrometer. YAG + P47 consists of three alternative transmission detector screens (some with a central hole), selected by also sliding on an axis normal to the page. G.P. is a glass plate on which an optical image of the microdiffraction patterns forms and on which masks can be placed. The CCD camers is used to record CBED patterns; the photomultipliers (PMT) are used for fast STEM imaging.

ends, which could be moved into the path of the beam. Light losses in the fiber-optics were found to be unacceptable, and improved performance was obtained with the lens-coupled CCD system shown in Fig. 9.4. Here the entire column is narrowed at the microdiffraction detector level to allow large numerical aperture lenses to be placed outside the chamber but close to YAG and P47 detector screens. Screens are mounted on sliding rectangular plates at two levels, as shown. The upper screens are transmission phosphors, normal to the beam (YAG, P47), some of which contain a hole to allow passage of the beam into the spectrometer. Above the screen lies a set of mirrors inclined at 45° to the beam, allowing the screens to be imaged by the cameras on either side. (The mirrors may be rotated about an axis out of the page.) The screens and mirrors move independently along an axis normal to the page. The photomultiplier on the right may be used to form a scanning image using the fast (P47) detector, which can respond at video rates, or the entire microdiffraction patterns (including the central disk) may be recorded with extended exposures using the CCD camera on the left and a slower, more efficient phosphor (such as P20). A PEELS system may be fitted, without restriction on the speed of its detector, since this is not used for bright-field imaging. Alternatively, an image of the microdiffraction pattern may be formed on the glass plate shown and masks inserted. At the lower level, a reflection screen in the shape of an elliptical annulus is fitted for high-angle dark-field imaging. The size of the diffraction pattern on this may be varied using the post-specimen lenses. This screen may also be moved along an axis normal to the page. For the study of very thin samples, where filtering is not needed, this system seems ideal.

A third system has been described by Daberkow and Herrman (1988). Here the desired variable shape of the detector is obtained by reading out different portions of a photodiode array. This would appear to be the ideal arrangement, if it can be combined with an imaging energy filter.

The only system to offer two-dimensional elastic energy filtering of nanodiffraction patterns appears to be that of McMullan et al. (1990). This system is therefore useful for the study of thicker samples than those previously described. Here a CCD camera is used to record the PEELS spectrum formed by a quadrant spectrometer. Three quadrupole lenses are also used to focus the spectrum onto the CCD camera. If a slit is fitted to the spectrometer before the quadrupoles, these lenses can be used to form an energy-filtered two-dimensional image of the microdiffraction pattern through the spectrometer (Egerton, 1986; Krivanek and Ahn, 1986; Zaluzek and Staruaa, 1988). The slit is conjugate to the specimen. The beam may also be deflected to one side onto a YAP–PMT detector to form the bright-field STEM image. Figure 9.5 shows results from this system, as previously discussed in Section 5.4.

## 9.4. ELECTRON SOURCES

In this section we briefly review the factors which limit the performance of the various electron sources used for microdiffraction, and compare them. Reviews of electron-gun designs for electron microscopes can be found in many texts, including Reimer (1984), Spence (1988a), Wells (1974), and Grivet (1972). For field-emission sources, the book by Gomer (1961) and the review articles by Dyke (1956) and Crewe et al. (1968) can be recommended. An extensive literature exists on electron field-emission microscopy [for an excellent review, see Swanson

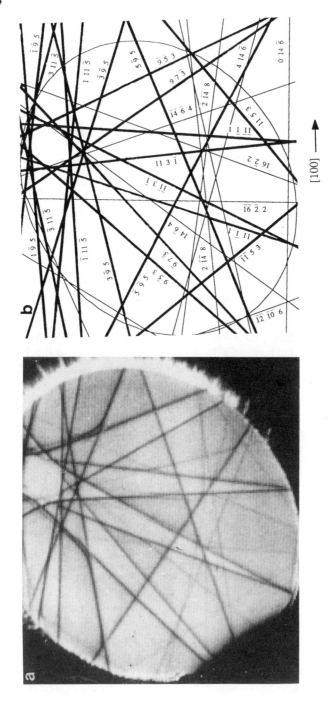

*Figure 9.5.* (a) Two-dimensional elastically filtered nanodiffraction pattern from silicon (Pike *et al.*, 1987) with probe size about 1 nm; 100 kV. The imaging spectrometer uses a quadrant design followed by three quadrupole lenses and a CCD detector. (b) Correctly indexed HOLZ lines, assuming an accelerating voltage of 96.4 kV and a lattice parameter of 0.5431.

and Bell (1973)], and this forms the background in physics to subsequent work on the development of field-emission guns for electron microscopy. The various sources most suitable for microdiffraction have also been reviewed by Brown (1981).

For microdiffraction, the primary requirements are high source brightness and emission stability. However, the requirements on stability of emission are relaxed considerably if area detectors, such as film, image plates, or CCD/YAG systems, are used, both because the recording time is reduced and because each pixel is detected simultaneously. For scanned readout systems, in which the pattern is scanned over an aperture, the demands on source stability may be severe.

Figure 9.6 shows a simplified ray diagram for a typical thermionic filament fitted to a triode gun. Since any axially symmetric system of electrodes forms a lens, the Wehnelt–anode system images the tip onto the plane D. [For a comprehesive treatment of electron optics, see Hawkes and Kasper (1989).] Electrons leaving the tip possess a Maxwellian distribution of velocities, including radial velocity components. A "crossover" is therefore formed as shown, and this forms the minimum cross section of the beam. It is a demagnified image of this crossover which is used to form the probe in microdiffraction experiments. Virtual rays may be extended back from the crossover to define a virtual source, which may lie behind the tip.

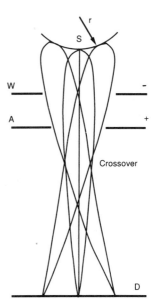

*Figure 9.6.* Simplified diagram of a thermal electron source $S$: W is the Wehnelt (biased negative with respect to S), A is the anode (positive), $r$ is the tip radius, and D is a detector screen.

The brightness $\beta$ of a source is defined as the emission current density $J$ per unit solid angle. For small angles $\alpha$ this is given by

$$\beta = J/\pi\alpha^2 \tag{9.1}$$

It may be shown that the brightness is a constant for all points along the optic axis, from tip to detector, even if lens aberrations and aperture stops are permitted. This follows from the electron-optical analog of the Helmholtz–Lagrange theorem which, in simplified form, gives the product of area, energy, and local solid angle as constant at all points along the beam path. Thus the effect of a demagnifying lens, for example, is to increase current density $J$ (proportional to $M^2$), and increase the angular aperture $\alpha$ (proportional to $M^2$), leaving brightness constant. (In fact, brightness conservation applies more generally and can be derived from Liouville's theorem. This requires that a group of particles occupy the same element of velocity–volume phase space as they travel, so that the density of particles in phase space does not change.) By applying Eq. (9.1) at the sample level, we see that the probe current density $j$ can only be increased by increasing the illumination aperture $\theta_c = \alpha$ (provided that it remains filled with electrons) or by increasing the intrinsic source brightness. There exists, however, an upper limit on brightness, due to the inevitable presence of some transverse momentum $p_t$ at the source. This momentum depends on temperature. If the axial momentum (which depends on both temperature and accelerating voltage) is $p_z$, then $\alpha = p_t/p_z$, and the maximum brightness obtainable becomes

$$\beta_{max} = J_s E_r/(\pi kT) \tag{9.2}$$

where $E_r$ is the relativistic accelerating voltage ($E_r \approx E + E^2$ if $E$ is given in MeV). Thus brightness increases with accelerating voltage, as the electron emission becomes more forward-directed. This can also be understood from Fig. 9.5. At higher voltages, the transverse kinetic energy (approximately $kT$) is reduced relative to the forward energy $eE$. Hence the size of the crossover is reduced and the current density increased [see, for example, Moss (1968)]. For a thermal emitter, the Richardson–Dushman equation then gives the maximum current density $j_T$ emitted by a uniform metal surface at temperature $T$ as

$$J_T = AT^2 \exp(-\phi/kT) \tag{9.3a}$$

where $A$ is a materials-dependent constant ($A \approx 120 \text{ Å K}^{-2} \text{cm}^{-2}$ for tungsten), Boltzman's constant $k = 1.38 \times 10^{-23}$ J K$^{-1}$, and $\phi$ is the work function of the emitter (see Table 9.1). At high temperatures, electrons in the tail of the Fermi distribution acquire enough energy to overcome the work function and are emitted into the vacuum. Tungsten is used for thermal emitters because this emission occurs before the metal melts. Lanthanum hexaboride tips are used because their lower work function leads to higher brightness. They require somewhat better vacuum conditions than tungsten thermal emitters (to reduce damage from positive-ion collisions), but are not nearly as demanding in this regard as cold field-emission tips. Their energy spread is less than tungsten thermal emitters (see Table 9.1). Thermal guns are normally run in the "autobias" mode. Here the filament is connected via a variable bias resistor to the Wehnelt, which is in turn connected to the negative high-voltage terminal. The variable resistor forms the gun bias or beam-current control. This circuit can then be analyzed by analogy with a triode vacuum tube. Any temporary increase in beam current causes an increase in the negative Wehnelt bias, which tends to decrease the beam current. An increase in bias resistance (less beam current) reduces the energy spread of the source, which varies between about 2.4 and 0.7 V FWHM [see Spence (1988a) for experimental measurements].

Equation (8.3) gives an approximation to the probe size as a function of illumination aperture $\theta_c$ as it is limited by various aberrations. The current into a probe whose geometric size is $d_s$ is

$$I_p = \beta \pi \theta_c^2 (d_s/2)^2 \pi \tag{9.3b}$$

If we consider only the spherical aberration term $d_{sa}$ and the diffraction limit term $d_d$ in Eq. (8.3), and use the optimum illumination aperture given by Eq. (8.6), it is possible to combine these expressions with Eq. (9.3b) to obtain an expression for probe current as a function of probe size $d_0$ for given $\lambda$, brightness $\beta$, and $C_s$. The conclusions which can be drawn from such curves (shown in Fig. 9.6 for a multilens system) will be discussed following our summary of field-emission sources. For CBED from undeformed, defect-free material it might be argued that the optimum illumination aperture is $\theta_c = \theta_B$, since this extracts the maximum rocking-curve information.

Schottky pointed out that the energy barrier to thermal emission (the work function $\phi$) could be reduced by the application of an external field, resulting in increased emission. At still higher fields, electron tunneling through this barrier becomes possible, as first explained by Fowler and Nordheim (1928). The intermediate range, in which both processes operate, is known as thermionic field emission (T-F emission) [see Shimoyama and Maruse (1984) for a comparison of these modes].

Table 9.1. Electron Sources Compared at 100 kV

| | Thermionic emission | | | Field emission | | |
|---|---|---|---|---|---|---|
| | Hairpin | Point | LaB$_6$ | Thermal | Cold | W/ZrO[a] |
| Brightness (A/cm$^2$sr) | $5 \times 10^5$ at 100 kV | $1 \times 10^6$ at 100 kV | $5 \times 10^6$ at 100 kV | $10^8$ at 100 kV | $10^8$–$10^9$ at 100 kV | $5 \times 10^8$ |
| Crossover diameters | ~20 $\mu$ | ~5 $\mu$ | ~10 $\mu$ | ~200 Å | 50–100 Å | 150 Å |
| Typical life (hours) | ~30 | ~10 | ~500 | 500–1000 | ~$10^3$ | 5000 |
| Operating temperature (°K) | 2800 | 2900 | 1800–1900 | 1000–1800 | ambient | 1800 |
| Typical vacuum (torr) | ~$10^{-5}$ | ~$10^{-5}$ | ~$10^{-6}$ | $10^{-8}$–$10^{-9}$ | $10^{-9}$–$10^{-10}$ | ~$10^{-8}$ |
| Emission current ($\mu$A) | ~100 | ~10 | ~50 | 50–100 | 5–20 | |

| | | | | | |
|---|---|---|---|---|---|
| Current density (A/cm$^2$) | ~3 | ~6 | 20–30 | $10^5$–$10^6$ | $10^4$–$10^6$ |
| Current density ($\mu$Å/sr) | ~$10^5$ | ~$10^5$ | ~$10^5$ | ~$10^2$ | ~10 |
| Current fluctuation (short term) | good | good | good | a little worse | a little worse |
| Current stability (long term) | good | good | good | good | a little worse |
| Current Efficiency | 100% | 100% | 100% | 100% | ~5% |
| Work function (eV) | 4.5 (mean value) | 4.5 (mean value) | 2.7 (mean value) | 4.7 for (100) | 4.35 for (310) |
| Energy spread (eV) | 1–2 | 0.5–2 | 2–3 | ~0.5 | ~0.2 |
| Melting point (°C) | 3370 | 3370 | 2200 | 3370 | 3370 |

[a] Approximate values provided by L. Swanson for 25 kV

Cold field-emission tips depend on the temperature-independent process of quantum-mechanical tunneling. The tip is made very sharp, but is not heated (except for cleaning, and sharpening by "flashing"), and electron emission commences abruptly once the electric field at the tip exceeds a certain value $E_c$. This tip field is given very approximately by $E = V/5r$, where $r$ is the radius of the tip and $V$ the potential between the tip and the first extraction anode. (There is no Wehnelt.) Since $E_c \approx 1$ volt per angstrom, emission occurs at voltages between a few hundred and several thousand volts, depending on tip sharpness and crystallographic orientation. [Different crystal faces have different work functions—(310) and (111) oriented W tips are often used because of their low work functions.] A field-emission current of $1-10\,\mu\text{A}$ is typically produced from the tip into a semiangle of less than a radian. Additional accelerating anodes are required to bring the beam up to the full kinetic energy required. Since these and the anode all act as lenses (with aberrations), careful optical design is required. An additional complication arises because the position and size of the crossover depends on the accelerating voltages (which in turn depend on tip sharpness). A virtual source is be formed behind the tip, whose aberrations may be calculated (Liebl, 1989; Scheinfein *et al.*, 1992). An example of early gun design can be found in Crewe *et al.* (1968). The design considerations involved in combining magnetic lenses with a field-emission gun are described in Venables and Cox (1987). Such a magnetic gun lens has now been incorporated into the Vacuum Generators HB501S STEM.

Since tip sharpness is so critical, ultrahigh vacuum conditions (better than $10^{-11}$ torr) are required to prevent positively charged ions, focused along the field lines, from destroying the tip. Emission is generally less stable than for thermal emitters, except under the highest vacuum conditions, since gas molecules adsorbed on the surface alter the work function. A great deal of research has been done on the processes of "build-up" and "dulling," in which the tip is heated and shaped under an intense field, causing surface tungsten atoms to migrate. The interested reader is referred to the field ion microscopy literature (Tsong, 1990).

From the uncertainty principle, we might expect field emission to commence once the uncertainty in axial position coordinate $\Delta z$ for an electron at the tip surface exceeds the width of the barrier potential. If the extraction field is $E$, then the width of a triangular barrier is approximately $\Delta z = \phi/eE$, where $\phi$ is the work function in electron-volts. The positional uncertainty is $\Delta z = h/4\pi\,\Delta p$, with $\Delta p$ the momentum uncertainty and $h$ Planck's constant. Using plane-waves states at the Fermi level, $\Delta p = (2m\phi)^{1/2}$. Thus the onset of field emission is expected

INSTRUMENTATION AND EXPERIMENTAL TECHNIQUE

at
$$E = E_c = 4\pi(2m)^{1/2}\phi^{3/2}/he \tag{9.4}$$

This agrees approximately with the results of the more accurate calculation, first given by Fowler and Nordheim (1928) [see also Everhart (1967)]. They obtained an expression (the Fowler–Nordheim equation) for the emission current density as

$$J_F = 6.2 \times 10^6 \left(\frac{\mu}{\varphi}\right)^{1/2} (\mu + \varphi)^{-1} E^2 \exp\left(\frac{-6.8 \times 10^7 \phi^{3/2}}{E}\right) \tag{9.5}$$

where $\mu$ is the Fermi energy for the metal, $E$ is given in volts per cm, while $\mu$ and $\phi$ are given in electron-volts. The argument of the exexponential equals 2.3 when Eq. (9.4) is satisfied. Plots of $\ln(J_F/V^2)$ versus $1/V$ are found to be linear. Equation (9.5) shows that current densities of $10^6$ A cm$^{-2}$ can be obtained using field emission, compared with perhaps 1 A cm$^{-2}$ for a thermal source. The total current for the field-emission source is a few microamps, however, compared to perhaps 50 microamps for a thermal source. The source size is much smaller (perhaps 1–10 nm) and the brightness about four orders of magnitude greater.

We now consider the current in the probe formed by the probe-forming lens in the presence of aberrations and apertures. As an indication of modern performance, the Vacuum Generators HB 601 field-emission STEM instrument generates a current of about 1 nA into a probe diameter at the sample of about one nm. There exists a "crossover" probe size, however, below which the total current into the probe will be greater using a field-emission gun, and above which the current will be greater using a thermal source. This can be understood by plotting Eq. (9.3b) (see subsequent discussion) for realistic values of $C_s$, $\lambda$, and $\beta$, or by the following argument. First, for a thermal emitter, the current in a probe of geometric diameter $d_s$ is given by

$$I_T = j\pi d_s^2/4 = \pi^2 \beta d_s^2 \theta_c^2/4 \tag{9.6}$$

If the contributions to probe broadening of spherical aberration, chromatic aberration, and diffraction are added in quadrature [see Eq. (8.3)], and the optimum value of $\theta_c$ used which produces the least broadening in Eq. (9.6), the total probe current becomes

$$I_T = 0.6 J_T |e| E_r C_s^{-2/3} d_s^{8/3}/(kT) \tag{9.7}$$

For a field-emission tip we assume a very small source, and allow broadening only due to spherical aberration, as given following Eq. (8.3). Then the probe diameter is $d_s = 0.5 C_s \theta_c^3$. Let the tip radius be $r$ and the emission semiangle for the marginal ray be $\alpha_1$. Then $\alpha_1 = M\,\theta_c$, with $M$ the magnification, and the total current into the probe is

$$I_F = \pi r^2 \alpha_1^2 J_F \tag{9.8}$$

where $J_F$ is the tip-emission current density. Expressing $\theta_c$ in terms of $d_s$, we find

$$I_F = \pi r^2 J_F M^2 (2 d_s)^{2/3} C_s^{-2/3} \tag{9.9}$$

Equations (9.9) and (9.7) therefore yield the ratio of the probe currents for field emission and thermal emission as

$$R = \frac{I_F}{I_T} = 8.31 M^2 \left(\frac{J_F}{J_T}\right)\left(\frac{kT}{eE_r}\right)\left(\frac{r}{d_s}\right)^2 \tag{9.10}$$

This very simplified analysis shows that the field-emission gun provides a better performance if $d_s/r$ is small, i.e., the probe size should be comparable to the emission area of the tip. Typically, $J_F/J_T = 10^6$, $M = 1$ (this is independent of $d_s$, since we have assumed the probe size to be aberration limited), and $(kT/eE_r) = 10^{-5}$. Then $R \approx 80 (r/d_s)^2$, so that the thermal filament will give greater total current for probe sizes greater than ten times the field-emission tip radius. However, we have made many simplifying assumptions (e.g., that the lenses have the same aberration coefficients, that only a single lens is used, neglect of chromatic aberration, etc.). In practice, the crossover point in the performance where $R = 1$ is found to be in the range of 10–100 nm for cold and T–F guns. A more realistic analysis of this problem can be found in Cosslett (1972). Figure 9.7 shows more detailed calculated values of probe current against probe size for the 100-kV Vacuum Generators HB501 cold field-emission STEM instrument. The optimum focus of Eq. (8.7) has been used, and aberration contributions added in quadrature as in Eq. (8.3). A total of four lenses are used—a newly designed magnetic gun lens (Venables and Cox, 1987), two condenser lenses, and the objective lens. The total emission current is taken to be 10 $\mu$A—the probe current is approximately proportional to this. The gun brightness assumed is that which has been measured for the HB501. This is $4.6 \times 10^9$ A cm$^{-2}$ sr$^{-1}$ (Mory et al., 1987).

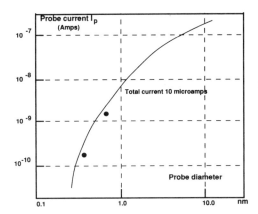

Figure 9.7. Calculated probe current as a function of probe size for the Vacuum Generators HB501 STEM operating at a total emission current of 10 μA (Venables and Cox, 1987). Experimental measurements are also shown (Mory et al., 1987).

Table 9.1 summarizes the properties of the electron sources currently in use. The entries for the zirconiated tungsten Schottky source (provided by L. Swanson) are given for 25 kV, since these sources have so far only been used in SEM instruments at lower voltages; the values at 100 kV will be slightly higher. More details of these promising sources, which tolerate poorer vacuum conditions than cold field-emission tips, can be found in Samoto et al. (1985), Swanson (1991), and Speidel and Brauchle (1987).

The contributions from sample defects, bending, and thickness variation are minimized when using the smallest probes, so we conclude that field-emission sources are the most suitable for electron microdiffraction studies, provided that contamination levels are acceptable and that the emission is sufficiently stable. This conclusion is based partly on the experimental finding that most modern fine-grained materials contain a high density of defects, in which case the quality of CBED patterns improves considerably as the probe is made smaller than 10 nm in diameter. Manufacturers data for the zirconiated tungsten sources, the T-F sources, and the cold field-emission sources should be carefully compared with regard to emission stability. Practical experience with a T-F gun (Philips EM400 ST) shows a great improvement in the quality of CBED patterns recorded on film from 5-nm-diameter areas by comparison with patterns obtained using a thermal source, particularly from partially ordered materials such as the oxide superconductors. [Reviews of the performance of this machine are given by Carpenter (1980) and Bentley (1981).] However, experience also shows that some field-emission sources may not be sufficiently stable for quantitative intensity measurement using a serially scanned readout, and that a cold stage is needed to minimize contamination. Area detectors, such as the

CCD–YAG combination or the image plate, largely eliminate this problem of emission stability but require an imaging energy filter.

It is interesting to note that the use of a field-emission source is also beneficial for the study of diffuse scattering, for which point diffraction patterns must be used with a very small illumination aperture. Then the high brightness of the field-emission source will produce a more intense pattern than a conventional source, other things being equal.

# References

Allen, L. J. and C. J. Rossouw (1989). *Phys. Rev.* **B39**, 8313.
Angelini, P. and J. Bentley (1984). In *Analytical Electron Microscopy*, eds. D. Williams and D. Joy, San Francisco Press, San Francisco, p. 93.
Autrata, R., P. Schauer, J. Kvapil, and J. Kvapil (1978). *J. Phys. E* **11**, 707.
Avilov, A. S., S. A. Semiletov, and V. V. Storozenko (1989). *Sov. Phys. Crystallogr.* **34**, 110.
Ayer, R. (1989). *J. Electron Microsc.* **13**, 16.
Bangert, U. and P. Charsley (1989). *Philos. Mag.* **A59**, 629.
Bates, R. H. T. and J. M. Rodenburg (1989). *Ultramicroscopy* **31**, 303.
Bentley, J. (1981). *Norelco Rep.* **28**, 11.
Berger, S. D., D. Imeson, and R. H. Milne (1987). *Ultramicroscopy* **21**, 293.
Bevington, R. (1969). *Data Reduction and Error Analysis for the Physical Sciences*, McGraw-Hill, New York.
Bird, D. (1990). *Acta Crystallogr.* **A46**, 208.
Bird, D. (1992). *Acta Crystallogr.* **A48**, 555.
Bird, D. and R. James (1988). *Ultramicroscopy* **26**, 31.
Bird, D. and Q. A. King (1990). *Acta Crystallogr.* **A46**, 202.
Bird, D., R. James, and Q. King (1989). *Phys. Rev. Lett.* **63**, 1118.
Bird, D., R. James, and A. R. Preston (1987). *Phys. Rev. Lett.* **59**, 1216.
Bird, D. M. and A. G. Wright (1989). *Acta Crystallogr.* **A45**, 104.
Bithell, E. G. and W. M. Stobbs (1989). *J. Microsc.* **153**, 39.
Boe, N. and K. Gjønnes (1989). *Micron Microscopica Acta* **22**, 113.
Britton, E. G. and W. M. Stobbs (1987). *Ultramicroscopy* **21**, 1.
Brown, L. M. (1981). *J. Phys. F* **11**, 1.
Brown, L. M., A. J. Craven, L. Jones, A. Griffith, W. M. Stobbs, and C. J. Wilson (1976). In *Scanning Electron Microscopy 1976*, ed. O. Johari, I.I.T.R. Institute, Chicago, p. 353.
Brown, L. M., J. M. Rodenburg, and W. T. Pike (1988). EUREM 1988, Vol 93, Institute of Physics, London, p. 3.
Buerger, M. J. (1956) *Elementary Crystallography*, Wiley, New York.
Buseck, P., J. M. Cowely, and L. Eyring (1989). *High Resolution Electron Microscopy and Related Techniques*, Oxford University Press, New York.
Buxton, B. and J. E. Loveluck (1977). *J. Phys. C.* **10**, 3941.
Buxton, B. F. (1976). *Proc. R. Soc. London* **A350**, 335.
Buxton, B. F. and P. T. Tremewan (1980). *Acta Crystallogr.* **A36**, 304.
Buxton, B. F., J. A. Eades, J. W. Steeds, and G. M. Rackham (1976). *Philos. Trans. R. Soc. London, Ser. A.* **281**, 171.
Carpenter, R. (1980). *Norelco Rep.* **27**, 20.
Carpenter, R. W. and J. C. H. Spence (1982). *Acta Crystallogr.* **A38**, 55.

Chadi, J. (1989). Personal communication.
Chang, S.-L. (1987). *Cryst. Rev.* **1**, 87–189.
Cheng, Y. and R. Wang (1989). *Phys. Status Solidi* **B152**, 33.
Cherns, D. and A. R. Preston (1986). In *Proceedings of the 11th International Congress on Electron Microscopy*, ed. T. Imura, The Japanese Society of Electron Microscopy, Kyoto, p. 207.
Cherns, D. and A. Preston (1989). *J. Electron Microsc.* **13**, 111.
Cherns, D., C. J. Kiely, and A. Preston (1988). *Ultramicroscopy* **24**, 355.
Chew, N. G. and A. G. Cullis (1987). *Ultramicroscopy* **23**, 175.
Chou, C. T., L. J. Zhao, and T. Ko (1989). *Philos. Mag.* **A59**, 1221.
Christenson, K. K. and J. A. Eades (1987). *Ultramicroscopy* **21**, 101.
Cockayne, D. J. H. and D. R. McKenzie (1988). *Acta Crystallogr.* **A44**, 870.
Cosslett, V. E. (1972). *Optik* **36**, 1972.
Cowley, J. M. (1967). "Crystal Structure Determination by Electron Diffraction," in *Progress in Materials Science,* Pergamon Press, Oxford.
Cowley, J. M. (1969). *Acta Crystallogr.* **A25**, 129.
Cowley, J. M. (1976). *Ultramicroscopy* **1**, 255.
Cowley, J. M. (1978). *Adv. Electron. Electron Phys.* **46**, 1.
Cowley, J. M. (1979). *Ultramicroscopy* **4**, 435.
Cowley, J. M. (1981a). *Ultramicroscopy* **7**, 19.
Cowley, J. M. (1981b). *Diffraction Physics,* North-Holland, New York.
Cowley, J. M. (1981c). "Electron Microdiffraction and Microscopy of Amorphous Materials," in *Diffraction Studies on Non-crystalline Substances,* eds. I. Hargittai and W. J. Orville-Thomas, Elsevier, New York, pp. 849–891.
Cowley, J. M. (1985). *Ultramicroscopy* **18**, 11.
Cowley, J. M. (1986). *J. Electron Micros. Tech.* **3**, 25.
Cowley, J. M. ed. (1992a). *Techniques of Transmission Electron Diffraction,* Oxford University Press, New York.
Cowley, J. M. (1992b). *Ultramicroscopy* **41**, 335.
Cowley, J. M. and M. M. Disko (1980). *Ultramicroscopy* **5**, 469.
Cowley, J. M. and A. F. Moodie (1957). *Proc. Phys. Soc.* **B70**, 486.
Cowley, J. M. and A. F. Moodie (1960). *Proc. Phys. Soc.* **B76**, 378.
Cowley, J. M. and H.-J. Ou (1989). *J. Microsc. Technol.* **11**, 143.
Cowley, J. M. and J. C. H. Spence (1979). *Ultramicroscopy* **3**, 433.
Cowley, J. M. and J. C. H. Spence (1981). *Ultramicroscopy* **6**, 359.
Cowley, J. M. and D. J. Walker (1981) *Ultramicroscopy* **6**, 71.
Cowley, J., M. Osman, and P. Humble (1984). *Ultramicroscopy* **15**, 311.
Crewe, A. V. (1985). In *Scanning Electron Microscopy,* 1985, ed. O. Johari, I.I.T.R. Institute, Chicago, p. 467.
Crewe, A. V., D. Eggenberger, J. Wall, and L. M. Welter (1968). *Rev. Sci. Instrum.* **39**, 576.
Daberkow, I. and K. H. Herrman (1988). EUREM 1988, Vol. 93, Institute of Physics, London, p. 125.
Daberkow, I., K.-H. Herrmann, L. Liu, and R. W. D (1991). *Ultramicroscopy* **38**, 215.
Dowell, W. C. T. and P. Goodman (1976). *Optik* **45**, 93.
Downs, J. W., F. K. Ross, and G. V. Gibbs (1985). *Acta Crystallogr.* **B41**, 425.
Doyle, P. A. and P. S. Turner (1968). *Acta Crystallogr.* **A24**, 390.
Dravid, V., J. Sutliff, A. Westwood, M. Notis, and C. Lyman (1990). *Philos. Mag.* **61**, 417.
Dyke, W. (1956). "Electron field emission," in *Advances in Electronics and Electron Physics,* New York, Academic Press.
Eades, J. (1988c). *Ultramicroscopy* **24**, 143.

Eades, J., M. Shannon, and B. Buxton (1983). In *Scanning Electron Microscopy*, ed. O. Johari, I.I.T.R. Institute, Chicago, p. 83.
Eades, J. A. (1980). *Ultramicroscopy* **5**, 71.
Eades, J. A. (1984). *J. Electron Microsc. Tech.* **1**, 229.
Eades, J. A. (1988a). "Glide Planes and Screw Axes in CBED: The Standard Procedure", in *Microbeam Analysis 1988*, ed. D. Newbury, San Francisco Press, San Francisco, p. 75.
Eades, J. A. (1988b). *Proceedings of the 9th European Congress on Electron Microscopy*, (Inst. Phys. Conf. Ser., Vol 93, Institute of Physics, London, p. 3.
Eades, J. A. (1991). Personal communication.
Eagelsham, D. J. (1989). *J. Electron Microsc. Tech.* **13**, 67.
Eagelsham, D. J., C. J. Humphreys, N. Alford, W. J. Clegg, M. A. Harmer, and J. D. Birchall (1987) In *EMAG 1987*, ed. L. M. Brown, Institute of Physics, Inst. Phys. Conf. Ser., Vol. 90, London, 295.
Eagelsham, D. J., D. M. Maher, H. L. Fraser, C. J. Humphreys, and J. C. Bean (1989). *Appl. Phys. Lett.* **54**, 222.
Ecob, R., M. P. Shaw, A. J. Porter, and B. Ralph (1981a). *Philos. Mag.* **A44**, 1117.
Ecob, R. C. (1986). *Scr. Metall.* **20**, 1001.
Ecob, R. C., R. A. Ricks, and A. J. Porter (1982). *Scr. Metall.* **16**, 1085.
Ecob, R. C., M. P. Shaw, A. J. Porter, and B. Ralph (1981b). *Philos. Mag.* **A44**, 1135.
Egerton, R. F. (1986). *Electron Energy-Loss Spectroscopy in the Electron Microscope*, Plenum Press, New York.
Everhart, T. E. (1967). *J. Appl. Phys.* **38**, 4944.
Fink, H. W., H. Schmid, H. J. Kreuzer, and A. Wierzbicki (1991). *Phys. Rev. Lett.* **67**, 1543.
Fournier, D., G. L'Esperance, and G. Saint-Jacques (1989). *J. Electron Microsc. Tech.* **13**, 123.
Fowler, R. H. and L. Nordheim (1928). *Proc. R. Soc. London* **A119**, 173.
Fox, A. G. and R. M. Fisher (1988). *Aust. J. Phys.* **41**, 461.
Fraser, H. (1983). *J. Microsc. Spectrosc. Electron.* **8**, 431.
Fraser, H., D. M. Maher, C. J. Humprheys, C. J. D. Hetherington, R. V. Knoell, and J. C. Bean (1985). *Microsc. Semicond*, **76**, 307.
Freeman, L. A., A. Howie, A. B. Mistry, and P. H. Gaskell (1976). *The Structure of Non-crystalline Materials*, Taylor and Francis, London.
Friedman, S. L. and J. M. Rodenburg (1992). *J. Phys. D*, in press.
Fues, E. (1949). *Z. Phys.* **125**, 531.
Fukahara, A. (1966). *J. Phys. Soc. Jpn.* **21**, 2645.
Fung, K. K. (1984). *Ultramicroscopy* **12**, 243.
Fung, K. K. (1985). *Ultramicroscopy* **17**, 81.
Gajdardziska-Josifovska, M., M. McCartney, J. K. Weiss, W. de Ruijter, and D. J. Smith (1992). *Ultramicroscopy*, in press.
Gjønnes, J. (1971). *Z. Naturforsch.* **27a**, 434.
Gjønnes, J. and R. Høier (1969). *Acta Crystallogr.* **A25**, 595.
Gjønnes, J. and R. Høier (1971). *Acta Crystallogr.* **A27**, 313.
Gjønnes, J. and A. F. Moodie (1965). *Acta Crystallogr.* **19**, 65.
Gjønnes, J. and J. Taftø (1976). *Nucl. Instrum. Methods* **132**, 141.
Gjønnes, J. and D. Watanabe (1966). *Acta Crystallogr.* **21**, 297.
Gjønnes, K. (1985). *Ultramicroscopy* **17**, 133.
Gjønnes, K., J. Gjønnes, J. Zuo, and J. C. H. Spence (1981). *Acta Crystallogr.* **A44**, 810.
Gomer, R. (1961). *Field Emission and Field Ionization*, Harvard University, Cambridge, Mass.

Gong, H. and F. W. Schapink (1991). *Ultramicroscopy* **35**, 171.
Goodhew, P. F. (1975). *Specimen Preparation in Materials Science*, North-Holland/Elsevier, New York.
Goodman, P. (1971). *Acta Crystallogr.* **A27**, 140.
Goodman, P. (1975). *Acta Crystallogr.* **A31**, 793.
Goodman, P. (1976). *Acta Crystallogr.* **A32**, 793.
Goodman, P. (1978). *EMAG 1978,* Inst. Phys. Conf. Ser., Vol. 41, Institute of Physics, London, p. 116.
Goodman, P. (1984). *Acta Crystallogr.* **A40**, 635.
Goodman, P. and A. W. Johnson (1977). *Acta Crystallogr.* **A33**, 997.
Goodman, P. and G. Lehmpfuhl (1964). *Z. Naturforsch.* **19a**, 818.
Goodman, P. and G. Lehmpfuhl (1967). *Acta Crystallogr.* **22**, 14.
Goodman, P. and G. Lehmpfuhl (1968). *Acta Crystallogr.* **24**, 339.
Graczyk, J. F., K. N. Tu, B. Y. Tsaur and J. W. Mayer (1982). *J. Appl. Phys.* **53**, 6772.
Graham, R. A. Ourmazd and J. C. H. Spence (1987). Unpublished work.
Green, A. E. and J. A. Eades (1986). In *Proceedings of the 44th Annual EMSA Meeting,* ed. G. W. Bailey, San Francisco Press, San Francisco, p. 624.
Grivet, P. (1972). *Electron Optics,* Pergamon Press, New York.
Haine, M. E. and T. Mulvey (1952). *J. Opt. Soc. Am.* **42**, 763.
Hall, C. R. and P. B. Hirsch (1965). *Proc. R. Soc. London* **A286**, 158.
Hall, S. (1991). Crystallography Center, University of Western Australia, Nedlands, Australia.
Hanszen, K. J. (1982). "Holography in Electron Microscopy," in *Advances in Electronics and Electron Physics,* Academic Press, New York.
Hashimoto, H., A. Howie, and M. J. Whelan (1962). *Proc. R. Soc. London* **A269**, 80.
Hawkes, P. W. and E. Kasper (1989). *Principles of Electron Optics,* Academic Press, New York.
Herrman, K. H. (1983). In *Quantitative Electron Microscopy,* 25th Scottish Universities Summer School in Physics, p. 63.
Herrman, K. H. (1990). In *Proceedings of the International Congress on Electron Microscopy,* ed. R. Fisher, San Francisco Press, San Francisco, p. 112.
Hewat, E. A. and C. J. Humphreys (1974). *High-Voltage Electron Microscopy,* Academic Press, New York.
Higgs, A. and O. Krivanek (1981). In *Proceedings of the 39th Annual EMSA Meeting,* ed. G. Bailey, San Francisco Press, San Francisco, p. 346.
Hirsch, P. S., A. Howie, R. B. Nicholson, D. W. Pashley, and M. J. Whelan (1977). *Electron Microscopy of Thin Crystals.* Robert E. Krieger Publ. Co. Inc., Florida.
Hohenberg, P. and W. Kohn (1964). *Phys. Rev. B* **136**, 864.
Høier, R. (1969). *Acta Crystallogr.* **A25**, 516.
Høier, R. (1972). *Phys. Status Solidi* **11**, 597.
Høier, R. and B. Andersson (1974) *Acta Crystallogr.* **A30**, 93.
Høier, R. and K. Marthinsen (1983). *Acta Crystallogr.* **A39**, 854.
Høier, R., J. M. Zuo, K. Marthinsen, and J. C. H. Spence (1988). *Ultramicroscopy* **26**, 25.
Honjo, G. and K. Mihama (1954). *J. Phys. Soc. Jpn.* **9**, 184.
Howe, J. M., M. Sarikaya, and R. Gronsky (1986). *Acta Crystallogr.* **A42**, 368.
Howie, A. (1966). *Philos. Mag.* **14**, 223.
Howie, A. (1988). "Highly Disordered Materials," in *High Resolution Electron Microscopy,* eds. P. Buseck, J. M. Cowley, and L. Eyring, Oxford University Press, New York.
Howie, A. and U. Valdre (1967). *Philos. Mag.* **15**, 777.
Hummer, K. and H. W. Billy (1986). *Acta Crystallogr.* **A42**, 127.

Humphreys, C. J. (1979). *Rep. Prog. Phys.* **42**, 1825.
Humphreys, C. J. and P. B. Hirsch (1968). *Philos. Mag.* **18**, 115.
Humphreys, C. J. and J. C. H. Spence (1979). In *EMSA 1979*, ed. G. W. Bailey, Claitors, Louisiana, p. 554.
Humphreys, C. J. D. Eagelsham, D. Maher, and H. L. Fraser (1988a). *Ultramicroscopy* **26**, 13.
Humphreys, C. J. D. M. Maher, H. L. Fraser, and D. J. Eagelsham (1988b). *Philos. Mag.* **A58**, 787.
Hurley, A. and A. F. Moodie (1980). *Acta Crystallogr.* **A36**, 737.
Hussein, A. and H. Wagenfeld (1978). In *Electron Diffraction 1927–1978*, ed. P. Dobson, Inst. Phys. Conf. Ser. Vol. 41, Institute of Physics, London, p. 47.
Ibers, J. A. (1958). *Acta Crystallogr.* **11**, 178.
Ichimiya, A. and R. Uyeda (1977). *Z. Naturforsch.* **32a**, 750.
Ishikawa, A., H. Suda, A. Fukami, T. Oikawa and Y. Arai (1990). *Proceedings of the 12th International Congress on Electron Microscopy*, San Francisco Press, San Francisco, p. 474.
Ishizuka, K. (1982). *Ultramicroscopy* **9**, 255.
Ishizuka, K. and J. Taftø (1984). *Acta Crystallogr.* **B40**, 332.
Isoda, S., K. Saitoh, S. Moriguchi, and T. Kobayashi (1991). *Ultramicroscopy* **35**, 329.
Jackson, A. G. (1991). *Handbook of Crystallography*, Springer, New York.
Jesson, D. E. and J. W. Steeds (1990) *Philos. Mag.* **A61**, 385.
Johnson, A. W. (1976). Personal communication.
Johnson, A. W. S. (1972). *Acta Crystallogr.* **A28**, 89.
Jones, P., G. M. Rackham, and J. M. Steeds (1977). *Proc. R. Soc. London* **A354**, 197.
Jordan, I. K., C. J. Rossouw, and R. Vincent (1991). *Ultramicroscopy* **35**, 237.
Kambe, K. (1957). *J. Phys. Soc. Jpn.* **12**, 1.
Karle, J. (1989). *Phys. Today* **42**, 22.
Kaufman, M. J., D. D. Pearson, and H. L. Fraser (1986). *Philos. Mag.* **A54**, 79.
Kelly, A. and G. W. Groves (1970). *Crystallography and Crystal Defects*, Longmans, London.
Kelly, P. M., A. Jostons, R. G. Blake, and J. G. Napier (1975). *Phys. Status Solidi* **A31**, 771.
Kim, Y., J. C. H. Spence, N. Long, W. Bergholz, and M. O'Keeffe (1987). *J. Appl. Phys.* **62**, 419.
Kitamura, N. (1966). *J. Appl. Crystallogr.* **37**, 2187.
Kittel, C. (1976). *Introduction to Solid State Physics*, Wiley, New York.
Kogiso, M. and H. Takahashi (1977). *J. Phys. Soc. Jpn.* **42**, 223.
Kolby, P. and J. Taftø (1991). *Micron Microsc. Acta* **22**, 151.
Kondo, Y., T. Ito, and Y. Harada (1984). *Jpn. J. Appl. Phys.* **23**, L178.
Konnert, J., D'Antonio, J. Cowley, A. Higgs, and H.-J. Ou (1989). *Ultramicroscopy* **30**, 371.
Kossel, W. and G. Möllenstedt (1942). *Ann. Phys.* **42**, 287.
Krakow, W. and L. Howland (1976). *Ultramicroscopy* **2**, 53.
Krakow, W. and M. O'Keefe (1989). *Computer Simulation of Electron Microscope Diffraction and Images*, The Minerals, Metals and Materials Society, New York.
Kreutle, M. and G. Meyer-Ehmson (1969). *Phys. Status Solidi* **35**, K17.
Krivanek, O. L. and C. Ahn (1986). *Proceedings of the International Congress Electron Microscopy*, Marusen, Tokyo.
Krivanek, O. L., A. J. Gubbens, and N. Dellby (1991). *J. Microsc. Microanal. Microstructures*, **2**, 315.
Krivy, I. and B. Gruber (1976). *Acta Crystallogr.* **A32**, 297.

Lally, J., C. J. Humphreys, A. J. Metherell, and R. M. Fisher (1972). *Philos. Mag.* **25**, 321.
Lanio, S. (1986). *Optik* **73**, 99.
Le Page, Y. (1987). *J. Appl. Crystallogr.* **20**, 264.
Le Page, Y. (1992). *Microsc. Res. Techn.* **21**, 158.
Lehmpfuhl, G. and A. Reissland (1968). *Z. Naturforsch.* **23a**, 544.
Lewis, A. L., R. E. Villagrana, and A. J. F. Metherall (1978). *Acta Crystallogr.* **A34**, 138.
Lichte, H. (1991). Personal communication.
Liebl, H. (1989). *Optik.* **83**, 129.
Lin, J. A. and J. M. Cowley (1986a). *Ultramicroscopy* **19**, 31.
Lin, J. A. and J. M. Cowley (1986b). *Ultramicroscopy* **19**, 179.
Lin, Y. P., D. M. Bird, and R. Vincent (1989). *Ultramicroscopy* **27**, 233.
Lin, Y. P., A. R. Preston, and R. Vincent (1987). *EMAG 1987*, Inst. Phys. Conf. Ser., Vol. 90, Institute of Physics, London, p. 115.
Liu, J. and J. M. Cowley (1990). *Ultramicroscopy* **34**, 119.
Loane, R., P. Xu, and J. Silcox (1991). *Acta Crystallogr.* **A47**, 267.
Lodge, E. and J. M. Cowley (1984). *Ultramicroscopy* **13**, 215.
Loretto, M. H. (1984). *Electron Beam Analysis of Materials*, Chapman and Hall, London.
Lundqvist, S. and N. H. March (1983). *Theory of the Inhomogeneous Electron Gas*, Plenum Press, New York.
Lynch, J. P., E. Lesage, H. Dexpert, and E. Freund (1981). *EMAG 1981*, Institute of Physics, London, p. 67.
Ma, Y., J. Gjønnes, and J. Taftø (1991). *Micron Microscopica Acta* **22**, 163.
MacGillavry, C. H. (1940). *Physica* **7**, 329.
Maher, D. M., H. Fraser, C. J. Humprheys, R. V. Knoell, and J. C. Bean (1987). *Appl. Phys. Lett.* **50**, 574.
Malacara, D. (1978). *Optical Shop Testing*, Wiley, New York.
Mansfield, J. (1989). *J. Microsc. Technol.* **13**, 3.
Marthinsen, K. and R. Høier (1986). *Acta Crystallogr.* **A42**, 484.
Marthinsen, K. and R. Høier (1988). *Acta Crystallogr.* **A44**, 558.
Marthinsen, K., R. Hoier, and L. Bakken (1990). In *Proceedings of the 12th International Congress on Electron Microscopy*, San Francisco Press, San Francisco, p. 492.
Marthinsen, K., H. Matsuhata, R. Høier, and J. Gjønnes (1988). *Aust. J. Phys.* **41**, 449.
Marthinsen, M. and R. Høier (1989). In *EMSA*, ed. G. Bailey, San Francisco Press, San Francisco, p. 484.
Maslen, E. N. (1988). *Acta Crystallogr.* **A44**, 33.
Matsuhata, H. and J. W. Steeds (1987). *Philos. Mag.* **55**, 39.
Matsuhata, H., Y. Tomokiyo, Y. Watanabe, and T. Eguchi (1984). *Acta Crystallogr.* **A36**, 686.
Matsumura, S., Y. Tomokiyo, and K. Oki (1989). *J. Electron Microsc. Technol.* **12**, 262.
Mayer, J., J. C. H. Spence, and G. Mobus (1991). *Proc. 49th Annual EMSA Meeting*, ed. G. Bailey, San Francisco Press, San Francisco, p. 786.
McKie, D. and C. McKie (1986). *Essentials of Crystallography*, Blackwell Scientific, Oxford.
McMullan, D. and S. D. Berger (1988). *Ultramicroscopy* **25**, 349.
McMullan, D., J. M. Rodenburg, and W. T. Pike (1990). In *Proceedings of the 12th International Congress on Electron Microscopy (Seattle)*, ed. R. Fisher, San Francisco Press, San Francisco, p. 15.
Menzel-Kopp, C. (1962). *J. Phys. Soc. Jpn.* **B-II 17**, 76.
Metherall, A. J. F. (1975). *Electron Microscopy and Materials Science, Third Course of the International School of Electron Microscopy*, Commission of Euro Communities, Directorate General, "Scientific and Technical Information," Luxembourg.

Miyake, S. (1940). *Proc. Phys. Math. Soc. Jpn.* **22**, 666.
Miyake, S. (1955). *Acta Crystallogr.* **8**, 335.
Monosmith, W. B. and J. M. Cowley (1983). *Ultramicroscopy* **12**, 51.
Moodie, A. F. (1972). *Z. Naturforsch.* **27a**, 437.
Moodie, A. F. (1979). *Chem. Scripta*, **14**, 21.
Mooney, P. E., G. Y. Fan, C. E. Meyer, K. V. Truong, D. B. Bui, and O. L. Krivanek (1990). In *12th International Congress for Electron Microscopy*, ed. R. Fisher, San Francisco Press, San Francisco, p. 164.
Mori, N., T. Oikawa, T. Katoh, J. Miyahara, and Y. Harada (1988). *Ultramicroscopy* **25**, 195.
Morton, G. A. and E. G. Ramberg (1939). *Phys. Rev.* **56**, 705.
Mory, C., C. Colliex, and J. M. Cowley. (1987). *Ultramicroscopy* **21**, 171.
Moss, H. (1968). *Advances in Electronics Electron Physics*, Academic Press, New York.
Mott, N. F. (1930). *Proc. R. Soc. London* **A127**, 658.
Muddle, B. C. (1985). *Metals Forum* **8**, 93.
Munch, J. (1974). *Optik* **43**, 79.
Nagata, F. and A. Fukuhara (1967). *Jpn. J. Appl. Phys.* **6**, 1233.
Ou, H. J. and J. M. Cowley (1988). *Proc. 46th Annual EMSA Meeting*, ed. G. Bailey, San Francisco Press, San Francisco, p. 882.
Pan, M., J. Cowley, and J. Barry (1989). *Ultramicroscopy* **30**, 385.
Pan, M., J. Cowley, and R. Garcia (1987). *Micron* **18**, 165.
Pearson, W. B. (1967). *A Handbook of Lattice Spacings for Metals and Alloys*, Pergamon Press, Oxford.
Peng, L. M. and J. M. Cowley (1988). *Acta Crystallogr.* **A44**, 1.
Perovic, D., G. Weatherly, and D. Houghton (1992). *Philos. Mag.*, in press.
Pfister, H. (1953). *Ann. Phys.* **11**, 239.
Pike, W. T., L. M. Brown, R. Kubiak, S. M. Newstead, A. R. Powell, E. H. Parker, and T. E. Whall (1991). *J. Cryst. Growth* **111**, 925.
Pike, W. T., J. M. Rodenburg, and L. M. Brown (1987). In *EMAG 1987*, ed. L. M. Brown, Inst. Phys. Conf. Ser. Vol. 90, Institute of Physics, London, p. 127.
Pogany, A. P. and P. S. Turner (1968). *Acta Crystallogr.* **A24**, 103.
Polk, D. E. (1971). *J. Non-Cryst. Solids* **5**, 365.
Pond, R. and D. S. Viachavas (1983). *Proc. R. Soc. London* **A386**, 95.
Portier, R. and D. Gratias (1981). *EMAG 1981*, Inst. Phys. Conf. Ser., Vol. 61, Institute of Physics, London, p. 275.
Press, W. H., B. Flannery, S. Teukolsky, and W. Vetterling (1986). *Numerical Recipes*, Cambridge University Press, New York.
Preston, A. R. and Cherns (1986). Inst. Phys. Conf. Ser., Vol. 78, Institute of Physics, London, p. 41.
Qian, W., J. Spence, and J. M. Zuo (1992). *Acta Crystallogr.* **A**, in press.
Rackham, G. M. and J. A. Eades (1977). *Optik* **47**, 227.
Radi, G. (1970). *Acta Crystallogr.* **A26**, 41.
Raghavan, M., J. Y. Koo, and R. Petrovie-Luton (1983). *J. Metals* **35**, 44.
Raimes, S. (1961). *The Wave Mechanics of Electrons in Metals*, North-Holland, Amsterdam.
Randle, V., I. Barker, and B. Ralph (1989). *J. Microsc. Technol.* **13**, 51.
Reimer, L. (1984). *Transmission Electron Microscopy*, Springer-Verlag, New York.
Reimer, L., I. Fromm, and I. Naundorf (1990). *Ultramicroscopy* **32**, 80.
Rez, P. (1978). D.Phil. Thesis, University of Oxford.
Rez, P. (1979). In *Proc. 37th EMSA 1979*, ed. G. Bailey, Claitors, Baton Rouge, p. 438.
Roberts, P., J. Chapman, and A. MacLeod (1982). *Ultramicroscopy* **8**, 385.

Rodenberg, J. M. and I. A. Rauf (1989). *EMAG 89*, Institute of Physics, London, p. 119.
Rodenburg, J. M. (1988). *Ultramicroscopy* **25**, 329.
Ronchi, V. (1964). *Appl. Opt.* **3**, 437.
Rudee, M. L. and A. Howie (1972). *Philos. Mag.* **25**, 1001.
Samoto, N., R. Shimizu, H. Hashimoto, N. Tamura, K. Gamo, and S. Namba (1985). *Jpn. J. Appl. Phys.* **24**, 766.
Sands, D. E. (1969). *Introduction to Crystallography,* Benjamin, Reading, Mass.
Saxton, O., M. A. O'Keefe, D. J. Cockayne, and M. Wilkens (1983). *Ultramicroscopy* **12**, 75.
Schapink, F. W., S. K. E. Forghany, and R. P. Caron (1986). *Philos. Mag.* **A53**, 717.
Scheinfein, M., W. Qian, and J. Spence (1991). *J. Appl. Phys.,* in press.
Sears, V. F. and S. A. Shelley (1991). *Acta Crystallogr.* **A47**, 441.
Sellar, J., D. Imeson, and C. Humphreys (1980). *Acta Crystallogr.* **A36**, 482.
Shimoyama, H. and S. Maruse (1984). *Ultramicroscopy* **15**, 239.
Shindo, D., K. Hiraga, T. Oikawa, and N. Mori (1990). *J. Electron Microsc.* **39**, 449.
Shinohara, K. (1932). *Sci. Pap. Inst. Phys. Chem. Res. Tokyo* **20**, 39.
Shirley, C. G. and R. M. Fisher (1979). *Philos. Mag.* **39**, 91.
Shishido, T. and M. Tanaka (1976). *Phys. Status Solidi A* **38**, 453.
Sinclair, J. E. *J. Appl. Phys.* **42**, 5321.
Smart, D. J. and C. J. Humphreys (1978). *The Crystal Potential in Electron Diffraction and in Band Theory,* Inst. Phys. Conf. Ser. Vol. 41 Institute of Physics, London.
Smith, P. J. and G. Lehmpfuhl (1975) *Acta Crystallogr.* **A31**, S220.
Spackman, M. A. and E. N. Malsen (1986). *J. Phys. Chem.* **90**, 2020.
Speer, S., J. C. H. Spence, and E. Ihrig (1990). *Acta Crystallogr.* **A46**, 763.
Speidel, R. and P. Brauchle (1987). *Optik* **77**, 46.
Spence, J. and J. Mayer (1991). In *Proc. 49th Annual EMSA Meeting,* ed. G. Bailey, San Francisco Press, San Francisco, p. 786.
Spence, J. and W. Qian (1991). *Phys. Rev. B* **45**, 10271.
Spence, J. C. H. (1977). *Proc. 35th Annual EMSA, Meeting,* ed. G. Bailey, Claitors, Baton Rouge, p. 178.
Spence, J. C. H. (1978a). *Acta Crystallogr.* **A34**, 112.
Spence, J. C. H. (1978b). In *Scanning Electron Microscopy 1978,* ed. O. Johari, I.I.T.R. Institute, Chicago, p. 61.
Spence, J. C. H. (1978c). In *Proceedings of the 9th International Congress on Electron Microscopy,* ed. J. Sturgess, vol. 1, Microscopical Society of Canada, Toronto, p. 554.
Spence, J. C. H. (1988a). *Experimental High Resolution Electron Microscopy,* Oxord University Press, New York.
Spence, J. C. H. (1988b). "Inelastic electron scattering," in *High Resolution Transmission Electron Microscopy,* eds. P. Buseck, J. H. Cowley, and L. Eyring, Oxford University Press, New York.
Spence, J. C. H. (1992a). "Accurate Structure Factor Measurement by Electron Diffraction," in *Techniques of Transmission Electron Diffraction,* Oxford University Press, New York. J. M. Cowley, Ed. Vol. 1.
Spence, J. C. H. (1992b). "Electron Channelling," in *Techniques of Electron Diffraction,* Oxford University Press, New York.
Spence, J. C. H. (1992c). *Acta Crystallogr.* **A**, in press.
Spence, J. C. H. (1992d) *Optik.* (to be published).
Spence, J. C. H. and R. W. Carpenter (1986). "Electron Microdiffraction," in *Elements of Analytical Electron Microscopy.* eds. D. Joy, A. Romig, H. Goldstein, Plenum Press, New York.
Spence, J. C. H. and J. M. Cowley (1978) *Optik* **50**, 129.

Spence, J. C. H. and M. Katz (1979). Unpublished work.
Spence, J. C. H. and J. Lynch (1982). *Ultramicroscopy* **9**, 267.
Spence, J. C. H. and J. M. Zuo (1988). *Rev. Sci. Instrum.* **59**, 2102.
Steeds, J. and J. Mansfield (1984). *Convergent Beam Electron Diffraction of Alloy Phases*, Adam Hilger, Ltd., Bristol, U.K.
Steeds, J. W. (1979). In *Introduction to Analytic Electron Microscopy*, eds. J. Hren, J. Goldstein, and D. Joy, Plenum Press, New York.
Steeds, J. W. and N. S. Evans (1980). In *Proc. 38th Annual EMSA Meeting*, ed. G. Bailey, Claitors, Baton Rouge, p. 188.
Steeds, J. W. and R. Vincent (1983). *J. Appl. Cryst.* **16**, 317.
Stoter, L. (1981). *J. Mater. Sci.* **16**, 1039.
Stout, G. H. and L. H. Jensen (1968). *X-ray Structure Determination*, Macmillan, London.
Sung, C. and D. B. Williams (1991). *J. Electron Microsc. Technol.* **17**, 95.
Swanson, W. (1991). In *Proc. 49th Annual EMSA Meeting*, ed. G. Bailey, San Francisco Press, San Francisco.
Swanson, L. W. and A. E. Bell (1973). "Recent Advances in Field Electron Microscopy of Metals," in *Advances in Electronics and Electron Physics*, Academic Press, New York.
Taftø, J. (1983). *Phys. Rev. Lett.* **51**, 654.
Taftø, J. (1987). *Acta Crystallogr.* **A43**, 208.
Taftø, J. and J. Gjønnes (1985). *Ultramicroscopy* **17**, 329.
Taftø, J. and T. H. Metzger (1985). *J. Appl. Cryst.* **18**, 110.
Taftø, J. and J. C. H. Spence (1982). *J. Appl. Cryst.* **15**, 60.
Tanaka, M. (1986). *J. Electron Microsc.* **35**, 314.
Tanaka, M. (1989). *J. Microsc. Technol.* **13**, 27.
Tanaka, M. and M. Terauchi (1985a). *Convergent-Beam Electron Diffraction*, JOEL, Ltd., Tokyo.
Tanaka, M. and M. Terauchi (1985b). *J. Electron Microsc.* **34**, 52.
Tanaka, M. and K. Tsuda (1990). *Proceedings of the 12th International Congress on Electron Microscopy*, San Francisco Press, San Francisco, p. 518.
Tanaka, M., R. Saito, X. Ueno, and Y. Harada (1980). *J. Electron Microsc.* **29**, 408.
Tanaka, M., H. Sekii, and T. Nagasawa (1983). *Acta Crystallogr.* **A39**, 825.
Tanaka, M., M. Terauchi, and T. Kaneyama (1988). *Convergent Beam Electron Diffraction II*, JOEL, Ltd. Tokyo.
Tanaka, M., M. Terauchi, and H. Sekii (1987). *Ultramicroscopy* **21**, 21.
Tietz, G. (1991). TVIPS Inc. Herbstrasse 7, D-8035 Gauting, West Germany.
Terasaki, O., D. Watanabe, and J. Gjønnes (1979). *Acta Crystallogr.* **A35**, 895.
Terauchi, M. and M. Tanaka (1985). *J. Electron Microsc.* **34**, 347.
Thomas, L. E., C. J. Shirley, J. S. Lally, and R. M. Fisher (1974). *High Voltage Electron Microscopy*, Academic Press, New York.
Tomokiyo, Y. and T. Kuroiwa (1990). *Proceedings of the 12th International Congress on Electron Microscopy* (Seattle), San Francisco Press, San Francisco, p. 526.
Tonomura, A. (1987). *Rev. Mod. Phys.* **59**, 639.
Treacy, M. J., J. M. Gibson, and A. Howie (1985). *Philos. Mag.* **A51**, 389.
Treacy, M. M. J., A. Howie, and C. J. Wilson (1978). *Philos. Mag.* **A38**, 569.
Tsong, T. T. (1990). *Atom Probe Field Ion Microscopy*, Cambridge University Press, New York.
Unwin, N. and R. Henderson (1975). *J. Mol. Biol.* **94**, 425.
Uyeda, R. (1968). *Acta Crystallogr.* **A24**, 175.
Vainshtein, B. K. (1964). *Structure Analysis by Electron Diffraction*, Pergamon Press, Oxford.

Valentine, R. C. (1966). "Response of Photographic Emulsions to Electrons," in *Advances in Optical and Electron Microscopy*, Academic Press, New York.
Venables, J. A. and G. Cox (1987). *Ultramicroscopy* **21**, 33.
Vincent, R. (1989). *J. Electron Microsc.* **13**, 40.
Vincent, R. and T. Bielicki (1981). *EMAG 1981*, Inst. Phys. Proc. Vol. 61, Institute of Physics, London, p. 491.
Vincent, R. and D. M. Bird (1986). *Philos. Mag.* **A53**, L35.
Vincent, R., D. Bird, and J. W. Steeds (1984). *Philos. Mag.* **50**, 765.
Vincent, R., D. Cherns, S. J. Bailey, and H. Morkoc (1987). *Philos. Mag. Lett.* **56**, 1.
Vincent, R., B. Krause, and J. W. Steeds (1986). *Proceedings of the 11th International Congress on Electron Microscopy* (Kyoto), Maruzen, Tokyo.
Vincent, R., A. Preston, and M. King (1988). *Philos. Mag.* **24**, 409.
Voss, R., G. Lehmpfuhl, and P. J. Smith (1980). *Z. Naturforsch.* **35**a, 973.
Wang, R. and J. Wen (1989). *Acta Crystallogr.* **A45**, 428.
Watari, F. and J. Cowley (1981). *Surf. Sci.* **105**, 240.
Watenabe, D., R. Uyeda, and M. Kogiso (1968). *Acta Crystallogr.* **A24**, 249.
Weickenmeier, A. and H. Kohl (1991). *Acta Crystallogr.* **A47**, 590.
Wells, O. C. (1974). *Scanning Electron Microscopy*, McGraw-Hill, New York.
Wen, J., R. Wang, and G. Lu (1989). *Acta Crystallogr.* **A45**, 422.
Williams, D. B. (1987). *Practical Analytical Electron Microscopy in Materials Science*, Philips, Mahwah, New Jersey.
Wolberg, J. R. (1967). *Prediction Analysis*, Van Nostrand, Princeton, p. 60, Eq. 3.10.40.
Xu, P., R. F. Loane, and John Silcox (1991). *Ultramicroscopy*, **38**, 127.
Yamamoto, N. (1982). Personal communication.
Yoshioka, H. (1957). *J. Phys. Soc. Jpn.* **12**, 618.
Yoshioka, H. and Y. Kainuma (1962). *J. Phys. Soc. Jpn., Suppl.* B2, 134.
Zachariasen, W. (1932). *J. Am. Chem. Soc.* **54**, 3841.
Zaluzek, N. and M. G. Staruaa (1988). *Proc. 46th Annual EMSA. Meeting*, ed. G. Bailey, San Francisco Press, San Francisco, p. 662.
Zhu, J. and J. M. Cowley (1982). *Acta Crystallogr.* **A38**, 718.
Zhu, J. and J. M. Cowley (1983). *J. Appl. Crystallogr.* **16**, 171.
Zhu, J. and J. M. Cowley (1985). *Ultramicroscopy* **18**, 419.
Zhu, J., L. M. Peng, and J. M. Cowley (1987). *J. Microsc. Technol.* **7**, 177.
Zhu, Y. A. Moodenbaugh, and J. Tafto (1990). *Physica C* **167**, 363.
Zou, H., X. Yao, and R. Wang (1991) *Acta Crystallogr.* **A47**, 363.
Zou, H., X. Yao, and R. Wang (1991). *Acta Crystallogr.* **A47**, 490.
Zuo, J., J. C. Spence, J. Downs, and J. Mayer (1992). *Acta Crystallogr.* **A**, in press.
Zuo, J. M. (1991). *Acta Crystallogr.* **A47**, 87.
Zuo, J. M. (1992a). *Ultramicroscopy*, **41**, 211.
Zuo, J. M. (1992b). *Acta Crystallogr.* (to be published).
Zuo, J. M. (1993). *Acta Crystallogr.* **A** (to be published).
Zuo, J. M. and J. C. H. Spence (1991). *Ultramicroscopy* **35**, 185.
Zuo, J. M. and J. C. H. Spence. *Philos. Mag.* (to be published).
Zuo, J. M., J. Foley, M. O'Keefe, and J. C. H. Spence (1989a). "On the Accurate Measurement of Structure Factors in Ceramics by Electron Diffraction," in *Metal-Ceramic Interfaces*, Pergamon Press, New York.
Zuo, J. M., R. Hoier, and J. C. H. Spence (1989b). *Acta Crystallogr.* **A45**, 839.
Zuo, J. M., J. C. H. Spence, and R. Hoier (1989c). *Phys. Rev. Lett.* **62**, 547.
Zuo, J. M., J. C. H. Spence, and M. O'Keefe (1988). *Phys. Rev. Lett.* **61**, 353.
Zvyagin, B. B. (1967). *Electron Diffraction Analysis of Clay Minerals*, Plenum Press, New York.

# Appendix 1. Useful Relationships in Dynamical Theory

The following is a compilation of frequently used results in dynamical theory:

$$U_g = \frac{2m|e|U_g}{h^2} = \frac{K\sigma V_g}{\pi} = \frac{\gamma F_g^B}{\pi W}$$

$$|U_g| = |K|/\xi_g$$

$\lambda$ (relativistic) $= hc/W\beta$  where $W = mc^2 = \gamma m_0 c^2 = 511 + E_0$, with $E_0$ in kV

$$\gamma = (1-\beta^2)^{-1/2} = \frac{W}{m_0 c^2} \quad \text{where } \beta = v/c$$

$$\sigma = \left(\frac{\pi}{\lambda E}\right)\frac{2}{1+\sqrt{(1-\beta^2)}} = \frac{2\pi m e \lambda}{h^2}$$

$$V_g = \frac{h^2 F_g^B}{8\pi \varepsilon_0 m_e |e| \Omega} = 47.878009 \frac{F_g^B}{\Omega} \quad \text{if the cell volume } \Omega \text{ is given in angstroms}^3$$

$$V_g = \pi/(\xi_g \sigma)$$

$|\mathbf{K}_0| = 1/\lambda \approx 511\gamma\beta/12.4$

$\mathbf{K}$ = wavevector corrected for mean inner potential, $|\mathbf{K}|^2 = |\mathbf{K}_0|^2 + U_0$

$h^2/m_0 \approx 300$ (eV, angstroms)

$hc = 12423$ (eV, angstroms)

$h/(m_0 c) = 0.02426$ angstrom

$S_g = -\lambda g^2/2$ in the zone-axis orientation

$|\mathbf{s}| = \sin\theta_B/\lambda \approx \Theta/2\lambda$ where $\Theta$ is the total scattering angle

$m_0 c^2 \approx 511$ kV

# Appendix 2. Electron Wavelengths, Physical Constants, etc.

The fundamental physical constants used in this book are listed below.

$c = 2.997924581(1.2) \times 10^8$ m s$^{-1}$

$e = 1.6021892(46) \times 10^{-19}$ C

$h = 6.626176(36) \times 10^{-34}$ J s

$m_e = 0.9109534(47) \times 10^{-30}$ kg

$\varepsilon_0 = 8.85418782(7) \times 10^{-12}$ F m$^{-1}$

In addition,

$1 \text{ eV} = 1.6021892 \times 10^{-19}$ J

$h = 4.14125 \times 10^{-15}$ eV s

The following table gives, as functions of the accelerating voltage $E$, the relativistic electron wavelength $\lambda$, wavenumber $\lambda^{-1}$, relativistic factor $\gamma = m/m_0 = (1 - \beta^2)^{-1/2}$, velocity $\beta = v/c$, and correction factor $v/v_{100}$ for $\xi_g$ relative to 100 kV.

| $E$ | $\lambda$ (Å) | $\lambda^{-1}$ (Å$^{-1}$) | $m/m_0$ | $v/c$ | $v/v_{100}$ |
|---|---|---|---|---|---|
| 1 V | 12.26 | 0.0815 | 1.0000020 | 0.0020 | 0.0036 |
| 10 V | 3.878 | 0.2579 | 1.0000196 | 0.0063 | 0.0114 |
| 100 V | 1.226 | 0.8154 | 1.0001957 | 0.0198 | 0.0361 |
| 500 V | 0.5483 | 1.824 | 1.0009785 | 0.0442 | 0.0806 |
| 1 kV | 0.3876 | 2.580 | 1.00196 | 0.0625 | 0.1139 |
| 2 kV | 0.2740 | 3.650 | 1.00391 | 0.0882 | 0.1609 |
| 3 kV | 0.2236 | 4.473 | 1.00587 | 0.1079 | 0.1968 |
| 4 kV | 0.1935 | 5.167 | 1.00783 | 0.1244 | 0.2269 |
| 5 kV | 0.1730 | 5.780 | 1.00978 | 0.1389 | 0.2533 |
| 6 kV | 0.1579 | 6.335 | 1.01174 | 0.1519 | 0.2771 |
| 7 kV | 0.1461 | 6.845 | 1.01370 | 0.1638 | 0.2989 |
| 8 kV | 0.1366 | 7.322 | 1.01566 | 0.1749 | 0.3190 |
| 9 kV | 0.1287 | 7.770 | 1.01761 | 0.1852 | 0.3379 |
| 10 kV | 0.1220 | 8.194 | 1.01957 | 0.1950 | 0.3557 |
| 20 kV | 0.0859 | 11.64 | 1.0391 | 0.2719 | 0.4959 |
| 30 kV | 0.0698 | 14.33 | 1.0587 | 0.3284 | 0.5990 |
| 40 kV | 0.0602 | 16.62 | 1.0783 | 0.3741 | 0.6823 |
| 50 kV | 0.0536 | 18.67 | 1.0978 | 0.4127 | 0.7528 |
| 60 kV | 0.0487 | 20.55 | 1.1174 | 0.4462 | 0.8139 |
| 70 kV | 0.0448 | 22.30 | 1.1370 | 0.4759 | 0.8680 |
| 80 kV | 0.0418 | 23.95 | 1.1566 | 0.5024 | 0.9164 |
| 90 kV | 0.0392 | 25.52 | 1.1761 | 0.5264 | 0.9602 |
| 100 kV | 0.0370 | 27.02 | 1.1957 | 0.5482 | 1.0000 |
| 200 kV | 0.0251 | 39.87 | 1.3914 | 0.6953 | 1.268 |
| 300 kV | 0.0197 | 50.80 | 1.5871 | 0.7765 | 1.416 |
| 400 kV | 0.0164 | 60.83 | 1.7828 | 0.8279 | 1.510 |
| 500 kV | 0.0142 | 70.36 | 1.9785 | 0.8629 | 1.574 |
| 600 kV | 0.0126 | 79.57 | 2.1742 | 0.8879 | 1.620 |
| 700 kV | 0.0113 | 88.56 | 2.3698 | 0.9066 | 1.654 |
| 800 kV | 0.0103 | 97.38 | 2.5655 | 0.9209 | 1.680 |
| 900 kV | 0.0094 | 106.1 | 2.7612 | 0.9321 | 1.700 |
| 1 MeV | 0.0087 | 114.7 | 2.9569 | 0.9411 | 1.717 |
| 2 MeV | 0.0050 | 198.3 | 4.9138 | 0.9791 | 1.786 |
| 4 MeV | 0.0028 | 361.5 | 8.8277 | 0.9936 | 1.812 |
| 6 MeV | 0.0019 | 523.5 | 12.742 | 0.9969 | 1.818 |
| 8 MeV | 0.0015 | 685.2 | 16.655 | 0.9982 | 1.821 |
| 10 MeV | 0.0012 | 846.8 | 20.569 | 0.9988 | 1.822 |

# Appendix 3. Crystallographic Data

## A3.1. THE RECIPROCAL LATTICE

We define real-space primitive lattice vectors **a**, **b**, and **c**. Then the reciprocal lattice is defined by the relations

$$\mathbf{a}^* = \frac{\mathbf{b} \times \mathbf{c}}{\Omega}, \quad \mathbf{b}^* = \frac{\mathbf{c} \times \mathbf{a}}{\Omega}, \quad \mathbf{c}^* = \frac{\mathbf{a} \times \mathbf{b}}{\Omega}$$

where $\Omega = \mathbf{a} \cdot (\mathbf{b} \times \mathbf{c})$ is the cell volume. Then $\mathbf{a}^* \cdot \mathbf{a} = \mathbf{b}^* \cdot \mathbf{b} = \mathbf{c}^* \cdot \mathbf{c} = 1$. Every real-space lattice point may be written $\mathbf{r} = n\mathbf{a} + m\mathbf{b} + p\mathbf{c}$, where $n$, $m$, and $p$ are integers. Any reciprocal space vector may be written

$$\mathbf{g} = h\mathbf{a}^* + k\mathbf{b}^* + l\mathbf{c}^*$$

with $h$, $k$, and $l$ integers. These vectors have the important properties that:

(1) Vector **g** in reciprocal space is normal to the plane in real space whose Miller indices are $(h, k, l)$.

(2) The spacing of the $(h, k, l)$ planes in real space is $1/(|\mathbf{g}|)$. (But see McKie and McKie (1986), for example, for a fuller discussion of the "fictitious planes" used to describe diffraction phenomena—these planes do not always contain atoms.)

## A3.2. THE SEVEN CRYSTAL SYSTEMS

| Crystal | Axial length and angles | Minimum symmetry elements |
|---|---|---|
| Cubic | Three equal axes at right angles: $a = b = c$, $\alpha = \beta = \gamma = 90°$ | Four, threefold rotation axes |
| Hexagonal | Two axes at 120°, third axis at right angles to both. $a = b \neq c$, $\alpha = \beta = 90°$, $\gamma = 120°$ | One, sixfold rotation (or rotation-inversion) axis |
| Trigonal (or rhombohedral) | Three equal axes, equally inclined: $a = b = c$, $\alpha = \beta = \gamma \neq 90°$ | One, threefold rotation (or rotation-inversion) axis |
| Tetragonal | Three axes at right angles: $a = b \neq c$, $\alpha = \beta = \gamma = 90°$ | One, fourfold rotation (or rotation-inversion) axis |
| Orthorhombic | Three orthogonal unequal axes: $a \neq b \neq c$, $\alpha = \gamma = 90° = \beta$ | Three, perpendicular twofold (or rotation-inversion) axes |
| Monoclinic | Three unequal axes, one pair not orthogonal: $a \neq b \neq c$, $\alpha = \gamma = 90° \neq \beta$ | One, twofold rotation (or rotation-inversion) axis |
| Triclinic | Three unequal axes, none at right angles: $a \neq b \neq c$, $\alpha \neq \beta \neq \gamma \neq 90°$ | None |

## A3.3. INTERPLANAR SPACINGS

The distance $d$ between adjacent planes with Miller indices $(hkl)$ may be found using the following expressions:

Cubic: $\quad \dfrac{1}{d^2} = \dfrac{h^2 + k^2 + l^2}{a^2}$

Tetragonal: $\quad \dfrac{1}{d^2} = \dfrac{h^2 + k^2}{a^2} + \dfrac{l^2}{c^2}$

Hexagonal: $\quad \dfrac{1}{d^2} = \dfrac{4}{3}\left(\dfrac{h^2 + hk + k^2}{a^2}\right) + \dfrac{l^2}{c^2}$

Rhombohedral: $\quad \dfrac{1}{d^2} = \dfrac{(h^2 + k^2 + l^2)\sin^2\alpha + 2(hk + kl + hl)(\cos^2\alpha - \cos\alpha)}{a^2(1 - 3\cos^2\alpha + 2\cos^3\alpha)}$

Orthorhombic: $\quad \dfrac{1}{d^2} = \dfrac{h^2}{a^2} + \dfrac{k^2}{b^2} + \dfrac{l^2}{c^2}$

Monoclinic: $\quad \dfrac{1}{d^2} = \dfrac{1}{\sin^2\beta}\left(\dfrac{h^2}{a^2} + \dfrac{k^2\sin^2\beta}{b^2} + \dfrac{l^2}{c^2} - \dfrac{2hl\cos\beta}{ac}\right)$

Triclinic; $\quad \dfrac{1}{d^2} = \dfrac{1}{V^2}(S_{11}h^2 + S_{22}k^2 + S_{33}l^2 + 2S_{12}hk + 2S_{23}kl + 2S_{13}hl)$

In the equation for triclinic crystals,

$V$ = volume of unit cell (see Section A3.7)

$S_{11} = b^2c^2 \sin^2\alpha$

$S_{22} = a^2c^2 \sin^2\beta$

$S_{33} = a^2b^2 \sin^2\gamma$

$S_{12} = abc^2(\cos\alpha \cos\beta - \cos\gamma)$

$S_{23} = a^2bc(\cos\beta \cos\gamma - \cos\alpha)$

$S_{13} = ab^2c(\cos\gamma \cos\alpha - \cos\beta)$

## A3.4. EXTINCTION CONDITIONS RESULTING FROM SCREW AND GLIDE SYMMETRY

### A3.4.1. Absent Reflections Resulting from the Bravais Lattice

For the various Bravais lattices, the following extinction conditions are imposed on $U_g$. (Additional extinctions may result from the presence of screw and/or glide symmetry elements.) See also Table A3.4.2.

| Lattice | Allowed reflections | Extinctions |
|---|---|---|
| Primitive P | All | None |
| Body centered I | $(h + k + l)$ even | $(h + k + l)$ odd |
| Face centered F | $h, k, l$ all odd or even | $h, k, l$ mixed |
| Base centered C | $h$ and $k$ both odd or even | $h$ and $k$ mixed |

## A3.4.2. Extinctions Due to Screw and Glide Symmetry Elements

| Symmetry element | | Affected reflection | Condition for absence of reflection |
|---|---|---|---|
| 2-fold screw ($2_1$) | $a$ | $h00$ | $h = 2n + 1 =$ odd |
| 4-fold screw ($4_2$) along | $b$ | $0k0$ | $k = 2n + 1$ |
| 6-fold screw ($6_3$) | $c$ | $00l$ | $l = 2n + 1$ |
| 3-fold screw ($3_1, 3_2$) | along $c$ | $00l$ | $l = 3n + 1, 3n + 2,$ i.e., not evenly divisible by 3 |
| 6-fold screw ($6_2, 6_4$) | | | |
| 4-fold screw ($4_1, 4_3$) along $a$ | | $h00$ | $h = 4n + 1, 2,$ or $3$ |
| $b$ | | $0k0$ | $k = 4n + 1, 2,$ or $3$ |
| $c$ | | $00l$ | $l = 4n + 1, 2,$ or $3$ |
| 6-fold screw ($6_1, 6_5$) along $c$ | | $00l$ | $l = 6n + 1, 2, 3, 4,$ or $5$ |
| Glide plane perpendicular to $a$ | | | |
| Translation $b/2$ ($b$ glide) | | $0kl$ | $k = 2n + 1$ |
| $c/2$ ($c$ glide) | | | $l = 2n + 1$ |
| $b/2 + c/2$ ($n$ glide) | | | $k = l = 2n + 1$ |
| $b/4 + c/4$ ($d$ glide) | | | $k + l = 4n + 1, 2,$ or $3$ |
| Glide plane perpendicular to $b$ | | | |
| Translation $a/2$ ($a$ glide) | | $h0l$ | $h = 2n + 1$ |
| $c/2$ ($c$ glide) | | | $l = 3n + 1$ |
| $a/2 + c/2$ ($n$ glide) | | | $h + l = 2n + 1$ |
| $a/4 + c/4$ ($d$ glide) | | | $h + l = 4n + 1, 2,$ or $3$ |
| Glide plane perpendicular to $c$ | | | |
| Translation $a/2$ ($a$ glide) | | $hk0$ | $h = 2n + 1$ |
| $b/2$ ($c$ glide) | | | $k = 2n + 1$ |
| $a/2 + b/2$ ($n$ glide) | | | $h + k = 2n + 1$ |
| $a/4 + b/4$ ($d$ glide) | | | $h + k = 4n + 1, 2,$ or $3$ |

## A3.5. SYMMETRIES IN ZONE-AXIS CBED PATTERNS

The table below shows the symmetries which will be observed in convergent-beam diffraction patterns from crystals of the 32 point groups in the electron beam directions indicated (Buxton et al. 1976)

| Electron Beam Direction | | | | | | | | |
|---|---|---|---|---|---|---|---|---|
| Crystal system | Point group | ⟨111⟩ | ⟨100⟩ | ⟨110⟩ | ⟨uv0⟩ | ⟨uuw⟩ | ⟨uū0w⟩ | ⟨uvtw⟩ |
| Cubic | $m3m$ | $6_R mm_R$ | $4mm1_R$ | $2mm1_R$ | $2_R mm_R$ | $2_R mm_R$ | $2_R$ | |
| | $\bar{4}3m$ | $3m$ | $4_R mm_R$ | $m1_R$ | $m_R$ | $m$ | 1 | |
| | 432 | $3m_R$ | $4m_R m_R$ | $2m_R m_R$ | $m_R$ | $m_R$ | 1 | |
| | $m3$ | $6_R$ | $2mm1_R$ | | $2_R mm_R$ | | $2_R$ | |
| | 23 | 3 | $2m_R m_R$ | | $m_R$ | | 1 | |
| Hexagonal | $6/mmm$ | [0001] | ⟨11$\bar{2}$0⟩ | ⟨1$\bar{1}$00⟩ | ⟨uv0⟩ | ⟨uutw⟩ | ⟨uū0w⟩ | ⟨uvtw⟩ |
| | | $6mm1_R$ | $2mm1_R$ | $2mm1_R$ | $2_R mm_R$ | $2_R mm_R$ | $2_R mm_R$ | $2_R$ |
| | $\bar{6}m2$ | $3m1_R$ | $m1_R$ | $2mm$ | $m$ | $m_R$ | $m$ | 1 |
| | $6mm$ | $6mm$ | $m1_R$ | $m1_R$ | $m_R$ | $m$ | $m$ | 1 |
| | 622 | $6m_R m_R$ | $2m_R m_R$ | $2m_R m_R$ | $2_R mm_R$ | $m_R$ | $m_R$ | 1 |
| | $6/m$ | $61_R$ | | | $m$ | | | $2_R$ |
| | $\bar{6}$ | $31_R$ | | | $m_R$ | | | 1 |
| | 6 | 6 | | | | | | 1 |
| Trigonal | $\bar{3}m$ | [0001] | ⟨11$\bar{2}$0⟩ | ⟨1$\bar{1}$00⟩ | ⟨uv0⟩ | ⟨uutw⟩ | ⟨uū0w⟩ | ⟨uvtw⟩ |
| | | $6_R mm_R$ | $21_R$ | | | | $2_R mm_R$ | $2_R$ |
| | $3m$ | $3m$ | $1_R$ | | | | $m$ | 1 |
| | 32 | $3m_R$ | 2 | | | | $m_R$ | 1 |
| | $\bar{3}$ | $6_R$ | | | | | $2_R$ | $2_R$ |
| | 3 | 3 | | | | | 1 | 1 |

(continued)

*Electron Beam Direction*

| Crystal system | Point group | ⟨111⟩ | ⟨100⟩ | ⟨110⟩ | ⟨uv0⟩ | ⟨uuw⟩ | ⟨uvw⟩ |
|---|---|---|---|---|---|---|---|
| Tetragonal | | [001] | ⟨100⟩ | ⟨110⟩ | [u0w] | [uv0] | [uuw] |
| | $4/mmm$ | $4mm1_R$ | $2mm1_R$ | $2mm1_R$ | $2_Rmm_R$ | $2_Rmm_R$ | $2_Rmm_R$ |
| | $\bar{4}2m$ | $4_Rmm_R$ | $2m_Rm_R$ | $m1_R$ | $m_R$ | $m_R$ | $m$ |
| | $4mm$ | $4mm$ | $m1_R$ | $m1_R$ | $m$ | $m_R$ | $m$ |
| | $422$ | $4m_Rm_R$ | $2m_Rm_R$ | $2m_Rm_R$ | $m_R$ | $2_Rmm_R$ | $m_R$ |
| | $4/m$ | $41_R$ | | | | $m_R$ | |
| | $\bar{4}$ | $4_R$ | | | | $m_R$ | |
| | $4$ | $4$ | | | | | |
| Orthorhombic | | [001] | ⟨100⟩ | ⟨uv0⟩ | [uv0] | [uvw] | |
| | $mmm$ | $2mm1_R$ | $2mm1_R$ | $2_Rmm_R$ | $2_Rmm_R$ | $2_R$ | |
| | $mm2$ | $2mm$ | $m1_R$ | $m$ | $m_R$ | $1$ | |
| | $222$ | $2m_Rm_R$ | $2m_Rm_R$ | $m_R$ | $m_R$ | $1$ | |
| Monoclinic | | [010] | [u0w] | | | ⟨uvw⟩ | |
| | $2/m$ | $21_R$ | $2_Rmm_R$ | | | $2_R$ | |
| | $m$ | $1_R$ | $m$ | | | $1$ | |
| | $2$ | $2$ | $m_R$ | | | $1$ | |
| Triclinic | | | | | | ⟨uvw⟩ | |
| | $\bar{1}$ | | | | | $2_R$ | |
| | $1$ | | | | | $1$ | |

## A3.6. HEIGHT H OF HOLZ PLANES FOR ZONE (u, v, w)

The information in this section is taken from Steeds (1979).

| Crystal class | $H$ in Eq. (2.13) |
|---|---|
| Monoclinic | $(u^2a^2 + v^2b^2 + w^2c^2 + 2uwac \cos \beta)^{-1/2}$ |
| Orthorhombic | $(u^2c^2 + v^2b^2 + w^2c^2)^{-1/2}$ |
| Hexagonal and Rhombohedral [with $D^2 = (\frac{2}{3})(c/a)^2$] | $[(3a^2/2)(u^2 + v^2 + t^2 + Dw^2)]^{-1/2}$ |
| Tetragonal | $[a^2(u^2 + v^2) + c^2w^2]^{-1/2}$ |
| Cubic | $a^{-1}(u^2 + v^2 + w^2)^{-1/2}$ |

## A3.7. THE USE OF A METRIC MATRIX FOR CRYSTALLOGRAPHIC CALCULATIONS

The crystallographic quantities required in a computer program can be derived most simply from a metric matrix **G** defined (for any crystal system) by Boisen and Gibbs (1985).

$$\mathbf{G} = \begin{pmatrix} \mathbf{a} \cdot \mathbf{a} & \mathbf{a} \cdot \mathbf{b} & \mathbf{a} \cdot \mathbf{c} \\ \mathbf{b} \cdot \mathbf{a} & \mathbf{b} \cdot \mathbf{b} & \mathbf{b} \cdot \mathbf{c} \\ \mathbf{c} \cdot \mathbf{a} & \mathbf{c} \cdot \mathbf{b} & \mathbf{c} \cdot \mathbf{c} \end{pmatrix} = \begin{pmatrix} a^2 & ab \cos \gamma & ac \cos \beta \\ ab \cos \gamma & b^2 & bc \cos \alpha \\ ac \cos \beta & bc \cos \alpha & c^2 \end{pmatrix}$$

The inverse metric matrix can be shown to be

$$\mathbf{G}^{-1} = \begin{pmatrix} \mathbf{a}^* \cdot \mathbf{a}^* & \mathbf{a}^* \cdot \mathbf{b}^* & \mathbf{a}^* \cdot \mathbf{c}^* \\ \mathbf{b}^* \cdot \mathbf{a}^* & \mathbf{b}^* \cdot \mathbf{b}^* & \mathbf{b}^* \cdot \mathbf{c}^* \\ \mathbf{c}^* \cdot \mathbf{a}^* & \mathbf{c}^* \cdot \mathbf{b}^* & \mathbf{c}^* \cdot \mathbf{c}^* \end{pmatrix}$$

For example, the $d$-spacing of $(hkl)$ planes is $d = 1/s$, where $s$ is the length of reciprocal lattice vector $\mathbf{g} = h\mathbf{a}^* + k\mathbf{b}^* + l\mathbf{c}^*$. In terms of the metric matrix,

$$s^2 = (hkl)\mathbf{G}^{-1} \begin{pmatrix} h \\ k \\ l \end{pmatrix}$$

The dot product of two real space vectors $\mathbf{r}_1 = n_1\mathbf{a} + m_1\mathbf{b} + p_1\mathbf{c}$ and

$\mathbf{r}_2 = n_2\mathbf{a} + m_2\mathbf{b} + p_2\mathbf{c}$ is

$$\mathbf{r}_1 \cdot \mathbf{r}_2 = (n_1 \; m_1 \; p_1)\mathbf{G}\begin{pmatrix} n_2 \\ m_2 \\ p_2 \end{pmatrix}$$

The dot product of two reciprocal space vectors $\mathbf{r}_1^* = h_1\mathbf{a}^* + k_1\mathbf{b}^* + l_1\mathbf{c}^*$ and $\mathbf{r}_2^* = h_2\mathbf{a}^* + k_2\mathbf{b}^* + l_2\mathbf{c}^*$ is

$$\mathbf{r}_1^* \cdot \mathbf{r}_2^* = (h_1 \; k_1 \; l_1)\mathbf{G}^{-1}\begin{pmatrix} h_2 \\ k_2 \\ l_2 \end{pmatrix}$$

The transformation between real space and reciprocal space for the vector $\mathbf{r} = n\mathbf{a} + m\mathbf{b} + p\mathbf{c} = h\mathbf{a}^* + k\mathbf{b}^* + l\mathbf{c}^*$ is given by

$$\begin{pmatrix} h \\ k \\ l \end{pmatrix} = \mathbf{G}\begin{pmatrix} n \\ m \\ p \end{pmatrix} \quad \text{or} \quad \begin{pmatrix} n \\ m \\ p \end{pmatrix} = \mathbf{G}^{-1}\begin{pmatrix} h \\ k \\ l \end{pmatrix}$$

The volume of the unit cell is given by

$$V = \sqrt{\det(\mathbf{G})} = abc(1 - \cos^2\alpha - \cos^2\beta - \cos^2\gamma + 2\cos\alpha\cos\beta\cos\gamma)^{1/2} = 1/V^*$$

# APPENDIX 4. INDEXED DIFFRACTION PATTERNS WITH HOLZ

The diagrams in Figures A4.1–A4.4 show indexed diffraction patterns for commonly encountered structures. By successive addition of the basis vectors given (in three dimensions) it is possible to construct the entire three-dimensional lattice as an aid to indexing lattice points in the outer HOLZ ring. The cross indicates one HOLZ point, from which the entire three-dimensional lattice may be constructed by translating the origin of the ZOLZ lattice to the cross.

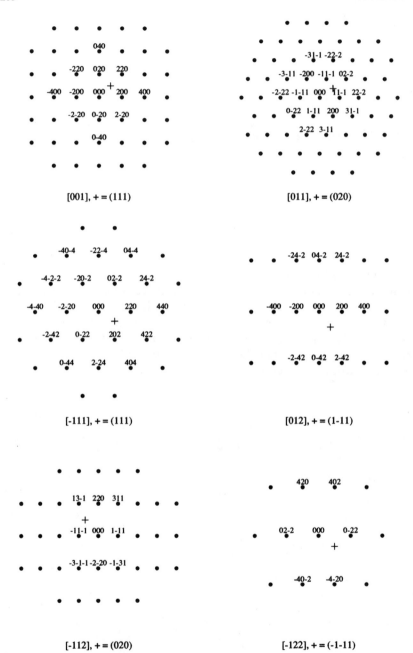

*Figure A4.1.* Twelve low-index zone-axis diffraction patterns for face-centered cubic crystals.

# INDEXED DIFFRACTION PATTERNS WITH HOLZ

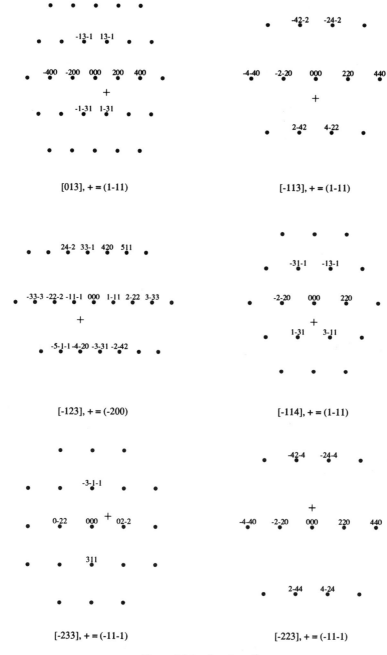

Figure A4.1.  (continued)

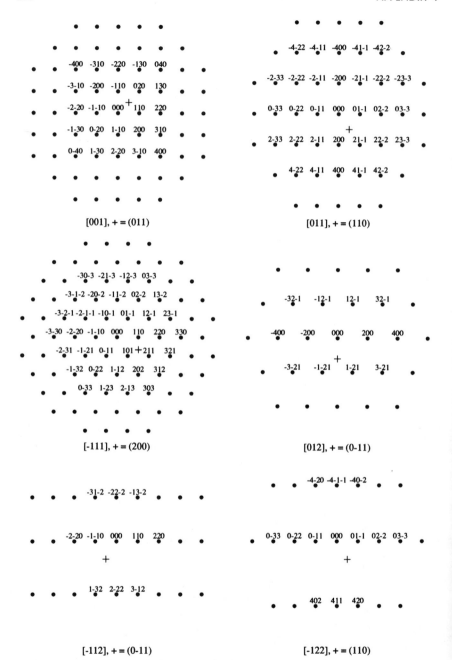

Figure A4.2. Twelve low-index zone-axis diffraction patterns for body-centered cubic crystals.

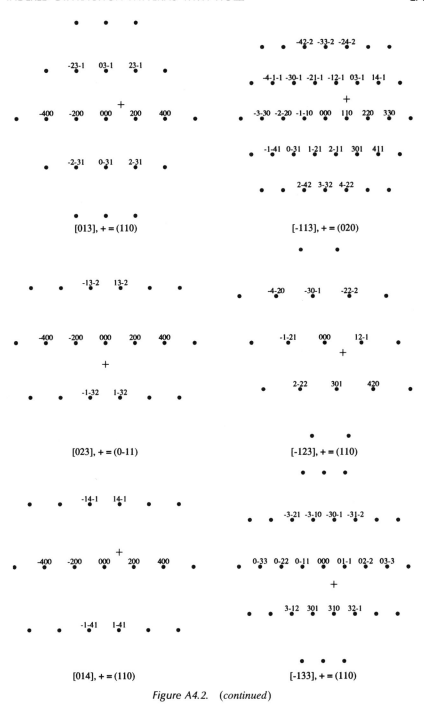

Figure A4.2. (continued)

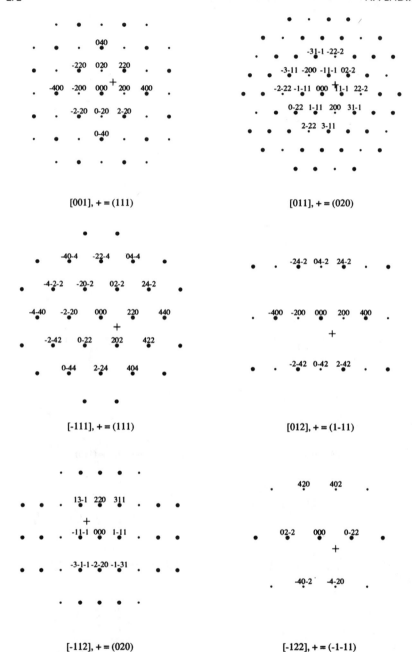

*Figure A4.3.* Twelve low-index zone-axis diffraction patterns for the diamond lattice. The reflections indicated by small dots are kinematically forbidden if the atoms are spherical.

# INDEXED DIFFRACTION PATTERNS WITH HOLZ

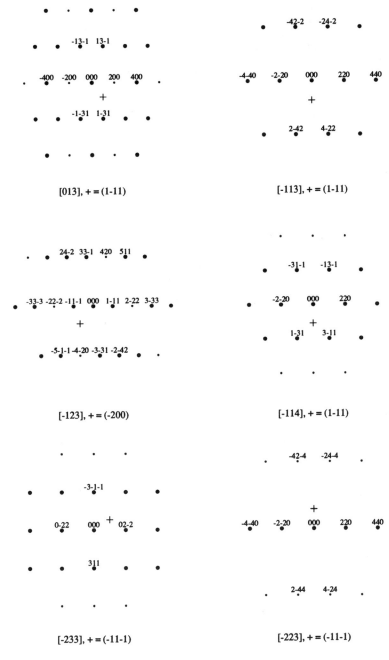

Figure A4.3. (continued)

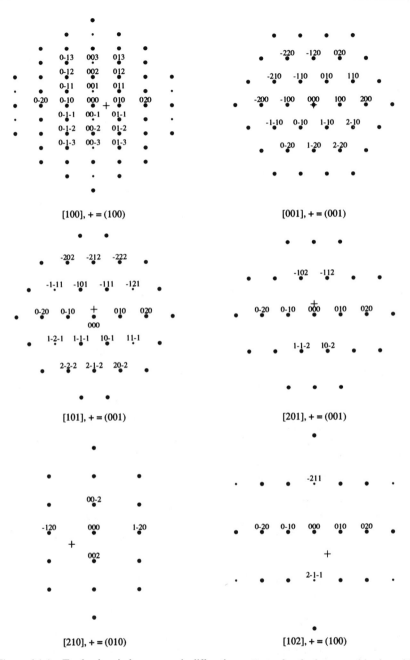

*Figure A4.4.* Twelve low-index zone-axis diffraction patterns for the hexagonal lattice with $c/a = 1.633$.

# INDEXED DIFFRACTION PATTERNS WITH HOLZ 275

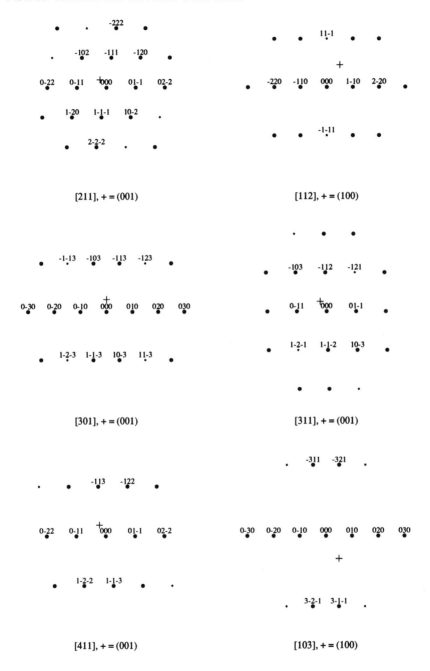

Figure A4.4. (continued)

# Appendix 5. Computer Programs

## A5.1. PLOTTING HOLZ LINES

The following FORTRAN program generates a pattern of HOLZ lines in the central disk of a CBED pattern for a given crystal structure, accelerating voltage, cell constants, and orientation. It is based on Eq. (2.15). A comparison of the results of this program with those of several others has been published (Eades, 1992). The results are based on the kinematic theory of electron diffraction and take no account of dynamical effects, which can cause small shifts in the lines. Because of this, the results of this program may be in error by several kilovolts if used to calibrate the accelerating voltage of an electron microscope using a HOLZ pattern recorded at a high-symmetry zone axis. As discussed in Sections 3.8 and 5.4, these "dynamical shifts" are minimized away from zone-axis orientations.

The program produces an output file in the "Postscript" page description language, suitable for most laser printers.

The program consists of a main program and several subroutines. They are divided into four groups according to their function:

1. HOLZ-line calculation and plotting: HOLZ, CODING, GETLIN, XXY, and YXY.
2. Automatic beam generation subroutines: BEAM_SETUP, BEAM_CHECK, EXCIT, and SORTIN.
3. Crystallographic calculation subroutines based on Appendix 3.7: METRIC, ANGLE, SCALE, CROSS, TRANS, SUM, and INVERT.
4. Postscript plotting subroutines: INITPS, WINDOW, NWPATH, SEGRAY, SETLNW, SHOW, SETDSH, CIRCLE, DRAWLN, FONT, TEXT, and TEXTP.

Test input data is also supplied—this should generate Fig. 2.11b.

```
C
C THIS PROGRAM PLOTS GEOMETRIC CBED AND HOLZ/KIKUCHI LINE PATTERNS
C WITH POSTSCRIPT GRAPHICS
C BY J. M. ZUO, SEPT., 1991
C
      PROGRAM HOLZ

      CHARACTER FILNAM*20,LISNAM*20,PSNAME*20,TITLE*80,FONAM*30
      CHARACTER DISKW*4,INDEXW*4,CENTER*4,TITLEW*4,AUTO*4,LINDEX*4
      CHARACTER DIR*1,POS*1,STRING*80,GREYL*4

      PARAMETER (NM=500)
      INTEGER   HKL(3,NM)
      PARAMETER (NJ=10)
      CHARACTER*4  LABEL(NJ)
      REAL      ZT(NJ),CX(NJ),AX(4,NJ),BX(4,NJ),AF(NJ)
      PARAMETER (NK=500)
      DIMENSION ITYPE(NK),XP(3,NK),DW(NK)

      REAL*8       GMX(3,3),GMXR(3,3)
      COMMON/METRIC/ GMX,GMXR

      REAL      CELL(6),TILT(3),ZONE(3),GH(3),GG(3),GL(3),GK(3),
     *          GHX(3),GGX(3),DSTAR(1000)
      INTEGER   INDZ(3),GX(3),GY(3),GZ(3)
      REAL*8    KV,PI,TWOPI,TEMP,CN,SN,S2,A(3,3),AT(3,3),BIGK,
     *          WAVEL,DS,RAD,RATIO,FIE,THETA,GAMMA,BIGKZ
      DATA      PI/3.141592654/,TWOPI/6.283185307/,
     *          NR,NW/10,12/,DELTA/1.0E-7/

      INTEGER   LDSTAT(NM),LESTAT(NM),LEVEL(NM)
      REAL      XJ(NM),YJ(NM),ZJ(NM),K0,C0,XDL(2,NM),YDL(2,NM),
     *          XEL(2,NM),YEL(2,NM),XLP(2),YLP(2),AMP2(NM),GREY(5)

      PRINT *,'ENTER INPUT DATA FILENAME'
      READ (5,100) FILNAM
100   FORMAT (80A)
      PRINT *,'ENTER OUTPUT LISTING FILENAME'
      READ (5,100) LISNAM
      PRINT *,'ENTER OUTPUT POSTSCRIPT FILENAME'
      READ (5,100) PSNAME
      OPEN (UNIT=NR,FILE=FILNAM,STATUS='OLD',FORM='FORMATTED')
      OPEN (UNIT=NW,FILE=LISNAM,STATUS='NEW',FORM='FORMATTED')
C
C TEXT: 80 CHARACTER TITLE
C
      READ (NR,100) TITLE
      WRITE (NW,120) TITLE
120   FORMAT (/,' OUTPUT LISTING OF BLOCH WAVE PROGRAM BY ZUO, V3.0',
     *        /,X,80A1)
C
C KV    INCIDENT ELECTRON ENERGY IN UNIT OF KeV
C INDZ(3) NEAREST ZONE AXIS INDEX
C TILT(3) INCIDENT BEAM DIRECTION AT THE CENTER OF ZERO DISK
C      THE INCIDENT BEAM DIRECTION IS DEFINED BY THE COORDINATES OF
C        THE CENTER OF THE LAUE CIRCLE IN UNITS OF THE RECIPROCAL LATTICE
C CL    CAMERA LENGTH
C
      READ (NR,*) KV,(INDZ(I),I=1,3),(TILT(J),J=1,3),CL
      TEMP=KV*(1.0+0.97846707E-03*KV)
      WAVEL=0.3878314/DSQRT(TEMP)
      WRITE (NW,140) KV,WAVEL,CL
140   FORMAT (/,' INCIDENT ELECTRON BEAM HIGH VOLTAGE ',F12.6,' KV',
     *        /,'                         WAVELENGTH ',F12.6,' A',
     *        /,'                         CAMERA LENGTH ',F12.6,' MM')
      WRITE (NW,160) (INDZ(I),I=1,3)
```

```
160     FORMAT (/,'                         NEAREST ZONE AXIS ',3I4)
        WRITE (NW,180) (TILT(J),J=1,3)
180     FORMAT (/,' THE INCIDENT BEAM DIRECTION : (',3F12.6,' )')
C
C CELL(6) CRYSTAL UNIT CELL PARAMETERS IN THE ORDER
C       a, b, c, alpha, beta, and gamma
C
        READ (NR,*) (CELL(I),I=1,6)
        WRITE (NW,220) (CELL(I),I=1,6)
220     FORMAT (/,' THE UNIT CELL IS DEFINED BY a = ',F9.5,' Angstrom',
       *          /,'                             b = ',F9.5,' Angstrom',
       *          /,'                             c = ',F9.5,' Angstrom',
       *          /,'                         alpha = ',F9.5,' Degree',
       *          /,'                          beta = ',F9.5,' Degree',
       *          /,'                         gamma = ',F9.5,' Degree')
        DO I=4,6
          CELL(I)=COS(PI*CELL(I)/180.0)
        END DO
        CALL METRIC(CELL)
        WRITE (NW,*) 'METRIC MATRIX'
        DO I=1,3
          WRITE (NW,210) (GMX(I,J),J=1,3)
        END DO
210     FORMAT (1X,3E15.7)
        WRITE (NW,*) 'INVERSE METRIC MATRIX G-1'
        DO I=1,3
          WRITE (NW,210) (GMXR(I,J),J=1,3)
        END DO
        DO I=1,3
          ZONE(I)=FLOAT(INDZ(I))
        END DO
        CALL SCALE(0,ZONE,TEMP)
        DO I=1,3
          ZONE(I)=ZONE(I)/TEMP
        END DO
C
C NATOMS  NUMBER OF ATOMS IN THE UNIT CELL
C NTYPE   NUMBER OF TYPES OF ATOM IN THE UNIT CELL
C
        READ (NR,*) NATOMS,NTYPE
        WRITE (NW,240) NATOMS,NTYPE
240     FORMAT (/,' THERE ARE TOTAL',I4,' ATOMS AND',I3,' SPECIES')
        WRITE (NW,260)
260     FORMAT (/,' THE ATOMS, THEIR X-RAY SCATTERING FACTOR PARAMETERS'
       *         /,' AND THEIR DEBYE-WALLER FACTOR')
        DO I=1,NTYPE
C
C LABEL       ATOMIC SYMBOL
C
        READ (NR,110) LABEL(I)
110     FORMAT (A4)
C
C ZT          ATOMIC NUMBER
C AX,BX,CX    9 GAUSSIAN FITTING PARAMETERS FROM ATOMIC X-RAY SCATTERING
C             FACTORS
C
        READ (NR,*) ZT(I),(AX(J,I),BX(J,I),J=1,4),CX(I)
        IF (ZT(I).EQ.0.0) THEN
          ZT(I)=CX(I)
          DO J=1,4
            ZT(I)=ZT(I)+AX(J,I)
          END DO
        END IF
        WRITE (NW,280) I,LABEL(I),ZT(I),(AX(J,I),BX(J,I),J=1,4),
       *               CX(I)
280     FORMAT (/,X,I2,2X,A4,' Z=',F6.2,/,9F8.4)
        END DO
```

```
C
C ITYPE          THE SEQUENTIAL NUMBER ASSIGNED TO THE ATOMS IN THE PREVIOUS
C               X-RAY SCATTERING FACTOR INPUT
C XP(3)          FRACTIONAL ATOMIC COORDINATES
C DW             DEBYE-WALLER FACTOR OF THE ATOM
C
       READ (NR,*) (ITYPE(I),(XP(J,I),J=1,3),DW(I),I=1,NATOMS)
       WRITE (NW,300)
       WRITE (NW,320) (ITYPE(I),(XP(J,I),J=1,3),DW(I),I=1,NATOMS)
300    FORMAT (/,' THE ATOMS TYPE, THEIR COORDINATES AND D-W FACTOR')
320    FORMAT (2(X,I2,3F9.5,F6.3))
       READ (NR,*) (GX(I),I=1,3),(GY(I),I=1,3),(GZ(I),I=1,3)
       WRITE (NW,200) (GX(I),I=1,3),(GY(I),I=1,3),(GZ(I),I=1,3)
200    FORMAT (/,' BASE G VECTOR ALONG X DIRECTION ',3I4,
      *           /,'                         Y DIRECTION ',3I4,
      *           /,'                         Z DIRECTION ',3I4)
       READ (NR,100) AUTO
       IF (AUTO(1:3).EQ.'YES') THEN
C
C IHOLZ NUMBER OF ZONES TO BE INCLUDED
C SGMAX THE MAXIMUM EXCITATION ERROR TO BE CONSIDERED
C
       READ (NR,*) IHOLZ,SGMAX
C
C SETUP THE BEAMS
C
       CALL BEAM_SETUP(NBEAMS,HKL,GX,GY,GZ,ZONE,IHOLZ,TILT,
      *                SGMAX,WAVEL,1)
       ELSE
         READ (NR,*) NBEAMS
         READ (NR,*) ((HKL(I,J),I=1,3),J=1,NBEAMS)
       END IF
       WRITE (NW,420) NBEAMS
420    FORMAT (/,' TOTAL NUMBER OF BEAMS INCLUDED',I5)
       WRITE (NW,440) ((HKL(I,J),I=1,3),J=1,NBEAMS)
440    FORMAT (1X,4(3I4,','),3I4)
C
C CONTROL PARAMETER
C CENTER(1:3) = 'YES' CENTER THE ZERO BEAM, OTHERWISE NO
C
       READ (NR,100) CENTER
C
C CONTROL PARAMETER
C DISKW(1:4) = 'DISK' DRAW CBED DISK ONLY
C              'LINE' DRAW HOLZ/KIKUCHI LINE ONLY
C              'BOTH' DRAW BOTH CBED DISK AND HOLZ/KIKUCHI LINES
C
       READ (NR,100) DISKW
       IF (DISKW(1:4).EQ.'DISK'.OR.DISKW(1:4).EQ.'BOTH') THEN
C
C DISKS: CBED DISK SIZE
C INDEXW = 'YES' INDEX CBED DISKS
C
         READ (NR,*) DISKS
         READ (NR,100) INDEXW
       END IF
C
C GREYL = 'YES' USE DIFFERENT LINE WIDTH TO SHOW HOLZ LINES
C FIVE LINE WIDTHES ARE USED TO REPRESENT INTENSITY FROM 1 TO 5
C IN LOG10 SCALE
C
       IF (DISKW(1:4).EQ.'BOTH'.OR.DISKW(1:4).EQ.'LINE') THEN
         READ (NR,100) GREYL
       END IF
C
C LINDEX = 'YES' INDEX THE LINES
```

```
C
      READ (NR,100) LINDEX
C
C *END OF INPUT*
C
      VOL=1.0-CELL(4)*CELL(4)-CELL(5)*CELL(5)-CELL(6)*CELL(6)+
     *    2.0*CELL(4)*CELL(5)*CELL(6)
      VOL=SQRT(VOL)*CELL(1)*CELL(2)*CELL(3)
      RATIO=(1+1.9569341E-03*KV)/(PI*VOL)
C
C CALCULATE THE U(000) AND K VECTOR HERE
C
      DO I=1,NTYPE
        AF(I)=0.0
        DO J=1,4
          AF(I)=AF(I)+AX(J,I)*BX(J,I)
        END DO
        AF(I)=0.023933754*AF(I)
      END DO
      U0=0.0
      DO I=1,NATOMS
        J=ITYPE(I)
        U0=U0+AF(J)
      END DO
      U0=U0*RATIO
      WRITE (NW,580) U0
580   FORMAT (/,' THE MEAN CRYSTAL POTENTIAL IS
      TEMP=1.0/(WAVEL*WAVEL)+U0
      BIGK=DSQRT(TEMP)
      CALL SCALE(1,TILT,TEMP)
      BIGKZ=DSQRT(BIGK**2-TEMP**2)
C
C FIND A NEW BASIS WITH ZONE AXIS AS THE Z-AXIS
C
      DO I=1,3
        GG(I)=GX(I)
        GH(I)=GY(I)
      END DO
      CALL SCALE(1,GG,TEMP)
      BASEL=TEMP
      CALL TRANS(GMX,ZONE,GL)
      CALL SCALE(1,GL,TEMP)
      DO I=1,3
        GL(I)=GL(I)*BASEL/TEMP
      END DO
      CALL ANGLE(1,GG,GH,TEMP)
      IF (ABS(TEMP).LT.1.0E-7) THEN
        CALL SCALE(1,GH,TEMP)
        DO I=1,3
          GH(I)=GH(I)*BASEL/TEMP
        END DO
      ELSE
        GK(1)=GL(2)*GG(3)-GL(3)*GG(2)
        GK(2)=GL(3)*GG(1)-GL(1)*GG(3)
        GK(3)=GL(1)*GG(2)-GL(2)*GG(1)
        CALL TRANS(GMX,GK,GH)
        CALL SCALE(1,GH,TEMP)
        DO I=1,3
          GH(I)=GH(I)*BASEL/TEMP
        END DO
      END IF
      WRITE (NW,600) GG,GH,GL
600   FORMAT (/,' THE NEW CORDINATES, X: ',3F12.8,
     *         /,'                     Y: ',3F12.8,
     *         /,'                     Z: ',3F12.8)
```

```
C
C FIND THE TRANSFORMATION MATRIX
C
      DO I=1,3
        A(I,1)=GG(I)
        A(I,2)=GH(I)
        A(I,3)=GL(I)
      END DO
      CALL INVERT(A,AT)
      WRITE (NW,602) ( (AT(I,J),J=1,3) ,I=1,3)
602   FORMAT (/,' THE PROJECTION MATRIX ',3(/,1X,3E12.5))
C
C   THE COORDINATES OF G IN THE NEW COORDINATES
C
      SCALER=CL*BASEL/BIGK
      WRITE (NW,605) SCALER
605   FORMAT(/,' THE PROJECTED LENGTH OF BASE VECTOR GX ',E12.5,' MM')
      DO I=2,NBEAMS
        GGX(1)=HKL(1,I)
        GGX(2)=HKL(2,I)
        GGX(3)=HKL(3,I)
        CALL TRANS(AT,GGX,GHX)
        XJ(I)=GHX(1)*SCALER
        YJ(I)=GHX(2)*SCALER
      END DO
C
C SETUP THE BOUNDARY
C
      CALL TRANS(AT,TILT,GHX)
      XC=GHX(1)*SCALER
      YC=GHX(2)*SCALER
      WRITE (NW,610) XC,YC
610   FORMAT (/,' THE CENTER OF THE LAUE CIRCLE ',2E12.5)
C
C DISPLAY WINDOW SIZE 150mmX150mm
C
      IF (CENTER(1:3).EQ.'YES') THEN
        XMN=-75.0+XC
        XMX=75.0+XC
        YMN=-75.0+YC
        YMX=75.0+YC
      ELSE
        XMN=-75.0
        XMX=75.0
        YMN=-75.0
        YMX=75.0
      END IF
C
C CALCULATE THE STRUCTURE FACTOR HERE. SETUP KIKUCHI LINE LEVEL
C
      WRITE (NW,460)
460   FORMAT(/,' hkl AND STRUCTURE FACTOR U(hkl) AMPLITUDE AND PHASE')
      DO I=2,NBEAMS
        GGX(1)=HKL(1,I)
        GGX(2)=HKL(2,I)
        GGX(3)=HKL(3,I)
        S2=0.25*SUM(GMXR,GGX,GGX)
        DO K=1,NTYPE
          AF(K)=CX(K)
          DO M=1,4
            AF(K)=AF(K)+AX(M,K)*EXP(-BX(M,K)*S2)
          END DO
          AF(K)=0.023933754*(ZT(K)-AF(K))/S2
        END DO
        SFRE=0.0
        SFIM=0.0
```

```
      DO M=1,NATOMS
        TEMP=TWOPI*(GGX(1)*XP(1,M)+GGX(2)*XP(2,M)+GGX(3)*XP(3,M))
        CN=DCOS(TEMP)
        SN=DSIN(TEMP)
        TEMP=EXP(-DW(M)*S2)
        SFRE=SFRE+CN*AF(ITYPE(M))*TEMP
        SFIM=SFIM+SN*AF(ITYPE(M))*TEMP
      END DO
      SFRE=SFRE*RATIO
      SFIM=SFIM*RATIO
C
C STRUCTURE FACTOR LISTING HERE
C
      TEMP1=SFRE
      TEMP2=SFIM
      TEMP1=SQRT(TEMP1*TEMP1+TEMP2*TEMP2)
      TEMP2=180.0*ASIN(TEMP2/TEMP1)/PI
      IF (SFRE.LT.0) TEMP2=SIGN(180.0-ABS(TEMP2),TEMP2)
      SFRE=TEMP1
      SFIM=TEMP2
      WRITE (NW,480) (IFIX(GGX(M)),M=1,3),SFRE,SFIM
480   FORMAT (1X,3I5,E15.6,F13.6)
      AMP2(I)=SFRE*SFRE
    END DO
    AMAX=0.0
    DO I=2,NBEAMS
      IF (AMAX.LT.AMP2(I)) AMAX=AMP2(I)
    END DO
    DO I=2,NBEAMS
      AMP=AMP2(I)/AMAX
      IF (AMP.GT.0) THEN
        AMP=LOG10(AMP)
      ELSE
        AMP=-100
      END IF
      LEVEL(I)=-NINT(AMP)
    END DO
C
C SETUP PLOTTING
C
    CALL INITPS(PSNAME)
    CALL WINDOW(XMN,XMX,YMN,YMX,1.29724,7.20276,3.79725,9.70276)
    CALL SETDSH(1)
    FONAM='Times-Roman'
    CALL FONT(FONAM,8)
    IF (DISKW(1:4).EQ.'DISK') GOTO 770
C
C PLOT K-LINES
C
    DO 500 I=2,NBEAMS
    GHX(1)=HKL(1,I)
    GHX(2)=HKL(2,I)
      GHX(3)=HKL(3,I)
    DS=ZONE(1)*GHX(1)+ZONE(2)*GHX(2)+ZONE(3)*GHX(3)
    TEMP=SUM(GMXR,GHX,GHX)
    C0=TEMP-DS*2.0*BIGKZ
    SN=SUM(GMXR,GHX,GH)
    CN=SUM(GMXR,GHX,GG)
    IF (DABS(SN).GT.1.0E-7) THEN
        SLOPP=-CN/SN
      C0=-C0/(2.0*SN)
    ELSE
      SLOPP=1.0E+20
      C0=C0/(2.0*CN)*SLOPP
    END IF
```

```
C
C   C0 AND K0 : PARAMETERS FOR LINE EQUATION Y=K0*X+C0
C
        K0=SLOPP
        C0=C0*SCALER
        CALL GETLIN(C0,K0,XMN,XMX,YMN,YMX,XLP,YLP,LSTAT)
        LDSTAT(I)=LSTAT
        IF (LSTAT.GT.0) THEN
           XDL(1,I)=XLP(1)
           XDL(2,I)=XLP(2)
           YDL(1,I)=YLP(1)
           YDL(2,I)=YLP(2)
           WRITE (NW,620) GHX
620        FORMAT (/,' FOR BEAM ',3F5.0)
           WRITE (NW,630) DS,CN,SN,K0,C0
630        FORMAT ( ' D* ',F12.8,' Gx and Gy',2E13.6,
     *             /,' LINE EQUATION : y = ',E12.5,'x + ',E12.5)
        END IF
C
C THE EXCESS LINE
C
        C0=C0+YJ(I)-SLOPP*XJ(I)
        CALL GETLIN(C0,K0,XMN,XMX,YMN,YMX,XLP,YLP,LSTAT)
        LESTAT(I)=LSTAT
        IF (LSTAT.GT.0) THEN
           XEL(1,I)=XLP(1)
           XEL(2,I)=XLP(2)
           YEL(1,I)=YLP(1)
           YEL(2,I)=YLP(2)
        END IF
500     CONTINUE
C
C PLOT HOLZ/KIKUCHI LINES
C
        IF (GREYL(1:3).EQ.'YES') THEN
           GREY(1)=1.0
           GREY(2)=0.75
           GREY(3)=0.5
           GREY(4)=0.25
           GREY(5)=0.0
        ELSE
           DO I=1,5
              GREY(I)=0.5
           END DO
        END IF
        DO L=5,1,-1
           CALL SETLNW(GREY(L))
           DO I=2,NBEAMS
              IF (LEVEL(I).EQ.L-1) THEN
                IF (LDSTAT(I).GT.0) THEN
                CALL DRAWLN(XDL(1,I),YDL(1,I),XDL(2,I),YDL(2,I))
                END IF
                IF (LESTAT(I).GT.0) THEN
                CALL DRAWLN(XEL(1,I),YEL(1,I),XEL(2,I),YEL(2,I))
                END IF
              END IF
           END DO
        END DO
C
C INDEX THE LINES HERE
C
        CALL NWPATH
        IF (LINDEX(1:3).EQ.'YES') THEN
           IAL=-1
           DO I=2,NBEAMS
```

```
            IF (LEVEL(I).LT.5) THEN
              GX(1)=HKL(1,I)
              GX(2)=HKL(2,I)
              GX(3)=HKL(3,I)
              CALL CODING(GX,STRING,L)
              IF (LDSTAT(I).GT.0.OR.LESTAT(I).GT.0) THEN
                IK=1
                IF (IAL.GT.0) IK=2
              IAL=-IAL
              END IF
              IF (LDSTAT(I).GT.0) THEN
                XT=XDL(IK,I)
              YT=YDL(IK,I)
              POS='L'
              IF (YT.EQ.YMN.OR.YT.EQ.YMX) THEN
                IF (YT.EQ.YMN) POS='R'
                CALL TEXTP(STRING,L,XT,YT,POS)
              ELSE
                    IF (XT.EQ.XMN) POS='R'
                    CALL TEXT(STRING,L,XT,YT,POS)
              END IF
              END IF
              IF (LESTAT(I).GT.0) THEN
                XT=XEL(IK,I)
              YT=YEL(IK,I)
              POS='L'
              IF (YT.EQ.YMN.OR.YT.EQ.YMX) THEN
                IF (YT.EQ.YMN) POS='R'
                CALL TEXTP(STRING,L,XT,YT,POS)
              ELSE
                    IF (XT.EQ.XMN) POS='R'
                    CALL TEXT(STRING,L,XT,YT,POS)
              END IF
              END IF
            END IF
          END DO
        END IF
C
C PLOT CBED DISKS HERE
C
770     IF (DISKW(1:4).EQ.'DISK'.OR.DISKW(1:4).EQ.'BOTH') THEN
          CALL SETLNW(0.5)
          CALL NWPATH
          DISKS=SCALER*DISKS
          LEVEL(1)=0
          DO I=1,NBEAMS
            XJ(I)=XJ(I)+XC
            YJ(I)=YJ(I)+YC
            IF (XJ(I).GT.XMN.AND.XJ(I).LT.XMX.AND.
     *          YJ(I).GT.YMN.AND.YJ(I).LT.YMX) THEN
              CALL CIRCLE(XJ(I),YJ(I),DISKS)
            END IF
          END DO
C
C INDEX HERE
C
        IF (INDEXW(1:3).EQ.'YES') THEN
          CALL NWPATH
          DO I=1,NBEAMS
            GX(1)=HKL(1,I)
            GX(2)=HKL(2,I)
            GX(3)=HKL(3,I)
            IF (XJ(I).GT.XMN.AND.XJ(I).LT.XMX.AND.
     *          YJ(I).GT.YMN.AND.YJ(I).LT.YMX) THEN
              CALL CODING(GX,STRING,L)
                CALL TEXT(STRING,L,XJ(I),YJ(I),'C')
            END IF
          END DO
        END IF
      END IF
```

APPENDIX 5

```
C
C CAPTIONS
C
      CALL WINDOW(0.0,21.59,0.0,27.94,0.0,8.5,0.0,11.0)
      CALL FONT(FONAM,16)
      STRING(1:8)='TITLE : '
      STRING(9:80)=TITLE(1:71)
      CALL TEXT(STRING,78,3.295,6.5,'L')
      ENCODE(20,800,STRING) INDZ(1),INDZ(2),INDZ(3)
  800 FORMAT ('ZONE : [',I3,',',I3,',',I3,']')
      CALL TEXT(STRING,20,3.295,5.85,'L')
      ENCODE(18,820,STRING) KV
  820 FORMAT ('H.V. : ',F8.3,' KV')
      CALL TEXT(STRING,18,3.295,5.2,'L')
      ENCODE(18,810,STRING) CL
  810 FORMAT ('C.L. : ',F8.2,' mm')
      CALL TEXT(STRING,18,3.295,4.55,'L')
      ENCODE(31,840,STRING) TILT(1),TILT(2),TILT(3)
  840 FORMAT ('TILT : ',3F8.4)
      CALL TEXT(STRING,31,3.295,3.90,'L')
      STRING(1:18)='INPUT DATA FILE : '
      STRING(19:38)=FILNAM(1:20)
      CALL FONT(FONAM,12)
      CALL TEXT(STRING,38,3.295,2.0,'L')
      CALL SHOW
         END

      SUBROUTINE CODING(K,STRING,L)
C
C ENCODE THE BEAM INDEX
C
      INTEGER K(3),L
      CHARACTER D*1,P*1,STRING*80
      ENCODE (12,100,STRING) K(1),K(2),K(3)
  100 FORMAT (I3,',',I3,',',I3,'&')
      L=11
      I=0
  150 I=I+1
      IF (STRING(I:I).EQ.' ') THEN
        L=L-1
        DO J=I+1,12
           STRING(J-1:J-1)=STRING(J:J)
        END DO
        I=I-1
      END IF
      IF (STRING(I:I).NE.'&') GOTO 150
      RETURN
      END

      SUBROUTINE GETLIN(C0,K0,XMN,XMX,YMN,YMX,XL,YL,LSTAT)
      REAL XP(4),YP(4),XL(2),YL(2)
      INTEGER ID(4)
C
C  LOCATE THE POINT OF INTERSECTION OF LINE WITH THE FRAME
C
      CALL YXY(YMN,XMN,XMX,C0,K0,XP(1),YP(1),ID(1))
      CALL XXY(XMX,YMN,YMX,C0,K0,XP(2),YP(2),ID(2))
      CALL YXY(YMX,XMN,XMX,C0,K0,XP(3),YP(3),ID(3))
      CALL XXY(XMN,YMN,YMX,C0,K0,XP(4),YP(4),ID(4))
C
      N=1
      DO 200 IP=1,4
        IF (ID(IP).EQ.0) GOTO 200
        XL(N)=XP(IP)
        YL(N)=YP(IP)
        N=N+1
```

```
200       CONTINUE
          LSTAT=1
          IF (N.LT.2) LSTAT=0
          RETURN
          END
C
          SUBROUTINE XXY(X1,Y1,Y2,C,S,X,Y,ID)
          Y=S*X1+C
          X=X1
          ID=1
          IF (Y.LT.Y1.OR.Y.GT.Y2) ID=0
          RETURN
          END
C
          SUBROUTINE YXY(Y1,X1,X2,C,S,X,Y,ID)
          IF (ABS(S).LT.1.0E-12) THEN
            ID=0
          ELSE
            X=(Y1-C)/S
            Y=Y1
            ID=1
            IF (X.LT.X1.OR.X.GT.X2) ID=0
          END IF
          RETURN
          END

          SUBROUTINE BEAM_SETUP(NBEAMS,INDEX,GX,GY,GZ,ZONE,ILZ,
         *                     TILT,SGMAX,WL,IFLAG)
C
C   SET UP THE BEAMS IN A ZONE AXIS INVOLVED IN THE DIFFRACTION
C
          PARAMETER (ND=2500)
          INTEGER       INDEX(3,1),INDEX1(3,ND),GX(3),GY(3),GZ(3)
          REAL          ZONE(3),KZ,TILT(3)
          REAL*8        GMX(3,3),GMXR(3,3)
          COMMON/METRIC/    GMX,GMXR

          TEMP2=SUM(GMXR,TILT,TILT)
          KZ=SQRT(1.0/(WL*WL)-TEMP2)
          NBEAMS=0
          DO I=1,ILZ
            IH=I-1
            CALL BEAM_CHECK(INDEX1,NB,IH,SGMAX,GX,GY,GZ,ZONE,TILT,KZ,IFLAG)
            DO J=1,NB
              NBEAMS=NBEAMS+1
              DO K=1,3
                INDEX(K,NBEAMS)=INDEX1(K,J)
              END DO
            END DO
          END DO
          RETURN
          END

          SUBROUTINE BEAM_CHECK(INDEX,NBEAMS,IHOLZ,SGMAX,GX,GY,GZ,
         *                     ZONE,TILT,KZ,IFLAG)
C
C   SELECT BEAM BY THEIR EXCITATION ERROR
C
          PARAMETER (NB=51)
          REAL          TILT(3),TEMP(3),DSTAR(1000),KZ,ZONE(3)
          INTEGER       GX(3),GY(3),GZ(3),MESH(NB,NB)
          INTEGER       H,K,INDEX(3,1)
          REAL*8        GMX(3,3),GMXR(3,3)
          COMMON/METRIC/    GMX,GMXR
```

```
      NC=NB/2
      DO J=1,NB
        DO I=1,NB
          MESH(I,J)=0
          H=I-NC-1
          K=J-NC-1
          DO M=1,3
            TEMP(M)=H*GX(M)+K*GY(M)+IHOLZ*GZ(M)
          END DO
          CALL EXCIT(TEMP,TILT,ZONE,KZ,SG)
          IF (ABS(SG).LT.SGMAX) MESH(I,J)=1
        END DO
      END DO
      M=0
      DO J=1,NB
        DO I=1,NB
          IF (MESH(I,J).EQ.1) THEN
            M=M+1
            DO N=1,3
              INDEX(N,M)=(I-NC-1)*GX(N)+(J-NC-1)*GY(N)+IHOLZ*GZ(N)
            END DO
          END IF
        END DO
      END DO
      NBEAMS=M
      IF (IFLAG.LT.0) RETURN
      DO I=1,NBEAMS
        DO J=1,3
          TEMP(J)=INDEX(J,I)
        END DO
        CALL SCALE(1,TEMP,TEMP1)
        DSTAR(I)=TEMP1
      END DO
      CALL SORTIN(INDEX,DSTAR,NBEAMS)
      RETURN
      END

      SUBROUTINE EXCIT(H,TILT,ZONE,KZ,SG)
      REAL          ZONE(3),KZ,TILT(3),H(3),TEMP(3)
      REAL*8        GMX(3,3),GMXR(3,3)
      COMMON/METRIC/     GMX,GMXR
      DO I=1,3
        TEMP(I)=2.0*TILT(I)+H(I)
      END DO
      SG=SUM(GMXR,TEMP,H)
      SG=2.0*KZ*(ZONE(1)*H(1)+ZONE(2)*H(2)+ZONE(3)*H(3))-SG
      RETURN
      END

      SUBROUTINE SORTIN(INDEX,DSTAR,NB)
      REAL DSTAR(1)
      INTEGER    INDEX(3,1),NEXT(3)
      DO I=1,NB-1
        DO K=1,3
          NEXT(K)=INDEX(K,I+1)
        END DO
        DEXT=DSTAR(I+1)
        DO J=I,1,-1
          IF (DEXT.GE.DSTAR(J)) GOTO 100
          DSTAR(J+1)=DSTAR(J)
          DO K=1,3
            INDEX(K,J+1)=INDEX(K,J)
          END DO
        END DO
100     DO K=1,3
          INDEX(K,J+1)=NEXT(K)
        END DO
        DSTAR(J+1)=DEXT
      END DO
      RETURN
      END
```

```
C
C FOLLOWINGS ARE A SET OF FORTRAN SUBROUTINES FOR TRANSLATION OF FORTRAN
  CALLS
C INTO POSTSCRIPT GRAPHICS LANGUAGE
C J. M. ZUO SEPT., 1991
C
      SUBROUTINE INITPS(FILNAM)
C
C INITIALIZE FORTRAN CALLS FOR POSTSCRIPT GRAPHICS
C
      CHARACTER FILNAM*20
      COMMON /GRAPH/ IPS,T11,T22,X0,Y0,ISTAT(5)
      IPS=20
      OPEN (UNIT=IPS,FILE=FILNAM,STATUS='NEW',FORM='FORMATTED')
      WRITE (IPS,100)
100   FORMAT (/,' % postscript - adobe',
     *         /,' % created by J. M. Zuo, ASU')
      ISTAT(3)=0
      ISTAT(4)=0
      ISTAT(5)=0
      RETURN
      END

      SUBROUTINE WINDOW(XMN,XMX,YMN,YMX,WXMN,WXMX,WYMN,WYMX)
C
C XMN,XMX,YMN,YMX: MINIMUM AND MAXIMUM X AND Y VALUES OF PLOTTING
C WXN,WXX,WYN,WYX: CORRESPONDING WINDOW ON PLOTTING PAGE IN INCHS
C PS HAS A DEFAULT COORDINATES WITH ORIGIN AT LEFT HAND LOW CORNER
C AND UNITS IN 1/72 INCHES
C THIS SUBROUTINE SETUPS THE TRANSFORMATION MATRIX BETWEEN PLOTTING
C AND PAGE COORDINATES
C
      REAL XMN,XMX,YMN,YMX,WXMN,WXMX,WYMN,WYMX
      COMMON /GRAPH/ IPS,T11,T22,X0,Y0,ISTAT(5)
      WXN=WXMN*72.0     !CHANGE TO POINTS
      WXX=WXMX*72.0
      WYN=WYMN*72.0
      WYX=WYMX*72.0
      T11=(WXX-WXN)/(XMX-XMN)
      X0=WXN-T11*XMN
      T22=(WYX-WYN)/(YMX-YMN)
      Y0=WYN-T22*YMN
      RETURN
      END

      SUBROUTINE NWPATH
C START A NEW GRAPHIC PATH
      COMMON /GRAPH/ IPS,T11,T22,X0,Y0,ISTAT(5)
      WRITE (IPS,*) 'newpath'
      RETURN
      END

      SUBROUTINE SEGRAY(GRAY)
C SET THE GREY, GREY=0.0 BLACK; 1.0 WHITE
      COMMON /GRAPH/ IPS,T11,T22,X0,Y0,ISTAT(5)
      WRITE (IPS,100) GRAY
100   FORMAT (1X,F5.3,' setgray')
      RETURN
      END

      SUBROUTINE SETLNW(W)
C SET THE LINE WIDTH IN UNITS OF POINT (1/72 INCHES)
      COMMON /GRAPH/ IPS,T11,T22,X0,Y0,ISTAT(5)
      WRITE (IPS,100) W
100   FORMAT (1X,F6.3,' setlinewidth')
      RETURN
      END
```

```
      SUBROUTINE SHOW
C PRINT THE PAGE
      COMMON /GRAPH/ IPS,T11,T22,X0,Y0,ISTAT(5)
      WRITE (IPS,*) 'showpage end'
      CLOSE (UNIT=IPS)
      RETURN
      END
      SUBROUTINE SETDSH(ID)
C SET THE LINE STYLE, ID=1 FULL LINE; 2-5 DASHED LINES
      COMMON /GRAPH/ IPS,T11,T22,X0,Y0,ISTAT(5)
      IF (ID.EQ.1) WRITE (IPS,*) ' 0 setdash'
      IF (ID.EQ.2) WRITE (IPS,*) '[20 5] 0 setdash'
      IF (ID.EQ.3) WRITE (IPS,*) '[10 5] 0 setdash'
      IF (ID.EQ.4) WRITE (IPS,*) '[5 5] 0 setdash'
      IF (ID.EQ.5) WRITE (IPS,*) '[6 4 2 4] 0 setdash'
      RETURN
      END

      SUBROUTINE CIRCLE(X,Y,R)
C DRAW A CIRCLE OF RADIUS R AT X,Y. R IS SCALED AS THE X COORDINATES
      REAL X,Y,R
      COMMON /GRAPH/ IPS,T11,T22,X0,Y0,ISTAT(5)
      IF (ISTAT(3).EQ.0) THEN
         WRITE (IPS,*) '/doCircle {0 360 arc stroke} def'
         ISTAT(3)=1
      END IF
      X1=T11*X+X0
      Y1=T22*Y+Y0
      R1=T11*R
      WRITE (IPS,150) X1,Y1,R1
150   FORMAT (1X,F6.2,1X,F6.2,1X,F6.2,' doCircle')
      RETURN
      END

      SUBROUTINE DRAWLN(X1,Y1,X2,Y2)
C DRAW A LINE FROM POINT (X1,Y1) TO POINT (X2,Y2)
      REAL X1,Y1,X2,Y2
      COMMON /GRAPH/ IPS,T11,T22,X0,Y0,ISTAT(5)
      XS=X1*T11+X0
      YS=Y1*T22+Y0
      XE=X2*T11+X0
      YE=Y2*T22+Y0
      WRITE (IPS,100) XS,YS,XE,YE
100   FORMAT (' newpath ',F6.2,1X,F6.2,' moveto ',F6.2,1X,F6.2,
     *                   ' lineto stroke')
      RETURN
      END

      SUBROUTINE FONT(FONAME,FONSIZ)
C SET THE FONT TYPE AND SIZE, FONAME: NAME OF THE FONT. FONSIZ: SIZE
C OF THE FONT, in units of point
C FOR AVAILABLE FONT, SEE YOUR POSTSCRIPT PRINTER.
C FOLLOWING ARE EXAMPLES:
C Courier, Courier-Bold, Times-Roman, Times-Bold, Times-BoldItalic
C Helvetica, Helvetica-Bold, Helvetica-BoldOblique
      CHARACTER FONAME*30
      INTEGER FONSIZ
      COMMON /GRAPH/ IPS,T11,T22,X0,Y0,ISTAT(5)
      WRITE (IPS,100) FONAME(1:30)
100   FORMAT (1X,'/',30A)
      WRITE (IPS,150) FONSIZ
150   FORMAT (' findfont ',I3,' scalefont setfont')
      RETURN
      END
```

```
      SUBROUTINE TEXT(STRING1,NS,X,Y,CTL)
C WRITE THE TEXT IN CHARACTER STRING 'STRING1' OF LENGTH NS AT
C POSITION X,Y HORIZONTALLY
C CTL = 'C', CENTER THE TEXT; 'R', RIGHTHAND SIDE;
C        'L', LEFTHAND SIDE; 'U' SUPERSCRIPT 5 POINT
C        'S', SUBSCRIPT 3 POINT
      CHARACTER STRING*80,STRING1*80,CTL*1,CHAR*1
      REAL X,Y
      INTEGER NS,N
      COMMON /GRAPH/ IPS,T11,T22,X0,Y0,ISTAT(5)
      X1=T11*X+X0
      Y1=T22*Y+Y0
      STRING=STRING1
      DO I=NS+1,1,-1
        CHAR(1:1)=STRING(I-1:I-1)
        STRING(I:I)=CHAR(1:1)
      END DO
      N=NS+2
      STRING(N:N)=')'
      STRING(1:1)='('
C
C LEFT HANDED
C
      IF (CTL.EQ.'L'.OR.CTL.EQ.'l') THEN
        WRITE (IPS,100) X1,Y1
100     FORMAT (1X,F6.2,1X,F6.2,' moveto')
        WRITE (IPS,260) STRING(1:N)
        WRITE (IPS,130)
130     FORMAT (' show')
      END IF
C
C RIGHT HANDED
C
      IF (CTL.EQ.'R'.OR.CTL.EQ.'r') THEN
        WRITE (IPS,260) STRING(1:N)
        WRITE (IPS,160) X1,Y1
160     FORMAT (' dup stringwidth pop ',F6.2,
     *          ' exch sub ',F6.2,' moveto show')
      END IF
C
C CENTERED
C
      IF (CTL.EQ.'C'.OR.CTL.EQ.'c') THEN
        WRITE (IPS,260) STRING(1:N)
        WRITE (IPS,210) X1,Y1
210     FORMAT (' dup stringwidth pop 2 div ',F6.2,
     *          ' exch sub ',F6.2,' moveto show')
      END IF
C
C SUPERSCRIPT 5 POINT
C
      IF (CTL.EQ.'U'.OR.CTL.EQ.'u') THEN
        WRITE (IPS,250)
250     FORMAT (' currentpoint 5 add moveto')
        WRITE (IPS,260) STRING(1:N)
260     FORMAT (1X,80A)
        WRITE (IPS,270)
270     FORMAT (' show',/,' currentpoint 5 sub moveto')
      END IF
C
C SUBSCRIPT 3 POINT
C
      IF (CTL.EQ.'S'.OR.CTL.EQ.'s') THEN
        WRITE (IPS,300)
300     FORMAT (' currentpoint 3 sub moveto')
        WRITE (IPS,260) STRING(1:N)
        WRITE (IPS,320)
320     FORMAT (' show',/,' currentpoint 3 add moveto')
      END IF
      RETURN
      END
```

```
      SUBROUTINE TEXTP(STRING1,NS,X,Y,CTL)
C WRITE THE TEXT IN CHARACTER STRING 'STRING1' OF LENGTH NS AT
C POSITION X,Y VERTICALLY
C CTL = 'C', CENTER THE TEXT; 'R', RIGHTHAND SIDE;
C        'L', LEFTHAND SIDE; 'U' SUPERSCRIPT 5 POINT
C        'S', SUBSCRIPT 3 POINT
      CHARACTER STRING*80,STRING1*80,CTL*1,CHAR*1
      REAL X,Y
      INTEGER NS,N
      COMMON /GRAPH/ IPS,T11,T22,X0,Y0,ISTAT(5)
      X1=T11*X+X0
      Y1=T22*Y+Y0
      WRITE(IPS,50) X1,Y1
50    FORMAT (1X,F6.2,1X,F6.2,' translate 90.0 rotate')
      STRING=STRING1
      DO I=NS+1,1,-1
         CHAR(1:1)=STRING(I-1:I-1)
         STRING(I:I)=CHAR(1:1)
      END DO
      N=NS+2
      STRING(N:N)=')'
      STRING(1:1)='('
C
C LEFT HANDED
C
      IF (CTL.EQ.'L'.OR.CTL.EQ.'l') THEN
         WRITE (IPS,100)
100      FORMAT (' 0 0 moveto')
         WRITE (IPS,260) STRING(1:N)
         WRITE (IPS,130)
130      FORMAT (' show')
      END IF
C
C RIGHT HANDED
C
      IF (CTL.EQ.'R'.OR.CTL.EQ.'r') THEN
         WRITE (IPS,260) STRING(1:N)
         WRITE (IPS,160)
160      FORMAT (' dup stringwidth pop neg 0.0 moveto show')
      END IF
C
C CENTERED
C
      IF (CTL.EQ.'C'.OR.CTL.EQ.'c') THEN
         WRITE (IPS,260) STRING(1:N)
         WRITE (IPS,210)
210      FORMAT (' dup stringwidth pop 2 div neg 0.0 moveto show')
      END IF
C
C CENTERED ON BOTH X AND Y
C
      IF (CTL.EQ.'M'.OR.CTL.EQ.'m') THEN
         WRITE (IPS,260) STRING(1:N)
         WRITE (IPS,220)
220      FORMAT (' dup stringwidth 2 div exch 2 div neg exch moveto show')
      END IF
C
C SUPERSCRIPT 5 POINT
C
      IF (CTL.EQ.'U'.OR.CTL.EQ.'u') THEN
         WRITE (IPS,250)
250      FORMAT (' currentpoint 5 add moveto')
         WRITE (IPS,260) STRING(1:N)
260      FORMAT (1X,80A)
         WRITE (IPS,270)
270      FORMAT (' show',/,' currentpoint 5 sub moveto')
      END IF
```

```
C
C SUBSCRIPT 3 POINT
C
      IF (CTL.EQ.'S'.OR.CTL.EQ.'s') THEN
        WRITE (IPS,300)
300     FORMAT (' currentpoint 3 sub moveto')
        WRITE (IPS,260) STRING(1:N)
        WRITE (IPS,320)
320     FORMAT (' show',/,' currentpoint 3 add moveto')
      END IF
      WRITE (IPS,400) -X1,-Y1
400   FORMAT (' 0 0 moveto -90.0 rotate ',F7.2,1X,F7.2,' translate')
      RETURN
      END

C
C FOLLOWINGS ARE A SET OF SUBROUTINES AND FUNCTIONS FOR STANDARD
C CRYSTALLOGRAPHIC CALCULATIONS
C
      SUBROUTINE METRIC(CELL)
C
C CALCULATE THE METRICAL MATRIX GMX AND ITS INVERSE GMXR
C CELL(6) CONTAINS A,B,C,ALPHA,BETA,GAMMA IN SEQUENCE, REAL*4 TYPE
C
      REAL       CELL(6)
      REAL*8          GMX(3,3),GMXR(3,3)
      COMMON/METRIC/ GMX,GMXR

      DO I=1,3
        GMX(I,I)=CELL(I)*CELL(I)
      END DO
      GMX(1,2)=CELL(1)*CELL(2)*CELL(6)
      GMX(2,1)=GMX(1,2)
      GMX(1,3)=CELL(1)*CELL(3)*CELL(5)
      GMX(3,1)=GMX(1,3)
      GMX(2,3)=CELL(2)*CELL(3)*CELL(4)
      GMX(3,2)=GMX(2,3)
      CALL INVERT(GMX,GMXR)
      RETURN
      END

      SUBROUTINE ANGLE(IFLAG,H,K,COSA)
C
C CALCULATE THE COSINE ANGLE BETWEEN TWO VECTORS H AND K
C IFLAG = 0, H AND K IN REAL SPACE, 1 IN RECIPROCAL SPACE
C H(3),K(3) REAL*4 TYPE, COORDINATES OF THE VECTORS IN UNITS OF THE
C           REAL OR RECIPROCAL SPACE CELL
C COSA: COS(ANGLE), REAL*8 TYPE
C
      REAL CELL(6),H(3),K(3)
      REAL*8      GMX(3,3),GMXR(3,3),COSA,SH,SK
      COMMON/METRIC/ GMX,GMXR
      IF (IFLAG.EQ.0) THEN
        COSA=SUM(GMX,H,K)
        SH=SUM(GMX,H,H)
        SH=DSQRT(SH)
        SK=SUM(GMX,K,K)
        SK=DSQRT(SK)
        COSA=COSA/(SH*SK)
      ELSE
        COSA=SUM(GMXR,H,K)
        SH=SUM(GMXR,H,H)
        SH=DSQRT(SH)
        SK=SUM(GMXR,K,K)
        SK=DSQRT(SK)
        COSA=COSA/(SH*SK)
      END IF
      RETURN
      END
```

```
      SUBROUTINE SCALE(IFLAG,H,X)
C
C CALCULATE THE LENGTH OF VECTOR H, X THE RETURNED LENGTH, REAL*8
C
      REAL        CELL(6),H(3)
      REAL*8          GMX(3,3),GMXR(3,3),X
      COMMON/METRIC/  GMX,GMXR
      IF (IFLAG.EQ.0) THEN
         X=SUM(GMX,H,H)
         X=DSQRT(X)
      ELSE
         X=SUM(GMXR,H,H)
         X=DSQRT(X)
      END IF
      RETURN
      END

      SUBROUTINE CROSS(IFLAG,H,K,X)
C
C CALCULATE THE SINE OF THE ANGLE BETWEEN VECTORS H AND K
C X = SIN(ANGLE), REAL*8 TYPE
C
      REAL        CELL(6),H(3),K(3)
      REAL*8          GMX(3,3),GMXR(3,3),COSA,SH,SK,X,X1
      COMMON/METRIC/  GMX,GMXR
      IF (IFLAG.EQ.0) THEN
         SH=SUM(GMX,H,H)
         SH=DSQRT(SH)
         SK=SUM(GMX,K,K)
         SK=DSQRT(SK)
         X=SUM(GMX,H,K)
         X1=X/(SH*SK)
         X=DSQRT(1.0-X1*X1)
         X=X*SH*SK
      ELSE
         SH=SUM(GMXR,H,H)
         SH=DSQRT(SH)
         SK=SUM(GMXR,K,K)
         SK=DSQRT(SK)
         X=SUM(GMXR,H,K)
         X1=X/(SH*SK)
         X=DSQRT(1.0-X1*X1)
         X=X*SH*SK
      END IF
      RETURN
      END

      REAL*8 FUNCTION SUM(G,A,B)
C
C CALCULATE A*G*B, A AND B 3 DIMENSIONAL VECTORS (REAL*4), G 3X3 MATRIX
C
      REAL       A(3),B(3)
      REAL*8 G(3,3)
      SUM=0.0
      DO J=1,3
        DO I=1,3
           SUM=SUM+A(I)*G(I,J)*B(J)
        END DO
      END DO
      RETURN
      END
```

```
      SUBROUTINE TRANS(G,H,K)
C
C PERFORMS THE TRANSFORMATION OF A 3-DIMENSIONAL VECTOR, G IS THE
C TRANSFORMATION MATRIX. H IS THE VECTOR TO BE TRANSFORMED
C
      REAL H(3),K(3)
      REAL*8     G(3,3)
      DO I=1,3
        K(I)=0.0
        DO J=1,3
          K(I)=K(I)+G(I,J)*H(J)
        END DO
      END DO
      RETURN
      END

C
C INVERT: FIND THE INVERSE MATRIX OF A 3X3 MATRIX
C
      SUBROUTINE INVERT(A,AT)
      REAL*8 A(3,3),AT(3,3),DETA
      DETA=A(1,1)*A(2,2)*A(3,3)+A(2,1)*A(3,2)*A(1,3)+
     *     A(3,1)*A(1,2)*A(2,3)-A(1,3)*A(2,2)*A(3,1)-
     *     A(2,3)*A(3,2)*A(1,1)-A(1,2)*A(2,1)*A(3,3)
      IF (DABS(DETA).LT.1.0E-7) STOP 'FAIL'
      AT(1,1)=  (A(2,2)*A(3,3)-A(2,3)*A(3,2))/DETA
      AT(1,2)=-(A(1,2)*A(3,3)-A(1,3)*A(3,2))/DETA
      AT(1,3)=  (A(1,2)*A(2,3)-A(1,3)*A(2,2))/DETA
      AT(2,1)=-(A(2,1)*A(3,3)-A(2,3)*A(3,1))/DETA
      AT(2,2)=  (A(1,1)*A(3,3)-A(1,3)*A(3,1))/DETA
      AT(2,3)=-(A(1,1)*A(2,3)-A(1,3)*A(2,1))/DETA
      AT(3,1)=  (A(2,1)*A(3,2)-A(2,2)*A(3,1))/DETA
      AT(3,2)=-(A(1,1)*A(3,2)-A(1,2)*A(3,1))/DETA
      AT(3,3)=  (A(1,1)*A(2,2)-A(1,2)*A(2,1))/DETA
      RETURN
      END

C
C EXAMPLE INPUT DATA
C
SILICON
100.0 -2 2 1 0.0 0.0 0.0 8500.0
5.4307 5.4307 5.4307 90.0 90.0 90.0
8    1
SI
14  6.2915   2.4386   3.0353  32.3370   1.9891    0.6758   1.5410  81.6937  1.1407
1  0.0  0.0  0.0  0.46
1  0.5  0.5  0.0  0.46
1  0.5  0.0  0.5  0.46
1  0.0  0.5  0.5  0.46
1  0.25 0.25 0.25 0.46
1  .75  .75  .25  0.46
1  .75  .25  .75  0.46
1  .25  .75  .75  0.46
2 2 0
2 0 4
1 1 1
YES
2 0.6 5.0
YES_BEAM_CENTER
BOTH_DISK_AND_LINES
0.37
NO_DISK_INDEX
YES_GREY_LEVELS
YES_LINE_INDEX
```

## A5.2. BLOCH-WAVE DYNAMICAL PROGRAMS

Clear reviews of the Bloch-wave theory of dynamical electron diffraction can be found in Hirsch *et al.* (1977), Howie (1971), Humphreys (1979), Metherall (1975), and Reimer (1984), as discussed in Chapter 3. The exact incorporation of HOLZ reflections in this approach is described by Lewis *et al.* (1978). A recent review which emphasizes Bloch waves in real space and the atomic string approach can be found in Bird (1989), where the use of perturbation methods for the HOLZ is also reviewed. The papers by Ishizuka and Uyeda (1977) and Goodman and Moodie (1974) discuss the relationship between the various algorithms and the Schrödinger equation. Methods for computing CBED patterns from quasi-crystals are discussed by Cheng and Wang (1989).

The development of computer algorithms based on these expressions is reviewed in the books by Krakow and O'Keefe (1989) and Buseck *et al.* (1989). Many FORTRAN source programs useful for electron microscopy can be obtained at no charge by telephone (using a computer modem) from the Argonne Public Domain Software Library (Bitnet address: ZALUZEC@ANLMST, Username EMMPDL, Password EMMPDL; Telephone: USA 312-972-7919). 800 or 1200 Baud, 8 data, 1 stop bit, no parity. See also *EMSA Bulletin*, Vol. 16, No. 1, p. 42. A popular suite of programs (EMS) which compute both lattice images (multislice) and CBED patterns (Bloch wave) are described in the chapter by Stadelmann in Krakow and O'Keefe (1989), and in Stadelmann (1987). This program is installed in many laboratories around the world, operating on many computers including the Silicon Graphics and Vax types. Another popular commercial suite of programs for many types of diffraction calculations and atomic modeling is the Cerius

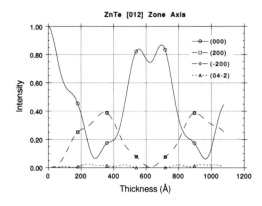

Figure A5.1. Thickness dependence of several beams for ZnTe at 200 kV. Incident beam along the [012] zone axis. ZOLZ reflections only have been included, without absorption. Bloch-wave method.

Crystals Module (from Cambridge Molecular Design, St John's Center, Cowley Rd., Cambridge, CB4 4WS, U.K.). The Refine/CB program discussed in Section 4.2 for automated structure-factor refinement is available from the authors (c/o Department of Physics, Arizona State University, Tempe, AZ 85287, U.S.A.), and is further discussed in Zuo and Spence (1991).

A listing of our Bloch-wave Fortran program TCBED follows, with sample input and output. An earlier version of this was listed in full in Zuo *et al.* (1989), and is available at no charge from the Argonne Public Domain Software Library by modem. New programs may be tested against the output given in Fig. A5.1, or against the data shown in Fig. 5.2, which agrees well with experiment at two thicknesses.

The TCBED program listed below uses the subroutines for automatic beam generation and crystallographic computations which have been listed in the previous section. The program also requires subroutines for the diagonalization of a complex general matrix. These can be found in the EISPACK package (Smith, 1976). These subroutines may be obtained from a public domain library at the electronic mail address NETLIB@RESEARCH.ATT.COM. For a listing of programs in this library (and for help), send a message to the above address with the words "send index."

```
PROGRAM TCBED
************************************************************************
* A FORTRAN PROGRAM FOR SIMULATING TRANSMISSION CONVERGENT BEAM        *
* ELECTRON DIFFRACTION PATTERNS WITH ABSORPTION BY BLOCH WAVE METHOD   *
* VERSION 2                                                            *
*                                                                      *
* THIS PROGRAM HAS THE FOLLOWING SPECIAL FEATURES:                     *
* 1) HOLZ INTERACTIONS ARE INCLUDED                                    *
* 2) ABSORPTION IS INCLUDED BY DIAGONALIZING A GENERAL COMPLEX MATRIX  *
* 3) THE DEFAULT ABSORPTION POTENTIAL IS CALCULATED BY "ATOM" PROGRAM  *
* 4) INDIVIDUAL STRUCTURE FACTORS AND ABSORPTION PARAMETERS CAN BE     *
*    ADJUSTED                                                        *
* 5) AUTOMATIC GENERATION OF BEAMS FOR SIMULATION                      *
*                                                                      *
* THIS PROGRAM ASSUMES THE FOLLOWING CONVENTIONS                       *
* 1) THE CRYSTALLOGRAPHIC SIGN CONVENTION  EXP(-IKR) FOR PLANE WAVE    *
* 2) ISOTROPIC DEBYE-WALLER FACTOR FOR EACH ATOM IN THE UNIT CELL      *
* 3) ANSTROM AS THE UNIT OF  LENGTH                                    *
* 4) INCIDENT BEAM DIRECTION SPECIFIED BY THE CENTER OF LAUE CIRCLE    *
* 5) Z AXIS IS ALONG THE ZONE AXIS DIRECTION, Z=0 FOR ENTRANCE         *
*    SURFACE AND Z=-t FOR EXIT SURFACE                              *
*                                                                      *
* OUTPUT FILE                                                          *
* OUTPUT DATA FILE IS UNFOMATED, TO READ THE FILE, USE                 *
* READ (UNIT) NP,(BINT(I),I=1,NP)                                      *
* NY=BINT(NP)        !NUMBER OF POINT IN Y-DIRECTION                   *
* NX=BINT(NP-1)      !NUMBER OF POINT IN X-DIRECTION                   *
* ND=BINT(NP-2)      !NUMBER OF THICKNESS                              *
* NB=BINT(NP-3)      !NUMBER OF OUTPUT BEAMS                           *
* TO GET   CBED INTENSITY FOR THE I1 REFLECTION, I2 THICKNESS, USE     *
```

```
*    DO J=1,NY                                                       *
*      DO I=1,NX                                                     *
*       IPOS=(J-1)*NX*ND*NB+(I-1)*ND*NB+(I1-1)*NB+I2                 *
*       CBED(I,J)=BINT(IPOS)                                         *
*      END DO                                                        *
*    END DO                                                          *
*                                                                    *
*  J.M. ZUO, NOV. 1989                                               *
**********************************************************************

      CHARACTER TEXT*80,STATUS*4

      PARAMETER  (NM=300)
      INTEGER          HKL(3,NM),HKLOUT(3,NM),HKLSR(3,NM),ID(NM)

      REAL       UGHR(NM,NM),UGHI(NM,NM),CCR(NM,NM),CCI(NM,NM),
     *           VR(NM),VI(NM),BINT(60000),UGSR(NM),UGSI(NM),
     *           ABSPR(NM),ABSPI(NM)
      COMPLEX          CVT(NM,NM),CINV(NM)

      PARAMETER  (NJ=10)
      CHARACTER*4      LABEL(NJ)
      REAL       ZT(NJ),CX(NJ),AX(4,NJ),BX(4,NJ),AF(NJ),AFI(NJ)
      PARAMETER  (NK=500)
      DIMENSION  ITYPE(NK),XP(3,NK),DW(NK)

      REAL*8           GMX(3,3),GMXR(3,3)
      COMMON/METRIC/ GMX,GMXR

      CHARACTER*2      ELNAME
      DOUBLE PRECISION MATM,SATM,KVATM,FEATM,FAATM
      INTEGER          STATM

      CHARACTER*2      ABSCTL
      REAL       KV,CELL(6),TILT0(3,3),ZONE(3),GH(3),GG(3)
      REAL       TILTX(3),TILTY(3),TILT(3),BIGK,DEPTH(5000)
      INTEGER          INDZ(3),GX(3),GY(3),GZ(3),NET(2,10000)
      REAL*8     PI,TWOPI,TEMP,CN,SN,SFRE,SFIM,S2
      COMPLEX          ZERO,ONE,C1,C2,C3
      DATA       PI/3.14159254/,TWOPI/6.283185307/,
     *           NR,NW,IW/10,12,14/,
     *           DELTA/1.0E-7/

100   FORMAT (80A)
110   FORMAT (A4)
120   FORMAT (/,' OUTPUT LISTING OF TCBED PROGRAM BY J. M. ZUO, V2.0',
     *         /,X,80A)
140   FORMAT (/,' INCIDENT ELECTRON BEAM HIGH VOLTAGE ',F12.5,' KV',
     *         /,'                           WAVELENGTH ',F12.6,' A')
160   FORMAT (/,'                         NEAREST ZONE AXIS ',3I4)
180   FORMAT (/,' 3 POINTS DEFINE A PARALLEL BOX CBED REGION',
     *         /,'                  POINT 1, THE ORIGIN ',3F10.5,
     *         /,'                  POINT 2, THE X-AXIS ',3F10.5,
     *         /,'                  POINT 3, THE Y-AXIS ',3F10.5,
     *         /,'                NUMBER OF SAMPLING POINTS IN X-DIR ',I4,
     *         /,'                                        IN Y-DIR ',I4)
200   FORMAT (/,' BASE G VECTOR ALONG X DIRECTION ',3I4,
     *         /,'                     Y DIRECTION ',3I4,
     *         /,'                     Z DIRECTION ',3I4)
220   FORMAT (/,' THE UNIT CELL IS DEFINED BY a = ',F9.5,' Angstrom',
     *         /,'                              b = ',F9.5,' Angstrom',
     *         /,'                              c = ',F9.5,' Angstrom',
     *         /,'                          alpha = ',F9.5,' Degree',
     *         /,'                           beta = ',F9.5,' Degree',
     *         /,'                          gamma = ',F9.5,' Degree')
240   FORMAT (/,' THERE ARE TOTAL',I4,' ATOMS AND',I3,' SPECIES')
260   FORMAT (/,' THE ATOMS, THEIR X-RAY SCATTERING FACTOR PARAMETERS'
     *         /,' AND THEIR DEBYE-WALLER FACTOR')
280   FORMAT (/,X,I2,2X,A4,' Z=',F6.2,/,9F8.4)
```

```
300   FORMAT (/,' THE ATOMS TYPE, THEIR COORDINATES AND D-W FACTOR')
320   FORMAT (2(X,I2,3F9.5,F6.3))
340   FORMAT (/,' STATUS = AUTO , BEAM INDEX (HKL) ARE GENERATED BY'
     *         ' PROGRAM BEAM_SETUP',
     *        /,' UP TO HOLZ ',I1,' AND MAXIMUM EXCITATION ERROR ',F5.2)
360   FORMAT (/,' NUMBER OF BEAMS EXCEEDS THE CAPACITY OF THIS PROGRAM')
380   FORMAT (/,' TOTAL',I4,' BEAMS ARE INCLUDED, THEY ARE')
400   FORMAT (5(X,3I4,','))
420   FORMAT (/,' BEAM INTENSITIES ARE CALCULATED FOR ',I3,' DIFFERENT'
     *         ' THICKNESSES, THEY ARE')
440   FORMAT (10(X,F8.3))
450   FORMAT (/,' BEAM INTENSITIES ARE CALCULATED FOR ',I3,' DIFFERENT'
     *         ' THICKNESSES WITH A INCREMENT',F10.5)
460   FORMAT (/,' STRUCTURE FACTOR AND ABSORPTION COEFFICIENTS IN Ugh',
     *         ' MATRIX',
     *       /,4X,'i',3X,'j',6X,'H',3X,'K',3X,'L',7X,'|Ugh|',3X,'AND',
     *        2X,'PHASE',9X,'|U''gh|',2X,'PHASE')
480   FORMAT (1X,2I4,' (',3I4,')',E15.5,F13.6,E15.5,F10.3)
500   FORMAT (/,' NUMBER OF BEAMS OUTPUTED ARE',I4,' THEY ARE')
520   FORMAT (/,' NUMBER OF REFINED REFLECTIONS ARE',I4,' THEY ARE')
540   FORMAT (' (',3I4,')',E15.5,F13.6,E15.5,F13.6)
560   FORMAT (/,' THE MEAN CRYSTAL ABSORPTION POTENTIAL IS ',F13.6)
580   FORMAT (/,' THE MEAN CRYSTAL POTENTIAL IS          ',F13.6)
600   FORMAT (/,' WARNING!!!, ERROR IN APSORPTION CAL. FOR ATOM ',A4,
     *         ' REFLECTION ',3I4)

      ZERO=CMPLX(0.0,0.0)
      ONE=CMPLX(1.0,0.0)
C
C START INPUT FROM CHANNEL 10 AND LISTING TO CHANNEL 12
C
C TEXT: 80 CHARACTER TITLE
C
      READ (NR,100) TEXT
      WRITE (NW,120) TEXT
C
C KV        INCIDENT ELECTRON ENERGY IN UNIT OF KeV
C INDZ(3)   NEAREST ZONE AXIS INDEX
C TILT0(3,3)    3 INCIDENT BEAM DIRECTIONS, TOGETHER THEY DEFINES A
PARALLEL
C             BOX REGION OF CBED. THE 1ST IS THE ORIGIN, THE 1ST AND 2ND IS
C             THE X SIDE AND 1ST AND 3RD IS THE Y SIDE OF THE
C             PARALLELOGRAM
C             THE INCIDENT BEAM DIRECTION IS DEFINED BY THE COORDINATES
C             OF THE LAUE CIRCLE IN UNITS OF CRYSTAL RECIPROCAL LATTICE
C NX,NY       NUMBER OF SAMPLING POINTS IN X AND Y DIRECTIONS
C
      READ (NR,*) KV,INDZ,((TILT0(I,J),J=1,3),I=1,3),NX,NY
      KVATM=KV
      WAVEL=0.3878314/SQRT(KV*(1.0+0.97846707E-03*KV))
      WRITE (NW,140) KV,WAVEL
      WRITE (NW,160) INDZ
      WRITE (NW,180) ((TILT0(I,J),J=1,3),I=1,3),NX,NY
C
C CELL(6) CRYSTAL UNIT CELL PARAMETERS IN THE ORDER OF
C         a, b, c, alpha, beta, and gamma
C
      READ (NR,*) CELL
      WRITE (NW,220) CELL
      DO I=4,6
        CELL(I)=COS(PI*CELL(I)/180.0)
      END DO
      CALL METRIC(CELL)
      DO I=1,3
        ZONE(I)=FLOAT(INDZ(I))
      END DO
      CALL SCALE(0,ZONE,TEMP1)
      DO I=1,3
        ZONE(I)=ZONE(I)/TEMP1
      END DO
```

```
C
C NATOMS    NUMBER OF ATOMS IN THE UNIT CELL
C NTYPE     NUMBER OF TYPES OF ATOMS IN THE UNIT CELL
C
      READ (NR,*) NATOMS,NTYPE
      WRITE (NW,240) NATOMS,NTYPE
      WRITE (NW,260)
      DO I=1,NTYPE
C
C LABEL            ATOMIC SYMBOL
C
      READ (NR,110) LABEL(I)
C
C ZT        ATOMIC NUMBER
C AX,BX,CX  9 GAUSSIAN FITTING PARAMETERS OF ATOMIC X-RAY SCATTERING
C           FACTORS
C
      READ (NR,*) ZT(I),(AX(J,I),BX(J,I),J=1,4),CX(I)
      IF (ZT(I).EQ.0) THEN
        ZT(I)=CX(I)
        DO J=1,4
          ZT(I)=AX(J,I)+ZT(I)
        END DO
      END IF
      WRITE (NW,280) I,LABEL(I),ZT(I),(AX(J,I),BX(J,I),J=1,4),
     *               CX(I)
      END DO
C
C ITYPE     THE SEQUENTIAL NUMBER ASSIGNED TO THE ATOMS IN PREVIOUS
C           X-RAY SCATTERING FACTOR INPUT
C XP(3)     FRACTIONAL ATOMIC COORDINATES
C DW        DEBYE-WALLER FACTOR OF THE ATOM
C
      READ (NR,*) (ITYPE(I),(XP(J,I),J=1,3),DW(I),I=1,NATOMS)
      WRITE (NW,300)
      WRITE (NW,320) (ITYPE(I),(XP(J,I),J=1,3),DW(I),I=1,NATOMS)
C
C STATUS    A PROGRAM FLOW CONTROL PARAMETER, IF STATUS =
C 'AUTO'    THEN
C              THE PROGRAM GENERATES THE REFLECTIONS NEEDED IN
C              THIS SIMULATION,
C           ELSE
C              USER ENTERS THE REFLECTIONS
C
      READ (NR,100) STATUS
      IF (STATUS(1:4).EQ.'AUTO'.OR.STATUS(1:4).EQ.'auto') THEN
C
C GX(3),GY(3),GZ(3)  THREE INDEPENDENT BASE G VECTORS FOR GENERATING
C                    THE ENTIRE G VECTORS IN THE ZONE AXIS
C                    GX AND GY FOR THE ZOLZ
C                    GZ FOR THE HOLZ
C
      READ (NR,*) GX,GY,GZ
      WRITE (NW,200) GX,GY,GZ
C
C IHOLZ     NUMBER OF HOLZ RINGS TO BE INCLUDED
C SGMAX     THE MAXIMUM EXCITATION ERRORS TO BE CONSIDERED
C           THIS PARAMETER DETERMINES WHICH REFLECTION SHOULD BE
C           INCLUDED, AND WHICH SHOULD NOT
C
      READ (NR,*) IHOLZ,SGMAX
      WRITE (NW,340) IHOLZ-1,SGMAX
      DO I=1,3
        TILT(I)=TILT0(1,I)+0.5*TILT0(2,I)+0.5*TILT0(3,I)
      END DO
      CALL BEAM_SETUP(NPEAMS,HKL,GX,GY,GZ,ZONE,IHOLZ,TILT,SGMAX,
     *                WAVEL,1)
      READ (NR,*) NBEAMS
      NPEAMS=NPEAMS-NBEAMS
      ELSE
```

```
C
C NBEAMS   NUMBER OF REFLECTIONS TO BE INCLUDED
C HKL(3,I)      H, K, L OF THE REFLECTION
C
      READ (NR,*) NPEAMS
      READ (NR,*) ((HKL(J,I),J=1,3),I=1,NPEAMS)
        READ (NR,*) NBEAMS
        NPEAMS=NPEAMS-NBEAMS
      END IF
      WRITE (NW,380) NBEAMS+NPEAMS
      WRITE (NW,400) ((HKL(J,I),J=1,3),I=1,NBEAMS+NPEAMS)
      IF (NBEAMS+NPEAMS.GT.NM) THEN
        WRITE (NW,360)
        STOP
      END IF
      WRITE (NW,402) NPEAMS
402   FORMAT (/,I5,' BEAMS ARE TREATED BY PERTURBATION')
C
C STATUS    A PROGRAM FLOW CONTROL PARAMETER, IF STATUS =
C 'STOP'    THEN
C              STOP
C           IF STATUS =
C 'AUTO'    THEN
C              PROGRAM GENERATES THE THICKNESS
C           ELSE
C              USER ENTERS THE THICKNESS
C
      READ (NR,100) STATUS
      IF (STATUS(1:4).EQ.'STOP'.OR.STATUS(1:4).EQ.'stop') STOP
C
C THICKNESS SETUP
C
      IF (STATUS(1:4).EQ.'AUTO'.OR.STATUS(1:4).EQ.'auto') THEN
C
C NDEPTH   NUMBER OF THICKNESSES TO BE CALCULATED
C TSTEP          THE THICKNESS STEP
C
        READ (NR,*) NDEPTH,TSTEP
        DO I=1,NDEPTH
          DEPTH(I)=I*TSTEP
        END DO
        WRITE (NW,450) NDEPTH,TSTEP
      ELSE
C
C DEPTH    THE THICKNESS TO BE CALCULATED
C
        READ (NR,*) NDEPTH,(DEPTH(I),I=1,NDEPTH)
        WRITE (NW,420) NDEPTH
        WRITE (NW,440) (DEPTH(I),I=1,NDEPTH)
      END IF
C
C NOUT         THE NUMBER OF REFLECTIONS WHOSE INTENSITY TO BE CALCULATED
C          IF NOUT = 0, CALCULATES FOR ALL REFLECTIONS
C
      READ (NR,*) NOUT
      IF (NOUT.EQ.0) THEN
        NOUT=NBEAMS
        DO I=1,NBEAMS
          ID(I)=I
          DO J=1,3
            HKLOUT(J,I)=HKL(J,I)
          END DO
        END DO
        GOTO 1000
      END IF
```

```
C
C HKLOUT(3,I)     THE H, K, L INDEX OF THE ith REFLECTION TO BE CALCULATED
C
      IF (NOUT.GT.0) THEN
        READ (NR,*) ((HKLOUT(J,I),J=1,3),I=1,NOUT)
      END IF
      DO I=1,NOUT
        DO K=1,NBEAMS
          DO J=1,3
            IF (HKL(J,K).NE.HKLOUT(J,I)) GOTO 900
          END DO
          ID(I)=K
          GOTO 950
900       CONTINUE
        END DO
950     CONTINUE
      END DO
1000  WRITE (NW,500) NOUT
      WRITE (NW,400) ((HKLOUT(J,I),J=1,3),I=1,NOUT)
C
C NRF           THE NUMBER OF REFLECTIONS WHOSE STRUCTURE FACTORS ARE TO
BE
C               ADJUSTED
C HKLSR         THE INDEX OF ADJUSTED REFLECTIONS
C TEMP1,TEMP2   THE AMPLITUDE (ANGSTROM**-2) AND PHASE (IN RADIANS) OF
C               ADJUSTED STRUCTURE FACTOR
C ABSPR,ABSPI   THE ABSORPTION COEFFICIENT RATIO (|Ug'/Ug|) AND PHASE
C               DIFFERENCE BETWEEN Ug' AND Ug (IN RADIANS) OF THE
C               ADJUSTED ABSORPTION PARAMETER
C
      READ (NR,*) NRF
      WRITE (NW,520) NRF
      IF (NRF.EQ.0) GOTO 1500
      DO I=1,NRF
        READ (NR,*) (HKLSR(J,I),J=1,3),UGSR(I),UGSI(I),ABSPR(I),ABSPI(I)
        WRITE (NW,540)
(HKLSR(J,I),J=1,3),UGSR(I),UGSI(I),ABSPR(I),ABSPI(I)
      END DO
C
C ABM           THE MEAN ABSORPTION
C
1500  READ (NR,*) ABM
      WRITE (NW,560) ABM
C
C OPTIONAL CONTROL PARAMETER:
C ABSCTL = 'NO'          NO DEFAULT ABSORPTION. IF ZERO DEBYE-WALLER
C                 FACTOR IS ENTERED, TO AVOID ERROR, ENTER 'NO'
C LISCTL = 'YES'         INTENSITY LISTING, DEFAULT NO
C
      READ (NR,100) ABSCTL
      IF (ABSCTL(1:2).EQ.' ') ABSCTL(1:2)='YE'
C
C * END OF INPUT *
C
      VOL=1.0-CELL(4)*CELL(4)-CELL(5)*CELL(5)-CELL(6)*CELL(6)+
     *        2.0*CELL(4)*CELL(5)*CELL(6)
      VOL=SQRT(VOL)*CELL(1)*CELL(2)*CELL(3)
      RATIO=(1+1.9569341E-03*KV)/(PI*VOL)
C
C GET U(000) AND K HERE
C
      DO I=1,NTYPE
        AF(I)=0.0
        DO J=1,4
          AF(I)=AF(I)+AX(J,I)*BX(J,I)
        END DO
        AF(I)=0.023933754*AF(I)
      END DO
```

```
      U0=0.0
      DO I=1,NATOMS
        J=ITYPE(I)
        U0=U0+AF(J)
      END DO
      U0=U0*RATIO
      WRITE (NW,580) U0
      BIGK=SQRT(1.0/(WAVEL*WAVEL)+U0)
C
C GET U(G,H) MATRIX HERE
C
      WRITE (NW,460)
      DO I=2,NBEAMS+NPEAMS
        L=I-1
        IF (L.GT.NBEAMS) L=NBEAMS
        DO J=1,L,1
          DO K=1,3
            GH(K)=HKL(K,I)-HKL(K,J)
          END DO
          S2=0.25*SUM(GMXR,GH,GH)
          SATM=DSQRT(S2)
          IF (S2.LT.DELTA) THEN
      WRITE(NW,*) 'ERROR!, SAME REFLECTION IS ENTERED TWICE FOR',I,J
            STOP
          END IF
          N=0
          DO K=1,NRF
            IF (GH(1).EQ.HKLSR(1,K).AND.GH(2).EQ.HKLSR(2,K).AND.
     *              GH(3).EQ.HKLSR(3,K)) THEN
            SFRE=UGSR(K)*COS(UGSI(K))
            SFIM=UGSR(K)*SIN(UGSI(K))
            AFRE=ABSPR(K)*UGSR(K)*COS(ABSPI(I))
            AFIM=ABSPR(K)*UGSR(K)*SIN(ABSPI(I))
            GOTO 1700
            END IF
            IF (GH(1).EQ.-HKLSR(1,K).AND.GH(2).EQ.-HKLSR(2,K).AND.
     *              GH(3).EQ.-HKLSR(3,K)) THEN
            SFRE=UGSR(K)*COS(UGSI(I))
            SFIM=-UGSI(K)*SIN(UGSI(I))
            AFRE=ABSPR(K)*UGSR(K)*COS(ABSPI(I))
            AFIM=-ABSPR(K)*UGSR(K)*SIN(ABSPI(I))
            GOTO 1700
            END IF
          END DO
          DO K=1,NTYPE
            AF(K)=CX(K)
            DO M=1,4
            AF(K)=AF(K)+AX(M,K)*EXP(-BX(M,K)*S2)
            END DO
            AF(K)=0.023933754*(ZT(K)-AF(K))/S2
          END DO
          SFRE=0.0
          SFIM=0.0
          AFRE=0.0
          AFIM=0.0
          DO M=1,NATOMS
            TEMP=TWOPI*(GH(1)*XP(1,M)+GH(2)*XP(2,M)+GH(3)*XP(3,M))
            CN=DCOS(TEMP)
            SN=DSIN(TEMP)
            TEMP=EXP(-DW(M)*S2)
            SFRE=SFRE+CN*AF(ITYPE(M))*TEMP
            SFIM=SFIM+SN*AF(ITYPE(M))*TEMP
            IF (ABSCTL(1:2).EQ.'YE'.AND.I.LE.NBEAMS) THEN
```

```
C
C  CALL ATOM SUBROUTINE FOR CALCULATION OF ABSORPTION POTENTIAL
C
                  ELNAME(1:2)=LABEL(ITYPE(M))(1:2)
                  MATM=DW(M)
                  CALL ATOM(ELNAME,MATM,SATM,KVATM,FEATM,FAATM,STATM)
                  IF (STATM.LT.0.0) THEN
                WRITE (NW,600) LABEL(ITYPE(M)),GH
                  END IF
                  AFRE=AFRE+CN*FAATM*TEMP
                  AFIM=AFIM+SN*FAATM*TEMP
              END IF
           END DO
           SFRE=SFRE*RATIO
           SFIM=SFIM*RATIO
           AFRE=AFRE*RATIO
           AFIM=AFIM*RATIO
1700       UGHR(I,J)=SFRE-AFIM
           UGHI(I,J)=SFIM+AFRE
           UGHR(J,I)=SFRE+AFIM
           UGHI(J,I)=-SFIM+AFRE
C
C  STRUCTURE FACTOR LISTING HERE
C
      IF (J.EQ.1) THEN
         TEMP1=SFRE
         TEMP2=SFIM
         TEMP1=SQRT(TEMP1*TEMP1+TEMP2*TEMP2)
         TEMP2=180.0*ASIN(TEMP2/TEMP1)/PI
         IF (SFRE.LT.0) TEMP2=SIGN(180.0-ABS(TEMP2),TEMP2)
         SFRE=TEMP1
         SFIM=TEMP2
         TEMP1=AFRE
         TEMP2=AFIM
         TEMP1=SQRT(TEMP1*TEMP1+TEMP2*TEMP2)
         IF (TEMP1.GT.DELTA) THEN
            TEMP2=180.0*ASIN(TEMP2/TEMP1)/PI
            IF (AFRE.LT.0) TEMP2=SIGN(180.0-ABS(TEMP2),TEMP2)
         ELSE
            TEMP2=0.0
         END IF
         WRITE (NW,480) I,J,(IFIX(GH(M)),M=1,3),SFRE,SFIM,TEMP1,TEMP2
      END IF
         END DO
      END DO
      OPEN (UNIT=7,FILE='BLOCH.TMP',STATUS='NEW',FORM='UNFORMATTED')
      DO J=1,NBEAMS
         WRITE (7) (UGHR(I,J),UGHI(I,J),I=1,NBEAMS)
      END DO
      CLOSE (UNIT=7)
C
C  SET UP BEAM DIRECTION HERE
C
      IF (NX.GT.1) THEN
        DO I=1,3
           TILTX(I)=(TILT0(2,I)-TILT0(1,I))/FLOAT(NX-1)
        END DO
      END IF
      IF (NY.GT.1) THEN
        DO I=1,3
           TILTY(I)=(TILT0(3,I)-TILT0(1,I))/FLOAT(NY-1)
        END DO
      END IF
      I1=0
      DO J=1,NY
        DO I=1,NX
           I1=I1+1
           NET(1,I1)=I-1
           NET(2,I1)=J-1
        END DO
      END DO
      NTILT=I1
```

```
C
C DO LOOP FOR CBED
C
      DO I1=1,NTILT
      DO I=1,3
        TILT(I)=NET(1,I1)*TILTX(I)+NET(2,I1)*TILTY(I)+TILT0(1,I)
      END DO
      IF (I1.GT.1) THEN
        OPEN (UNIT=7,FILE='BLOCH.TMP',FORM='UNFORMATTED',STATUS='OLD')
        DO J=1,NBEAMS
          READ (7) (UGHR(I,J),UGHI(I,J),I=1,NBEAMS)
        END DO
        CLOSE (UNIT=7)
      END IF
      UGHR(1,1)=0.0
      UGHI(1,1)=ABM*U0
C
C EXCITATION ERROR
C
      TEMP1=SUM(GMXR,TILT,TILT)
      BIGKZ=SQRT(BIGK*BIGK-TEMP1)
      DO I=2,NBEAMS+NPEAMS
        DO J=1,3
          GG(J)=HKL(J,I)
          GH(J)=2.0*TILT(J)+GG(J)
        END DO
        SG=SUM(GMXR,GH,GG)
        SG=2.0*BIGKZ*(ZONE(1)*GG(1)+ZONE(2)*GG(2)+ZONE(3)*GG(3))-SG
        UGHR(I,I)=SG
        UGHI(I,I)=ABM*U0
      END DO
C
C CALCULATES THE BETHE PERTURBATION POTENTIAL
C
      DO I=1,NBEAMS
        DO J=1,I-1,1
          TEMP1=0.0
          TEMP2=0.0
          DO K=NBEAMS+1,NBEAMS+NPEAMS
TEMP1=TEMP1+(UGHR(I,K)*UGHR(K,J)-UGHI(I,K)*UGHI(K,J))/UGHR(K,K)
TEMP2=TEMP2+(UGHR(I,K)*UGHI(K,J)+UGHI(I,K)*UGHR(K,J))/UGHR(K,K)
          END DO
          UGHR(I,J)=UGHR(I,J)-TEMP1
          UGHI(I,J)=UGHI(I,J)-TEMP2
          UGHR(J,I)=UGHR(J,I)-TEMP1
          UGHI(J,I)=UGHI(J,I)+TEMP2
        END DO
      END DO
      DO I=1,NBEAMS
        TEMP1=0.0
        DO K=NBEAMS+1,NBEAMS+NPEAMS
          TEMP1=TEMP1+(UGHR(I,K)**2+UGHI(I,K)**2)/UGHR(K,K)
        END DO
        UGHR(I,I)=UGHR(I,I)-TEMP1
      END DO
      CALL EISPACK(NM,NBEAMS,UGHR,UGHI,VR,VI,CCR,CCI,IERR)
      IF (IERR.NE.0) THEN
        WRITE (NW,*) 'EISPACK FAILS'
        STOP
      END IF
C
C OBTAIN THE FIRST COLUMN OF C-INVERSE MATRIX
C
      DO J=1,NBEAMS
        DO I=1,NBEAMS
          CVT(I,J)=CMPLX(CCR(I,J),CCI(I,J))
        END DO
      END DO
      CALL INVERSE(NM,NBEAMS,CVT,CINV,IERR)
      IF (IERR.EQ.0) THEN
        WRITE (NW,*) 'INVERSE FAILS'
        STOP
      END IF
```

```
C
C CALCULATE THE BEAM INTENSITY AND OUTPUT
C
      DO I=1,NDEPTH
        DO M=1,NOUT
          J=ID(M)
          SFRE=0.0
          SFIM=0.0
          DO K=1,NBEAMS
            C1=CMPLX(CCR(J,K),CCI(J,K))
C BIGKZ SHOULD BE NEGATIVE (ALONG NEGATIVE SURFACE NORMAL)
            TEMP=-PI*VR(K)*DEPTH(I)/BIGKZ
            CN1=DCOS(TEMP)
            SN1=DSIN(TEMP)
            C2=CMPLX(CN1,SN1)
            TEMP1=PI*VI(K)*DEPTH(I)/BIGKZ
            DECAY=EXP(-TEMP1)
            C1=C1*C2*CINV(K)*DECAY
            SFRE=SFRE+REAL(C1)
            SFIM=SFIM+AIMAG(C1)
          END DO
          IPOS=(I1-1)*NDEPTH*NOUT+(I-1)*NOUT+M
          BINT(IPOS)=SFRE*SFRE+SFIM*SFIM
        END DO
      END DO
      END DO
C
C OUPUT INTENSITY
C
      NP=NTILT*NDEPTH*NOUT+4
      BINT(NP-3)=NOUT
      BINT(NP-2)=NDEPTH
      BINT(NP-1)=NX
      BINT(NP)=NY
      WRITE (IW) NP,(BINT(J),J=1,NP)
      END

      SUBROUTINE INVERSE(NM,N,A,B,IERR)
C
C     TO CALCULATE THE FIRST ROW OF C-INVERSE BY GAUSSIAN ELIMINATION
C     AND BACK SUBSTITUTION
C       BY J.M. ZUO OCT. 1987
C
      COMPLEX A(NM,N),T,ZERO,B(N)
      INTEGER IPVT(100),KP1,JPVT,IERR,NM1
      IERR=100
      ZERO=CMPLX(0.0,0.0)
      IF (N.EQ.1) GOTO 35
      AMX=0.0
C
C     PIVOTING FOR THE FIRST ROW
C
      DO 5 J=1,N
        IF (CABS(A(1,J)).GT.AMX) THEN
          AMX=CABS(A(1,J))
          JPVT=J
        ELSE
        END IF
5     CONTINUE
      DO 10 I=1,N
        T=A(I,JPVT)
        A(I,JPVT)=A(I,N)
        A(I,N)=T
10    CONTINUE
C
C     INTERCHANGE FIRT ROW AND LAST ROW
```

```
C
      DO 12 I=1,N
        T=A(1,I)
        A(1,I)=A(N,I)
        A(N,I)=T
12    CONTINUE
C
C     GAUSSIAN ELIMINATION WITH PARTIAL PIVOTING
C
      NM1=N-1
      DO 35 K=1,NM1
        KP1=K+1
        M=K
        IF (KP1.EQ.NM1.OR.K.EQ.NM1) THEN
          DO 15 I=KP1,NM1
            IF(CABS(A(I,K)).GT.CABS(A(M,K))) M=I
15        CONTINUE
        ENDIF
        IPVT(K)=M
        T=A(M,K)
        A(M,K)=A(K,K)
        A(K,K)=T
        IF (CABS(T).LE.1.0E-07) GOTO 35
        DO 20 I=KP1,N
          A(I,K)=-A(I,K)/T
20      CONTINUE
        DO 30 J=KP1,N
          T=A(M,J)
          A(M,J)=A(K,J)
          A(K,J)=T
          IF (T.EQ.ZERO) GOTO 30
          DO 25 I=KP1,N
            A(I,J)=A(I,J)+A(I,K)*T
25        CONTINUE
30      CONTINUE
35    CONTINUE
      IF (CABS(A(N,N)).LE.1.0E-07) IERR=0
C
C     BACK SUBSTITUTION TO GET X
C
      B(N)=1.0/A(N,N)
      DO 50 KB=2,N
        K=N-KB+1
        KP1=K+1
        B(K)=ZERO
        DO 40 J=KP1,N
40        B(K)=B(K)-A(K,J)*B(J)
        B(K)=B(K)/A(K,K)
50    CONTINUE
      T=B(JPVT)
      B(JPVT)=B(N)
      B(N)=T
      RETURN
      END
      SUBROUTINE EISPACK(NM,N,AR,AI,WR,WI,VR,VI,IERR)
C
C     EIGENVALUES AND EIGENVECTORS OF A COMPLEX GENERAL MATRIX
C     CALLING EISPACK SUBROUTINE (B. T. SMITH, LECTURE NOTES IN COMPUTER
C     SCIENCE, VOL 6, SPRINGER-VERLAG 1976)
C     SUBROUTINES CBAL, CORTH, COMQR2, CBABK2 AND RELATED SUBROUTINES
C     ARE COPIED FROM NETLIB@RESEARCH.ATT.COM
C
C     NM ROW DIMENSION OF THW TWO-DIMENSIONAL MATRIX, MAXIMUM SET AT 300
C     N  ORDER OF THE MATRIX
C     AR,AI REAL AND IMAGINARY PART OF THE MATRIX TO BE DIAGONALIZED
C           AR,AI DESTROYED AFTER THE CALL
C     WR,WI RETURNED REAL AND IMAGINARY PART OF THE EIGENVALUES
C     VR,VI RETURNED REAL AND IMAGINARY PART OF THE EIGENVECTOR MATRIX
C     IERR ERROR FLAG, 0 SUCCESS, I HAS PROBLEM WITH ITH EIGENVALUES
```

```
C
      INTEGER NM,N,IS1,IS2
      REAL AR(NM,N),AI(NM,N),WR(N),WI(N),VR(NM,N),VI(NM,N),FV1(300)
      REAL FV2(300),FV3(300)
C
      CALL CBAL(NM,N,AR,AI,IS1,IS2,FV1)
      CALL CORTH(NM,N,IS1,IS2,AR,AI,FV2,FV3)
      CALL COMQR2(NM,N,IS1,IS2,FV2,FV3,AR,AI,WR,WI,VR,VI,IERR)
      IF (IERR.NE.0) RETURN
      CALL CBABK2(NM,N,IS1,IS2,FV1,N,VR,VI)
      RETURN
      END
c
c example input data
c
A TEST DATA
200.0
0 1 2
0.0 0.0 0.0
0.0 0.0 0.0
0.0 0.0 0.0
1 1
6.08112 6.08112 6.08112 90.0 90.0 90.0
8 2
ZN
0 14.0743 3.2655 7.0318 0.2333 5.1652 10.3163 2.4100 58.7097 1.3041
TE
0 19.9644 4.81742 19.0138 0.420885 6.14487 28.5284 2.5239 70.8403 4.3520
1 0.0  0.0  0.0  0.0
1 0.0  0.50 0.50 0.0
1 0.50 0.0  0.50 0.0
1 0.50 0.50 0.0  0.0
2 0.25 0.25 0.25 0.0
2 0.25 0.75 0.75 0.0
2 0.75 0.25 0.75 0.0
2 0.75 0.75 0.25 0.0
AUTO
2 0 0
0 4 -2
1 -1 1
1 8.0
23
AUTO
500 2.15
4
0 0 0, 2 0 0, -2 0 0, 0 4 -2
0
0.0
NO
```

## A5.3. MULTISLICE PROGRAMS

The use of the multislice algorithm for dynamical computations for perfect crystals has been reviewed extensively by Buseck *et al.* (1989). Additional reviews, emphasizing computational aspects, can be found in Cowley (1981), Self *et al.* (1983), Spence (1988), and Krakow and O'Keefe (1989). If sufficiently small slices are used (of thickness less than the periodicity of the crystal in the beam direction), the multislice algorithm treats three-dimensional scattering exactly, and so can be used for calculations of HOLZ intensities (Lynch, 1971). A problem arises, however, since, according to Bethe's original supposition, only those reciprocal lattice points which are close to the Ewald sphere make a significant contribution to the scattering, yet the formulation of the multislice algorithm (with very thin slices) automatically includes *all* reciprocal lattice points, regardless of their proximity to the sphere. Several modifications to the algorithm have been proposed to address this problem [see, for example, Van Dyck (1983) and Kilaas et al. (1987)].

In particular, if HOLZ reflections are to be included in a multislice program, the following points should be noted:

1. Many slices will be required per repeat period of the crystal in the beam direction. For greatest accuracy, the atomic planes should lie in the center of the slices. A common procedure is to take one slice per plane of atoms.
2. To save time, all the different phase gratings may be computed first, then stored for cyclical use during execution of the multislice.
3. Errors will be minimized if the number of beams used and the slice thickness $\Delta z$ are chosen such that $2\pi S_g \Delta z < 2\pi$ for all the reflections in the calculation.
4. Additional "termination" reflections will be seen which do not lie on the reciprocal lattice for the infinite crystal (Lynch, 1971). [Experimental evidence for these (which may also result from twinning) is given by Cherns (1974).]

The use of an artificial superlattice method for computing coherent CBED patterns was described by Spence (1978) and has since been extensively developed by the Cornell group (Loane et al., 1988). For recent reviews, see Kirkland et al. (1989) and Krakow and O'Keefe (1989). The method has been extended to allow inclusion of phonon excitations by Wang and Cowley (1989) and Loane et al. (1991).

Programs for computing single-beam images or CBED patterns from defects using the column approximation are listed in Head et al. (1973).

To summarize, three types of CBED patterns may broadly be distinguished: (1) Those obtained with a subnanometer probe on field-emission instruments; these may be obtained either from perfect crystals or from defects. (2) Patterns from defects or strained crystals recorded under incoherent conditions using a conventional electron source. (3) Patterns from perfect crystals obtained with conventional sources.

For the calculation of microdiffraction patterns obtained using a coherent, subnanometer-sized probe, the multislice artificial superlattice method (with the probe wavefunction in the first slice) appears to provide the most efficient algorithm, particularly for defects, or where overlapping orders and STEM imaging are involved. For perfect crystals this method will also be most efficient if the projected unit-cell dimensions normal to the beam are large and HOLZ interactions are not to be included. For coherent nanodiffraction patterns from small unit-cell perfect crystals, the Bloch-wave method will be more efficient. The efficiency of the multislice algorithm derives largely from its ability to use modern array processors for fast Fourier analysis.

CBED patterns from strained crystals obtained with a conventional electron source may be computed using the existing software developed for the simulation of diffraction-contrast TEM single-beam images, as described in Section 8.5.

For patterns obtained from small unit-cell crystals using conventional electron sources (LaB$_6$ and W), the Bloch-wave method will be most efficient, particularly for large thicknesses and high-index zone axes. Here the incorporation of HOLZ reflections is greatly facilitated. As an alternative in this case, and regardless of source size, a separate multislice computation can be performed for each (plane-wave) direction of incidence.

A comparison of the speed of these various algorithms for different structures (i.e., different densities of points in reciprocal space) and thicknesses would be most useful. The publication of CBED rocking curves for some standard conditions would also be useful in order to test

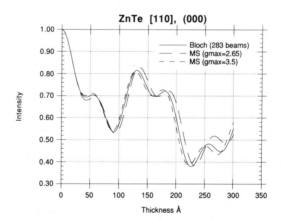

*Figure A5.2.* Thickness dependence of the (000) beam at 200 kV for acentric ZnTe along the [110] zone axis. ZOLZ reflections only have been included, without absorption. The Bloch-wave and multislice methods are compared for different numbers of beams (see text).

for errors in the programs. We suggest, therefore, that researchers may wish to repeat the calculations given in Section 5.2 for CdS, in Section 4.2 for MgO, in Section 5.1 for BeO, and in Fig. A5.2.

A multislice program which uses the fast Fourier transform (FFT) algorithm for convolution is listed below. A comparison between the multislice method and the Bloch-wave method is shown in Fig. A5.2. In this figure, "Bloch" and "MS" indicate Bloch-wave and multislice methods, respectively. The Bloch-wave calculations include 283 beams for ZnTe at 200 kV along the [110] zone axis, including ZOLZ reflections only. Two multislice calculations are performed, one with a maximum reciprocal lattice vector $|g|$ of gmax $=2.36$ (283 beams included), and the other with gmax $=3.5$ (1036 beams included), again using ZOLZ reflections only. The example data given should reproduce the multislice calculations for gmax $= 3.5$.

```
      PROGRAM MULTIS
C
C HIGH ENERGY ELECTRON DIFFRACTION BY COWLEY-MOODIE MULTISLICE METHOD
C VERSION 1
C BY J. M. ZUO, ASU, SEPT, 1989.
C
C THIS PROGRAM INCLUDES FOLLOWING FEATURES
C 1) FAST FOURIER TRANSFORM METHOD FOR CONVOLUTION
C 2) FLEXIBLE ARRAY SIZE, WHICH IS SET AT 131072. THIS CAN BE CHANGED BY
C    SET 'MAP = DESIRED (ALLOWED BY COMPUTER) SIZE' IN FILE MULTIS.FTN
C 3) SAMPLING NEED NOT TO BE SQUARE
C 4) HOLZ INCLUDED IF SLICE IS THIN ENOUGH
C 5) ABSORPTION IS NOT INCLUDED
C 6) NONCENTROCRYSTAL O.K.
C 7) LONG STACKING SEQUENCE IN BEAM DIRECTION IS O.K., NO LIMITATION ON
C    NUMBER OF DIFFERENT SLICES IN PRINCIPLE.
C
      IMPLICIT   INTEGER (H,I,J,K,L,M,N)

      INCLUDE       'MULTIS.FTN'
C
C FIVE INPUT/OUTPUT CHANNELS ARE USED IN THIS PROGRAM
C
C IN  = 8    FOR INPUT CONTROL DATA
C IW  = 9    FOR OUTPUT LISTING
C IPG = 10   FOR PHASE GRATING STORAGE AND READING
C IPR = 11   FOR PROPAGATOR STORAGE AND READING
C IMS = 12   FOR MULTISLICE RESULTS STORAGE
C
      PARAMETER  (MSL=20,MST=1000)
      REAL           KV,TILT(2)
      REAL*8         TEMP
      COMPLEX        BEAM(MSL,MST)
      INTEGER        INDZ(3),MSEQ(MST),MSOUT(MST),IB(MSL),
     *               HOUT(MSL),KOUT(MSL)
      CHARACTER TITLE*80,NAME1*6,COMND*80,MESSAGE*60
      CHARACTER*12   PGNAME(MSL),PRNAME,PVNAME,WAVE,WAVE0,BMNAME

100   FORMAT(80A)
110   FORMAT(/,' TRANSMISSION ELECTRON DIFFRACTION BY MULTISLICE METHOD',
     *       /,' WRITTEN BY J.M. ZUO, ASU, SEPT.,1989. VERSION 1',//,X,
     *        80A)
150   FORMAT(A8)
```

```
220   FORMAT(/,' INCIDENT ELETRON BEAM HIGH VOLTAGE ',F12.5,' KV',/,
     *                              WAVELENGTH ',F12.6,' A',/,
     *              '      INTERACTION CONSTANT (SIGMA) ',E12.5)
230   FORMAT(/,' NEAREST ZONE AXIS ',3I5,/
     *          ,' INCIDENT BEAM DIRECTION IN THE ZONE AXIS PLANE',2F10.5)
240   FORMAT(/,' THERE ARE ',I2,' DIFFERENT SLICES. FOR EACH SLICES,',
     *  /,' THERE ARE ',2I5,' SAMPLING POINTS IN X AND Y DIRECTION')
260   FORMAT(/,' THE PHASE GRATING RESULT IS STORED IN FOLLOWING FILES',
     8           '  FOR EACH SLICE:',/)
270   FORMAT(5(12A,2X))
275   FORMAT(/,' THE PROJECTED POTENTIAL IS STORED IN ',12A)
320   FORMAT(/,' THE 2-D CELL OF THE DIRECT SPACE IS DEFINED BY',/,
     *            '         a = ',F9.5,/,
     *            '         b = ',F9.5,/,
     *            '     gamma = ',F9.5)
340   FORMAT(/,' THE SLICE THICKNESS IS ',F8.5,' ANGSTROMS',/,
     *         ' THE ORIGN OF SLICE IS SHIFTED BY ',2F8.5,' FOR EACH'
     *         ' SLICE',/,
     *         ' THERE ARE',I2,' DIFFERENT ATOMIC SPECIES IN ALL SLICES'
     *         'THEY ARE')
360   FORMAT(X,I2,X,A8,'Z = ',F4.0,/,9F9.5,/)
380   FORMAT(/,' THE 2-D CELL OF THE RECIPROCAL SPACE IS DEFINED BY',/,
     *            '        a* = ',F9.5,/,
     *            '        b* = ',F9.5,/,
     *            '    gamma* = ',F9.5)
400   FORMAT(/,' THE PROPAGATOR IS STORED IN FILE ',A12)
420   FORMAT(/,' TOTAL',I4,' TIMES MULTISLICE CALCULATION ARE PERFORMED'
     *      /,' THE SLICE SEQUENCE IS AS FOLLOWING')
425   FORMAT(I3,29I2)
430   FORMAT(/,' TOTAL',I4,' MS CALCULATION RESULTS ARE OUTPUTED, THEY'
     *         ,' ARE NUMBER')
435   FORMAT(I5,14I4)
440   FORMAT(/,' THE MS RESULTS ARE STORED IN FILE ',A12)
460   FORMAT(/,' THE LAST SLICE IN FILE ',A12,' IS USED AS INITIAL WAVE')

      READ (IN,100) TITLE
      WRITE (IW,110) TITLE

C
C KV      HIGH VOLTAGE IN KILOVOLTS
C INDZ    ZONE AXIS INDICIES
C TILT    INCIDENT BEAM DIRECTION SPECIFIED BY THE FRACTIONAL COORDINATES
OF
C     A VECTOR IN THE RECIPROCAL PLANE OF THE ZONE AXIS.
C     IF ISCTL = 0, TILT IS SPECIFIED IN REFERENCE TO THE TWO BASE
C     VECTORS IN THE ZONE AS ENTERED LATE.
C
      READ (IN,*) KV,INDZ,TILT
      WAVEL=0.3878314/SQRT(KV*(1.0+0.97846707E-03*KV))
      SIGMA=0.020886416*WAVEL*(1.0+1.9569341E-03*KV)
      READ (IN,*) NSLICE,MESHX,MESHY
      WRITE (IW,220) KV,WAVEL,SIGMA
      WRITE (IW,230) INDZ,TILT
      WRITE (IW,240) NSLICE,MESHX,MESHY
C
C MSL = 20 IS THE MAXIMUM NUMBER OF SLICES ALLOWED IN THIS PROGRAM,
HOWEVER
C THIS CAN BE CHANGED WITHOUT ANY DIFFICULTY BY CHANGE THE VALUE OF MSL.
C
      IF (NSLICE.GT.MSL) THEN
         MESSAGE='EXCEEDING MAXIMUM ALLOWED NUMBER OF SLICES'
      PRINT *, MESSAGE
      END IF
      MT=MESHX*MESHY
      ISL=IN
      READ (ISL,*) (CELL2D(I),I=1,3)
      WRITE (IW,320) (CELL2D(I),I=1,3)
      CELL2D(3)=COS(3.141592654*CELL2D(3)/180.0)
```

```
      READ (ISL,*) DELTAZ,XSHIFT,YSHIFT
      READ (ISL,*) NTYPE
      WRITE (IW,340) DELTAZ,XSHIFT,YSHIFT,NTYPE
      DO I=1,NTYPE
        READ (ISL,150) LABEL(I)
        READ (ISL,*) ZT(I),(AX(J,I),BX(J,I),J=1,4),CX(I)
        IF (ZT(I).EQ.0) THEN
          ZT(I)=CX(I)
          DO J=1,4
            ZT(I)=AX(J,I)+ZT(I)
          END DO
        END IF
        WRITE (IW,360) I,LABEL(I),ZT(I),(AX(J,I),BX(J,I),J=1,4),
     *                 CX(I)
      END DO
C
C GET R-SPACE PARAMETERS
C
      VOL2D=CELL2D(1)*CELL2D(2)*SQRT(1-CELL2D(3)*CELL2D(3))
      CELL2R(1)=CELL2D(2)/VOL2D
      CELL2R(2)=CELL2D(1)/VOL2D
      CELL2R(3)=-CELL2D(3)
      GAMMA=180.0*ACOS(CELL2R(3))/3.141592654
      WRITE (IW,380) CELL2R(1),CELL2R(2),GAMMA
C
C DSAP MAXIMUM d(hkl) INCLUDED
C
      READ (IN,*) DSAP
C
C DSMAX MAXIMUM d(hkl) FOR STRUCTURE FACTOR LISTING
C
      READ (IN,*) DSMAX
C
C GET PGNAME
C
      READ (IN,100) NAME1
      CALL FNAME(PGNAME,NSLICE,NAME1)
      WRITE (IW,260)
      WRITE (IW,270) (PGNAME(I),I=1,NSLICE)
C
C SF AND PG CALCULATION
C
      DO I=1,NSLICE
        CALL SFMAP(ISL,I,PGNAME(I),DSAP)
        CALL SFLIST(DSMAX,I)
C
C THE OPERATION OF SFLIST DESTROYS THE PMAP
C
        OPEN (UNIT=IPG,FILE=PGNAME(I),STATUS='OLD',FORM='UNFORMATTED')
        READ (IPG) (PMAP(K),K=1,MT)
        CLOSE (UNIT=IPG)
        CALL OSHIFT
        CALL FFT(-1)
        CALL PGMAP(SIGMA,PGNAME(I))
      END DO
C
C CALCULATES THE PROPAGATOR HERE
C
      READ (IN,100) PRNAME
      WRITE (IW,400) PRNAME
      CALL PROPG(TILT,WAVEL,DELTAZ,XSHIFT,YSHIFT,PRNAME)
      READ (IN,*) MS,(MSEQ(I),I=1,MS)
      WRITE (IW,420) MS
      WRITE (IW,425) (MSEQ(I),I=1,MS)
C
C MSOUT SPECIFYS WHICH SLICE TO BE OUTPUTED, MOUT THE TOTAL NUMBER
C
      READ (IN,*) MOUT,(MSOUT(I),I=1,MOUT)
      WRITE (IW,430) MOUT
      WRITE (IW,435) (MSOUT(I),I=1,MOUT)
```

```
C
C WAVE THE FILE TO STORE THE MS RESULTS
C
      READ (IN,100) WAVE
      WRITE (IW,440) WAVE
      OPEN (UNIT=IMS,FILE=WAVE,STATUS='NEW',FORM='UNFORMATTED')
      JPG=MSEQ(1)
      TEMP=1.0/SQRT(FLOAT(MT))
      WRITE (IMS) MOUT,MESHX,MESHY
      WRITE (IMS) CELL2D
      OPEN (UNIT=IPG,FILE=PGNAME(JPG),STATUS='OLD',FORM='UNFORMATTED')
      DO J=1,MESHY
        J1=(J-1)*MESHX
        READ (IPG) (PMAP(J1+K),K=1,MESHX)
      END DO
      CLOSE (UNIT=IPG)
      DO I=1,MT
        PMAP(I)=TEMP*PMAP(I)
      END DO
C
C PREPARATION FOR PENDULUSUNG OUTPUT
C
      READ (IN,100) BMNAME
      READ (IN,*) MBOUT,(HOUT(I),KOUT(I),I=1,MBOUT)
      WRITE (IW,470) MBOUT
      WRITE (IW,475) (HOUT(I),KOUT(I),I=1,MBOUT)
  470 FORMAT (/,' TOTAL',I4,' BEAMS ARE OUTPUTED, THEY ARE')
  475 FORMAT ('         (H, K) ',2I5)
      DO I=1,MBOUT
        IF (HOUT(I).LT.0) HOUT(I)=MESHX+HOUT(I)
        IF (KOUT(I).LT.0) KOUT(I)=MESHY+KOUT(I)
        HOUT(I)=HOUT(I)+1
        KOUT(I)=KOUT(I)+1
        IB(I)=(KOUT(I)-1)*MESHX+HOUT(I)
      END DO
C
C MULTISLICE LOOP
C
      J=1
      I=1
 9999 CONTINUE
      CALL FFT(1)
      CALL MULTR(IPR,PRNAME)
C
C SELECTED SLICE OUTPUT
C
      IF (J.EQ.MSOUT(I)) THEN
        I=I+1
        CALL OUTPUT(MT)
      END IF
      DO I=1,MBOUT
        BEAM(I,J)=PMAP(IB(I))
      END DO
      CALL FFT(-1)
C
C EXIT HERE
C
      IF (J.EQ.MS) THEN
        CLOSE (UNIT=IMS)
        OPEN (UNIT=7,FILE=BMNAME,STATUS='NEW',FORM='UNFORMATTED')
        WRITE (7) MBOUT,J
        WRITE (7) ((BEAM(I,K),K=1,J),I=1,MBOUT)
        CLOSE (UNIT=7)
        CALL UTEST(J)
        STOP
      END IF
      J=J+1
      JPG=MSEQ(J)
      CALL MULT(IPG,PGNAME(JPG))
      GOTO 9999
      END
```

# COMPUTER PROGRAMS

```
      SUBROUTINE FNAME(PG,NSL,N1)
C
C GET PG FILENAME WHICH HAS A FORM XXXXXXIJ.PG, XXXXXX CHARACTER STRING
C GIVEN BY USER WITH MAXIMUM LENGTH 6, IJ THE SLICE'S NUMBER LIKE 1,2,
ETC.
C
      CHARACTER*12   PG(NSL)
      CHARACTER N1*6,N2*2,MESSAGE*60

100   FORMAT(I1)
200   FORMAT(I2)

      DO J=6,1,-1
        IF (N1(J:J).NE.' ') GOTO 300
      END DO
      MESSAGE='INPUT ERROR IN PG FILE NAME, DEFAULT IS ASSUMED'
      PRINT *,MESSAGE
      N1='PSLICE'
      J=6
300   DO I=1,NSL
        IF (I.LE.9) THEN
          ENCODE (1,100,N2) I
        ELSE
          ENCODE (2,200,N2) I
        END IF
        PG(I)=N1(1:J)//N2//'.PG'
      END DO
      RETURN
      END

      SUBROUTINE OUTPUT(MT)

      INCLUDE         'MULTIS.FTN'

      WRITE (IMS) (PMAP(I),I=1,MT)
      RETURN
      END

      SUBROUTINE UTEST(J)
      INCLUDE         'MULTIS.FTN'
      REAL*8     ONE,TEMPR,TEMPI
100   FORMAT(/,' THE TOTAL WAVE INTENSITY IN SLICE',I4,' IS',E16.6)
      MT=MESHX*MESHY
      ONE=0.0
      DO I=1,MT
        TEMPR=REAL(PMAP(I))
        TEMPI=AIMAG(PMAP(I))
        ONE=ONE+TEMPR*TEMPR+TEMPI*TEMPI
      END DO
      WRITE (IW,100) J,ONE
      RETURN
      END

      SUBROUTINE SFMAP(INS,IS,PGNAME,DSAP)
C
C CALCULATE THE STRUCTURE FACTOR
C
      IMPLICIT   INTEGER (H,I,J,K,L,M,N)

      INCLUDE         'MULTIS.FTN'

      PARAMETER  (MA=500)
      REAL       XPS(MA,2),OCC(MA),DW(MA)
      REAL*8         AF(MTYP),SFRE,SFIM,TEMP,CN,SN,VOL,TWOPI
      CHARACTER*12   PGNAME
      INTEGER        ITYPE(MA)
      DATA       DELTA/1.0E-6/
```

# APPENDIX 5

```
50      FORMAT(/,' SLICE NUMBER ',I2)
70      FORMAT(' THERE ARE ',I4,'ATOMS IN THIS SLICE, THEY ARE')
80      FORMAT(2(X,I2,2F10.6,2F7.4))

        TWOPI=6.283185307
        TEMP=CELL2D(3)*CELL2D(3)
        TEMP=DSQRT(1.0-TEMP)
        VOL=CELL2D(1)*CELL2D(2)*TEMP
        VOL=47.878009/VOL
        MX=MESHX/2
        MY=MESHY/2
        READ (INS,*)
  NBALL,(ITYPE(I),XPS(I,1),XPS(I,2),DW(I),OCC(I),I=1,NBALL)
        WRITE (IW,50) IS
        WRITE (IW,70) NBALL
        WRITE (IW,80) (ITYPE(I),XPS(I,1),XPS(I,2),DW(I),OCC(I),I=1,NBALL)
        NUMB=0
        DO J=2,MY+1
          K=J-MY-1
          JY=(J-1)*MESHX
          JP=-K+MY+1
          JP=(JP-1)*MESHX
          DO I=2,MESHX
            H=I-MX-1
            S2=H*H*CELL2R(1)*CELL2R(1)+K*K*CELL2R(2)*CELL2R(2)
       *       +2.0*H*K*CELL2R(1)*CELL2R(2)*CELL2R(3)
            IF (SQRT(S2).GT.DSAP) THEN
              SFRE=0.0
              SFIM=0.0
              GOTO 200
            END IF
            NUMB=NUMB+1
            S2=0.25*S2
            IF (S2.LT.DELTA) THEN
              DO M=1,NTYPE
                AF(M)=0.0
                DO N=1,4
                  AF(M)=AF(M)+AX(N,M)*BX(N,M)
                END DO
                AF(M)=0.023933754*VOL*AF(M)
              END DO
              GOTO 100
            END IF
            DO M=1,NTYPE
              AF(M)=CX(M)
              DO N=1,4
                AF(M)=AF(M)+AX(N,M)*EXP(-BX(N,M)*S2)
              END DO
              AF(M)=0.023933754*VOL*(ZT(M)-AF(M))/S2
            END DO
100         SFRE=0.0
            SFIM=0.0
            DO M=1,NBALL
              TEMP=TWOPI*(FLOAT(H)*XPS(M,1)+FLOAT(K)*XPS(M,2))
              CN=DCOS(TEMP)
              SN=DSIN(TEMP)
              SFRE=SFRE+AF(ITYPE(M))*CN*EXP(-DW(M)*S2)*OCC(M)
              SFIM=SFIM+AF(ITYPE(M))*SN*EXP(-DW(M)*S2)*OCC(M)
            END DO
200         IX=JY+I
            PMAP(IX)=CMPLX(SFRE,SFIM)
            IP=-H+MX+1
            IP=JP+IP
            PMAP(IP)=CMPLX(SFRE,-SFIM)
          END DO
        END DO
        IF (IS.EQ.1) WRITE (IW,250) 2*NUMB
250     FORMAT (/,' TOTAL',I5' BEAMS ARE INCLUDED')
```

```
C     CALL PAD0
      MT=MESHX*MESHY
      OPEN (UNIT=IPG,FILE=PGNAME,STATUS='NEW',FORM='UNFORMATTED')
      WRITE (IPG) (PMAP(K),K=1,MT)
      CLOSE (UNIT=IPG)
      RETURN
      END

      SUBROUTINE SFLIST(DSMAX,IS)

      INCLUDE         'MULTIS.FTN'

      PARAMETER (MS1=MAP/2,MS2=MS1/2,MS3=MS2/2)
      REAL       DS(MS1),TEMP(3),L(3)
      INTEGER*2  H(MS1),K(MS1),HEXT,KEXT
      COMPLEX         PEXT
      DATA       DELTA/1.0E-6/

      EQUIVALENCE    (PMAP(MS1+1),DS(1)),(PMAP(MS1+MS2+1),H(1)),
     *               (PMAP(MS1+MS2+MS3+1),K(1))

5     FORMAT(/,' STRUCTURE FACTOR LISTING FOR SLICE ',I2,
     *         ' UP TO S2 =',F8.5,/,
     *         5X,'H',4X,'K',6X,'REAL',9X,'IMAGE',
     *         10X,'AMP',7X,'PHASE')
10    FORMAT(X,2I4,X,E11.4,X,E11.4,2X,E11.4,X,F9.4)

      WRITE (IW,5) IS,DSMAX
      IMAX=IFIX(2.0*DSMAX/CELL2R(1))
      JMAX=IFIX(2.0*DSMAX/CELL2R(2))
      MX=MESHX/2
      MY=MESHY/2
      IF (IMAX.GT.MX) IMAX=MX
      IF (JMAX.GT.MY) JMAX=MY
      K1=0
      DO J=0,JMAX,1
        J1=J+MY+1
        J1=(J1-1)*MESHX
        DO I=-IMAX,IMAX,1
          I1=I+MX+1
          I1=J1+I1
          K1=K1+1
          PMAP(K1)=PMAP(I1)
        END DO
      END DO
      K1=0
      DO J=0,JMAX,1
        DO I=-IMAX,IMAX,1
          K1=K1+1
          H(K1)=I
          K(K1)=J
          DS(K1)=I*I*CELL2R(1)*CELL2R(1)+J*J*CELL2R(2)*CELL2R(2)
     *           +2.0*I*J*CELL2R(1)*CELL2R(2)*CELL2R(3)
          DS(K1)=0.5*SQRT(DS(K1))
        END DO
      END DO
C
C SORT IN THE ODER OF D DISTANCE
C
      DO I=1,K1-1
        DEXT=DS(I+1)
        PEXT=PMAP(I+1)
        HEXT=H(I+1)
        KEXT=K(I+1)
        DO J=I,1,-1
          IF (DEXT.GE.DS(J)) GOTO 1200
            DS(J+1)=DS(J)
            PMAP(J+1)=PMAP(J)
            H(J+1)=H(J)
            K(J+1)=K(J)
        END DO
```

```
1200      DS(J+1)=DEXT
          PMAP(J+1)=PEXT
          H(J+1)=HEXT
          K(J+1)=KEXT
        END DO
        TEMP(3)=0
        DO I=1,K1
          IF (DS(I).GT.DSMAX) GOTO 2000
          PRE=REAL(PMAP(I))
          PIM=AIMAG(PMAP(I))
          AMP=SQRT(PRE*PRE+PIM*PIM)
          IF (AMP.LT.DELTA) THEN
            AMP=0.0
            PHASE=0.0
          ELSE
            PHASE=180.0*ASIN(PIM/AMP)/3.141592654
            IF (PRE.LT.0.0) THEN
              DEXT=180.0-ABS(PHASE)
              PHASE=SIGN(DEXT,PHASE)
            END IF
          END IF
          WRITE(IW,10) H(I),K(I),PMAP(I),AMP,PHASE
        END DO
2000    RETURN
        END
        SUBROUTINE PGMAP(SIGMA,PGNAME)

        INCLUDE         'MULTIS.FTN'
        CHARACTER*12    PGNAME

        MT=MESHX*MESHY
        DO I=1,MT
          PR=REAL(PMAP(I))
          PI=AIMAG(PMAP(I))
          TEMP=-SIGMA*PR
          PI=EXP(SIGMA*PI)
          CN=COS(TEMP)*PI
          SN=SIN(TEMP)*PI
          PMAP(I)=CMPLX(CN,SN)
        END DO
        OPEN (UNIT=IPG,FILE=PGNAME,STATUS='NEW',FORM='UNFORMATTED')
        DO J=1,MESHY
          J1=(J-1)*MESHX
          WRITE (IPG) (PMAP(J1+K),K=1,MESHX)
        END DO
        CLOSE (UNIT=IPG)
        RETURN
        END

        SUBROUTINE PROPG(TILT,WAVEL,DELTAZ,X0,Y0,PRNAME)
C
C THIS PROGRAM CALCULATES THE FRESNEL PROPAGATOR IN THE RECIPROCAL
C SPACE BY FORMULA EXP(2*PI*i*Sg*Z), Sg IS THE EXCITATION ERROR OF
C RECIPROCAL LATTICE POINT G AND Z IS THE SLICE THICKNESS.
C
        INCLUDE         'MULTIS.FTN'
        REAL       TILT(2)
        CHARACTER*12    PRNAME

        TWOPI=6.2831853
        TWOKR=0.5*WAVEL
        MX=MESHX/2
        MY=MESHY/2
        A2=CELL2R(1)*CELL2R(1)
        B2=CELL2R(2)*CELL2R(2)
        ADOTB=CELL2R(1)*CELL2R(2)*CELL2R(3)
        DO J=2,MESHY
          K=J-MY-1
          J1=(J-1)*MESHX
```

```
      DO I=2,MESHX
        H=I-MX-1
        A1=2.0*TILT(1)+H
        B1=2.0*TILT(2)+K
        SG=-TWOKR*(A1*H*A2+B1*K*B2+(A1*K+B1*H)*ADOTB)
        TEMP=-TWOPI*(SG*DELTAZ+H*X0+K*Y0)
        CN=COS(TEMP)
        SN=SIN(TEMP)
        I1=J1+I
        PMAP(I1)=CMPLX(CN,SN)
      END DO
    END DO
    CALL PAD0
    CALL OSHIFT
    OPEN (UNIT=IPR,FILE=PRNAME,STATUS='NEW',FORM='UNFORMATTED')
    DO J=1,MESHY
      J1=(J-1)*MESHX
      WRITE (IPR) (PMAP(J1+K),K=1,MESHX)
    END DO
    CLOSE (UNIT=IPR)
    RETURN
    END

    SUBROUTINE PAD0

    INCLUDE          'MULTIS.FTN'
    COMPLEX          ZERO

    ZERO=CMPLX(0.0)
    DO I=1,MESHX
      PMAP(I)=ZERO
    END DO
    DO J=2,MESHY
      J1=(J-1)*MESHX+1
      PMAP(J1)=ZERO
    END DO
    RETURN
    END

    SUBROUTINE MULT(INN,NAME)

    INCLUDE          'MULTIS.FTN'
    PARAMETER (MD=MAP/32)
    COMPLEX          QMAP(MD)
    CHARACTER*12     NAME

    OPEN (UNIT=INN,FILE=NAME,STATUS='OLD',FORM='UNFORMATTED')
    K=0
    DO J=1,MESHY
      READ (INN) (QMAP(I),I=1,MESHX)
      DO I=1,MESHX
        K=K+1
        PMAP(K)=PMAP(K)*QMAP(I)
      END DO
    END DO
    CLOSE (UNIT=INN)
    RETURN
    END

    SUBROUTINE MULTR(INN,NAME)

    INCLUDE          'MULTIS.FTN'
    PARAMETER (MD=MAP/32)
    COMPLEX          QMAP(MD)
    CHARACTER*12     NAME
```

```
      OPEN (UNIT=INN,FILE=NAME,STATUS='OLD',FORM='UNFORMATTED')
      K=0
      MT=MESHX*MESHY
      RMT=1.0/FLOAT(MT)
      DO J=1,MESHY
        READ (INN) (QMAP(I),I=1,MESHX)
        DO I=1,MESHX
          K=K+1
          PMAP(K)=RMT*PMAP(K)*QMAP(I)
        END DO
      END DO
      CLOSE (UNIT=INN)
      RETURN
      END

      SUBROUTINE OSHIFT
      INCLUDE    'MULTIS.FTN'
      COMPLEX    TEMP
      MX=MESHX/2
      MY=MESHY/2
      DO J=1,MY
        J1=(J-1)*MESHX
        J2=(J+MY-1)*MESHX
        DO I=1,MX
          I1=J1+I
          I2=J2+I+MX
          TEMP=PMAP(I1)
          PMAP(I1)=PMAP(I2)
          PMAP(I2)=TEMP
        END DO
      END DO
      DO J=MY+1,MESHY
        J1=(J-1)*MESHX
        J2=(J-MY-1)*MESHX
        DO I=1,MX
          I1=J1+I
          I2=J2+I+MX
          TEMP=PMAP(I1)
          PMAP(I1)=PMAP(I2)
          PMAP(I2)=TEMP
        END DO
      END DO
      RETURN
      END

      SUBROUTINE FFT(ISIGN)
C
C FAST FOURIER TRANSFORM USING SUBROUTINE 'FOURN' FROM NUMERICAL RECIPES,
C BY WILLIAM H. PRESS et al., 1986, CAMBRIDGE UNIVERSITY PRESS
C
      INCLUDE         'MULTIS.FTN'
      REAL       DATA(2*MAP)
      INTEGER         NN(2)
      EQUIVALENCE     (PMAP(1),DATA(1))

      NN(1)=MESHX
      NN(2)=MESHY
      CALL FOURN(DATA,NN,2,ISIGN)
      RETURN
      END

      SUBROUTINE FOURN(DATA,NN,NDIM,ISIGN)

      REAL*8          WR,WI,WPR,WPI,WTEMP,THETA
      DIMENSION  NN(NDIM),DATA(*)

      NTOT=1
      DO 11 IDIM=1,NDIM
        NTOT=NTOT*NN(IDIM)
```

# COMPUTER PROGRAMS

```
11      CONTINUE
        NPREV=1
        DO 18 IDIM=1,NDIM
          N=NN(IDIM)
          NREM=NTOT/(N*NPREV)
          IP1=2*NPREV
          IP2=IP1*N
          IP3=IP2*NREM
          I2REV=1
          DO 14 I2=1,IP2,IP1
            IF (I2.LT.I2REV) THEN
              DO 13 I1=I2,I2+IP1-2,2
                DO 12 I3=I1,IP3,IP2
                  I3REV=I2REV+I3-I2
                  TEMPR=DATA(I3)
                  TEMPI=DATA(I3+1)
                  DATA(I3)=DATA(I3REV)
                  DATA(I3+1)=DATA(I3REV+1)
                  DATA(I3REV)=TEMPR
                  DATA(I3REV+1)=TEMPI
12              CONTINUE
13            CONTINUE
            END IF
            IBIT=IP2/2
1           IF ((IBIT.GE.IP1).AND.(I2REV.GT.IBIT)) THEN
              I2REV=I2REV-IBIT
              IBIT=IBIT/2
              GOTO 1
            END IF
            I2REV=I2REV+IBIT
14        CONTINUE
          IFP1=IP1
2         IF (IFP1.LT.IP2) THEN
            IFP2=2*IFP1
            THETA=ISIGN*6.28318530717959D0/(IFP2/IP1)
            WPR=-2.D0*DSIN(0.5D0*THETA)**2
            WPI=DSIN(THETA)
            WR=1.D0
            WI=0.D0
            DO 17 I3=1,IFP1,IP1
              DO 16 I1=I3,I3+IP1-2,2
                DO 15 I2=I1,IP3,IFP2
                  K1=I2
                  K2=K1+IFP1
                  TEMPR=SNGL(WR)*DATA(K2)-SNGL(WI)*DATA(K2+1)
                  TEMPI=SNGL(WR)*DATA(K2+1)+SNGL(WI)*DATA(K2)
                  DATA(K2)=DATA(K1)-TEMPR
                  DATA(K2+1)=DATA(K1+1)-TEMPI
                  DATA(K1)=DATA(K1)+TEMPR
                  DATA(K1+1)=DATA(K1+1)+TEMPI
15              CONTINUE
16            CONTINUE
              WTEMP=WR
              WR=WR*WPR-WI*WPI+WR
              WI=WI*WPR+WTEMP*WPI+WI
17          CONTINUE
            IFP1=IFP2
            GOTO 2
          END IF
          NPREV=N*NPREV
18      CONTINUE
        RETURN
        END
```

```
C
C COMMON DATA FILE 'MULTIS.FTN'
C
      PARAMETER  (MAP=131072)
      COMPLEX       PMAP(MAP)
      COMMON/MAP/   PMAP,MESHX,MESHY

      PARAMETER  (MTYP=10)
      REAL          ZT(MTYP),AX(4,MTYP),BX(4,MTYP),CX(MTYP),
     *              CELL2D(3),CELL2R(3)
      REAL*8        LABEL(MTYP)
      COMMON/SLICE/ CELL2D,CELL2R,NTYPE,LABEL,ZT,AX,BX,CX,DW

      DATA       IN,IW,IPG,IPR,IMS/8,9,10,11,12/
      COMMON/CHAN/   IN,IW,IPG,IPR,IMS

C
C EXAMPLE INPUT DATA
C
A ZNTE TEST DATA FOR VAN DYCK PROPOSAL
200.0 1 1 0 0.0 0.0
1 64 64
4.30 6.08112 90.0
1.075 0.0 0.0
2
ZN
0.0 14.0743 3.2655 7.0318 0.2333 5.1652 10.3163 2.4100 58.7097 1.3041
TE
0.0 19.9644 4.81742 19.0138 0.420885 6.14487 28.5284 2.5239 70.8403
4.3520
3.0
0.5
ZNTE
4
1 0.0 0.0   0.0  0.25
1 0.5 0.5   0.0  0.25
2 0.0 0.25  0.0  0.25
2 0.5 0.75  0.0  0.25
TEST.PRO
280
1 1 1 1 1 1 1 1 1 1   1 1 1 1 1 1 1 1 1 1   1 1 1 1 1 1 1 1 1 1
1 1 1 1 1 1 1 1 1 1   1 1 1 1 1 1 1 1 1 1   1 1 1 1 1 1 1 1 1 1
1 1 1 1 1 1 1 1 1 1   1 1 1 1 1 1 1 1 1 1   1 1 1 1 1 1 1 1 1 1
1 1 1 1 1 1 1 1 1 1   1 1 1 1 1 1 1 1 1 1   1 1 1 1 1 1 1 1 1 1
1 1 1 1 1 1 1 1 1 1   1 1 1 1 1 1 1 1 1 1   1 1 1 1 1 1 1 1 1 1
1 1 1 1 1 1 1 1 1 1   1 1 1 1 1 1 1 1 1 1   1 1 1 1 1 1 1 1 1 1
1 1 1 1 1 1 1 1 1 1   1 1 1 1 1 1 1 1 1 1   1 1 1 1 1 1 1 1 1 1
1 1 1 1 1 1 1 1 1 1   1 1 1 1 1 1 1 1 1 1   1 1 1 1 1 1 1 1 1 1
1 1 1 1 1 1 1 1 1 1   1 1 1 1 1 1 1 1 1 1   1 1 1 1 1 1 1 1 1 1
1 1 1 1 1 1 1 1 1 1
0
ZNTE.MS
ZNTE.PEN
6
0 0,  1 1,  -1 1,  1 -1,  0 4,  5 3
```

# Appendix 6. Crystal Structure Data

The following is a list of commonly occurring crystals, their structure, and their lattice constants. For more accurate data, consult, for example, Pearson (1967).

Table A6.1. Crystal Structures of the Elements[a]

| Atomic number | Element | Structure | Shortest distance between atoms (Å) | Cell constant |
|---|---|---|---|---|
| 1 | H | | | |
| 3 | Li | bcc | 3.04 | 3.491 |
| 4 | Be | cph | 2.24 | 2.27, 3.59 |
| 5 | B | trigonal | 1.71 | |
| 6 | C (graphite) | hexagonal | 1.42 | |
| | C (diamond) | cubic | 1.54 | 3.567 |
| 8 | O | | | |
| 9 | F | | | |
| 11 | Na | bcc | 3.71 | 4.225 |
| 12 | Mg | cph | 3.21 | 3.21, 5.21 |
| 13 | Al | fcc | 2.86 | 4.05 |
| 14 | Si | cubic (diamond) | 2.35 | 5.430 |
| 15 | P (black) | orthorhombic | 2.17 | |
| 16 | S | orthorhombic | 2.12 | |
| 17 | Cl | | | |
| 19 | K | bcc | 4.63 | 5.225 |
| 20 | Ca | fcc | 3.94 | 5.58 |
| 21 | Sc | fcc | 3.21 | 3.31, 5.27 |
| 22 | Ti | cph | 2.95 | 2.95, 4.68 |
| 23 | V | bcc | 2.63 | 3.03 |
| 24 | Cr | bcc | 2.50 | 2.88 |
| 25 | Mn | cubic | 2.24 | |
| 26 | Fe | bcc | 2.48 | 2.87 |

Table A6.1. (continued)

| Atomic number | Element | Structure | Shortest distance between atoms (Å) | Cell constant |
|---|---|---|---|---|
| 27 | Co | cph | 2.51 | 2.51 |
| 28 | Ni | fcc | 2.49 | 3.52 |
| 29 | Cu | fcc | 2.56 | 3.61 |
| 30 | Zn | cph | 2.66 | 2.66, 4.95 |
| 31 | Ga | orthorhombic | 2.44 | |
| 32 | Ge (diamond) | cubic | 2.45 | 5.658 |
| 33 | As | trigonal | 2.90 | |
| 34 | Se | hexagonal | 2.32 | |
| 35 | Br | | | |
| 37 | Rb | bcc | 4.90 | 5.585 |
| 38 | Sr | fcc | 4.30 | 6.08 |
| 39 | Y | cph | 3.62 | 3.65, 5.73 |
| 40 | Zr | cph | 3.20 | 3.23, 5.15 |
| 41 | Nb (Cb) | bcc | 2.86 | 3.30 |
| 42 | Mo | bcc | 2.72 | 3.15 |
| 43 | Tc | cph | 2.70 | 2.74, 4.40 |
| 44 | Ru | cph | 2.65 | 2.71, 4.28 |
| 45 | Rh | fcc | 2.69 | 3.80 |
| 46 | Pd | fcc | 2.75 | 3.89 |
| 47 | Ag | fcc | 2.89 | 4.09 |
| 48 | Cd | cph | 2.98 | 2.98, 5.62 |
| 49 | In | tetragonal | 3.25 | 3.25, 4.95 |
| 50 | Sn (white) | tetragonal | 3.02 | 6.49 |
| 51 | Sb | trigonal | 2.90 | |
| 52 | Te | hexagonal | 2.87 | |
| 53 | I | orthorhombic | 2.71 | |
| 55 | Cs | bcc | 5.25 | 6.045 |
| 56 | Ba | bcc | 4.35 | 5.02 |
| 57 | La | cph | 3.75 | 3.77 |
| 58 | Ce | fcc | 3.65 | 5.16 |
| 59 | Pr | hexagonal | 3.64 | 3.67 |
| 60 | Nd | hexagonal | 3.66 | 3.66 |
| 63 | Eu | bcc | 3.96 | 4.58 |
| 64 | Gd | cph | 3.56 | 3.63, 5.78 |
| 65 | Tb | cph | 3.53 | 3.60, 5.70 |
| 66 | Dy | cph | 3.50 | 3.59, 5.65 |
| 67 | Ho | cph | 3.52 | 3.58, 5.62 |
| 68 | Er | cph | 3.55 | 3.56, 5.59 |
| 69 | Tm | cph | 3.48 | 3.54, 5.56 |
| 70 | Yb | fcc | 3.87 | 5.48 |
| 71 | Lu | cph | 3.46 | 3.50, 5.55 |
| 72 | Hf | cph | 3.16 | 3.19, 5.05 |
| 73 | Ta | bcc | 2.86 | 3.30 |

Table A6.1. (continued)

| Atomic number | Element | Structure | Shortest distance between atoms (Å) | Cell constant |
|---|---|---|---|---|
| 74 | W | bcc | 2.74 | 3.16 |
| 75 | Re | cph | 2.74 | 2.76, 4.46 |
| 76 | Os | cph | 2.68 | 2.74, 4.32 |
| 77 | Ir | fcc | 2.71 | 3.84 |
| 78 | Pt | fcc | 2.77 | 3.92 |
| 79 | Au | fcc | 2.88 | 4.08 |
| 80 | Hg | trigonal | 3.00 | |
| 81 | Tl | cph | 3.41 | 3.46, 5.52 |
| 82 | Pb | fcc | 3.49 | 4.95 |
| 83 | Bi | trigonal | 3.11 | |
| 84 | Po | simple cubic | 3.35 | 3.34 |
| 89 | Ac | fcc | 3.75 | 5.31 |
| 90 | Th | fcc | 3.60 | 5.08 |
| 91 | Pa | bc tetragonal | 3.21 | 3.92, 32.4 |
| 92 | U | orthorhombic | 2.77 | |
| 93 | Np | orthorhombic | 2.60 | |
| 94 | Pu | monoclinic | 3.28 | |

<sup>a</sup> The cell constants for cph structures are given as $a$, $c$ in angstroms.

Table A6.2. Crystals with the Sodium Chloride Structure

| Crystal | $a_0$ (Å) | Crystal | $a_0$ (Å) |
|---|---|---|---|
| *Antimonides* | | *Carbides* | |
| CeSb | 6.40 | HfC | 4.46 |
| LaSb | 6.47 | NbC | 4.47 |
| ScSb | 5.86 | TaC | 4.45 |
| SnSb | 6.13 | TiC | 4.32 |
| ThSb | 6.32 | UC | 4.96 |
| USb | 6.19 | VC | 4.18 |
| *Arsenides* | | ZrC | 4.68 |
| CeAx | 6.06 | *Chlorides* | |
| LaAs | 6.13 | AgCl | 5.55 |
| ScAs | 5.49 | KCl | 6.29 |
| SnAs | 5.68 | LiCl | 5.13 |
| ThAs | 5.97 | NaCl | 5.64 |
| UAs | 5.77 | RbCl | 6.58 |
| *Borides* | | | |
| ZrB | 4.65 | | |

Table A6.2. (continued)

| Crystal | $a_0$ (Å) | Crystal | $a_0$ (Å) |
|---|---|---|---|
| *Bromides* | | *Cyanides* | |
| AgBr | 5.77 | KCN | 6.53 |
| KBr | 6.60 | NaCN | 5.89 |
| LiBr | 5.50 | RbCN | 6.82 |
| *Florides* | | *Phosphides* | |
| AgF | 4.92 | CeP | 5.90 |
| CsF | 6.01 | LaP | 6.01 |
| KF | 5.35 | ThP | 5.82 |
| LiF | 4.03 | UP | 5.59 |
| NaF | 4.62 | ZrP | 5.27 |
| RbF | 5.64 | *Selenides* | |
| *Hydrides* | | BaSe | 6.60 |
| CsH | 6.38 | CaSe | 5.91 |
| KH | 5.70 | CeSe | 5.98 |
| LiH | 4.08 | LaSe | 6.06 |
| HaH | 4.88 | MgSe | 5.45 |
| RbH | 6.04 | MnSe | 5.45 |
| *Iodides* | | PbSe | 6.12 |
| KI | 7.07 | SnSe | 6.02 |
| LiI | 6.00 | SrSe | 6.23 |
| NH$_4$I | 7.26 | ThSe | 5.87 |
| NaI | 6.47 | *Sulfides* | |
| RbI | 7.34 | BaS | 6.39 |
| *Nitrides* | | CaS | 5.69 |
| CeN | 5.01 | CeS | 5.78 |
| CrN | 4.14 | LaS | 5.84 |
| LaN | 5.30 | MgS | 5.20 |
| NbN | 4.70 | MnS | 4.45 |
| ScN | 4.44 | PbS | 5.94 |
| TiN | 4.23 | SrS | 6.02 |
| UN | 4.88 | ThS | 5.68 |
| VN | 4.13 | US | 5.48 |
| ZrN | 4.61 | ZrS | 5.25 |
| *Oxides* | | *Tellurides* | |
| BaO | 5.52 | BaTe | 6.99 |
| CaO | 4.81 | BiTe | 6.47 |
| CdO | 4.70 | CaTe | 6.34 |
| CoO | 4.27 | CeTe | 6.35 |
| FeO | 4.28–4.31 | LaTe | 6.41 |
| MgO | 4.21 | PbTe | 6.45 |
| MnO | 4.45 | SnTe | 6.31 |
| NbO | 4.21 | SrTe | 6.47 |
| NiO | 4.17 | UTe | 6.16 |
| SrO | 5.16 | | |
| TaO | 4.42–4.44 | | |
| TiO | 4.18 | | |

Table A6.3. Crystals with the Cesium Chloride Structure

| Crystal | $a_0$ (Å) |
|---|---|
| AgCd | 3.33 |
| AgMg | 3.28 |
| CuBe | 2.70 |
| CuZn | 2.94 |
| CsBr | 4.29 |
| CsCl | 4.12 |
| CsCN | 4.25 |
| CsI | 4.57 |
| TlBr | 3.97 |
| TlCl | 3.83 |
| TlCN | 3.82 |
| TlI | 4.20 |

Table A6.4. Crystals with the Sphalerite Structure

| Crystal | $a_0$ (Å) | Crystal | $a_0$ (Å) |
|---|---|---|---|
| AgI | 6.47 | GaSb | 6.12 |
| Bn | 3.62 | HgS | 5.85 |
| BeS | 4.85 | HgSe | 6.08 |
| BeSe | 5.07 | HgTe | 6.43 |
| BeTe | 5.54 | InAs | 6.04 |
| CdS | 5.82 | InP | 5.87 |
| CuBr | 5.69 | InSb | 6.48 |
| CuCl | 5.40 | SiC | 4.35 |
| CuF | 4.26 | ZnS | 5.41 |
| CuI | 6.04 | ZnSe | 5.67 |
| GaAs | 5.65 | ZnTe | 6.09 |
| GaP | 5.45 | | |

Table A6.5. Crystals with the Wurtzite Structure

| Crystal | $a_0$ (Å) | $c_0$ (Å) |
|---|---|---|
| AgI | 4.58 | 7.49 |
| AlN | 3.11 | 4.98 |
| BeO | 2.70 | 4.38 |
| CdS | 4.14 | 6.75 |
| CdSe | 4.30 | 7.02 |
| $NH_4F$ | 4.39 | 7.02 |
| SiC | 3.08 | 5.05 |
| TaN | 3.05 | 4.94 |
| ZnO | 3.25 | 5.21 |
| ZnS | 3.81 | 6.23 |
| ZnSe | 3.98 | 6.53 |
| ZnTe | 4.27 | 6.99 |

Table A6.6. Crystals with the Nickel Arsenide Structure

| Crystal | $a_0$ (Å) | $c_0$ (Å) | Crystal | $a_0$ (Å) | $c_0$ (Å) |
|---|---|---|---|---|---|
| CoS | 3.37 | 5.16 | MnTe | 4.14 | 6.70 |
| CoSb | 3.87 | 5.19 | NiAs | 3.60 | 5.01 |
| CoTe | 3.89 | 5.36 | NiSe | 3.66 | 5.36 |
| CrS | 3.45 | 5.75 | NiSn | 4.05 | 5.12 |
| CrSb | 4.11 | 5.44 | NiTe | 3.96 | 5.35 |
| CrSe | 3.68 | 6.02 | PtB[a] | 3.36 | 4.06 |
| FeS | 3.44 | 5.88 | PtBi | 4.31 | 5.49 |
| FeSb | 4.06 | 5.13 | PtSb | 4.13 | 5.47 |
| FeSe | 3.64 | 5.96 | PtSn | 4.10 | 5.43 |
| FeTe | 3.80 | 5.65 | VP | 3.18 | 6.22 |
| MnAs | 3.71 | 5.69 | VS | 3.36 | 5.98 |
| MnSb | 4.12 | 5.78 | VTe | 3.94 | 6.13 |

[a] Anti-nickel arsenide structure.

Table A6.7. Crystals with the Fluorite Structure

| Crystal | $a_0$ (Å) | Crystal | $a_0$ (Å) |
|---|---|---|---|
| $BaCl_2$ | 7.34 | $Li_2S$ | 5.71 |
| $BaF_2$ | 6.20 | $Li_2Se$ | 6.01 |
| $Be_2B$ | 4.67 | $Li_2Te$ | 6.50 |
| $Be_2C$ | 4.33 | $Na_2O$ | 5.55 |
| $CaF_2$ | 5.46 | $Na_2S$ | 6.53 |
| $CdF_2$ | 5.39 | $Na_2Se$ | 6.81 |
| $CeH_2$ | 5.59 | $Na_2Te$ | 7.31 |
| $CeO_2$ | 5.41 | $NbH_2$ | 4.56 |
| $CoSi_2$ | 5.36 | $NiSi_2$ | 5.39 |
| $HfO_2$ | 5.12 | $Rb_2O$ | 6.74 |
| $HgF_2$ | 5.54 | $Rb_2S$ | 7.65 |
| $K_2O$ | 6.44 | $SrCl_2$ | 6.98 |
| $K_2S$ | 7.39 | $SrF_2$ | 5.80 |
| $K_2Se$ | 7.68 | $ThO_2$ | 5.60 |
| $K_2Te$ | 8.15 | $UO_2$ | 5.47 |
| $Li_2O$ | 4.62 | $ZrO_2$ | 5.07 |

Table A6.8. Crystals with the Rutile Structure

| Crystal | $a_0$ (Å) | $c_0$ (Å) | Crystal | $a_0$ (Å) | $c_0$ (Å) |
|---|---|---|---|---|---|
| $CoF_2$ | 4.70 | 3.18 | $NbO_2$ | 4.77 | 2.96 |
| $FeF_2$ | 4.70 | 3.31 | $PbO_2$ | 4.95 | 3.38 |
| $GeO_2$ | 4.40 | 2.86 | $SnO_2$ | 4.74 | 3.19 |
| $MgF_2$ | 4.62 | 3.05 | $TaO_2$ | 4.71 | 3.06 |
| $MnF_2$ | 4.87 | 3.31 | $TiO_2$ | 4.59 | 2.96 |
| $MnO_2$ | 4.40 | 2.87 | $WO_2$ | 4.86 | 2.77 |
| $MoO_2$ | 4.86 | 2.79 | $ZnF_2$ | 4.70 | 3.13 |

# Appendix 7. A Bibliography of CBED Applications Indexed by Material

This appendix contains an alphabetical list of many of the materials which have been studied by CBED, with references to the original papers. The list is intended to prevent repetition of past work, to summarize the knowledge that has been obtained by CBED, and to assist with the indexing of patterns of these materials in future work. Only papers which name a material by chemical formula in the title have been included. Thus many papers on, for example, "CBED of quasi-crystals," or "CBED of oxide superconductors," etc., are not listed. Since systematic naming conventions are not always used ($YBa_2Cu_3O_7$, for instance, should actually be written $Ba_2YCu_3O_7$) it may be necessary to search under several entries. We have simplified many formulas (e.g., by deleting dopants, numbers, etc.) for the purposes of alphabetization.

The list is seen to reflect strongly the interests of the aircraft industry and the semiconductor industry, together with work on the new oxide superconductors.

Al    Cassada, W. A., Shiflet, G. J., and Poon, S. J. (1987). Quasicrystalline grain boundary precipitates in Al-alloy through solid–solid transformations, *J. Microsc.* **146**, 323–335.

Al    Cassada, W. A., Shiflet, G. J., and Starke, Jr., E. A. (1986a). Grain boundary precipitates with five-fold diffraction symmetry in an Al–Li–Cu alloy, *Scr. Metall.* **20**, 751–756.

Al    Chandrasekaran, M., Lin, Y. P., Vincent, R., and Staniek, G. (1988a). On a metastable rhombohedral Al–Fe–Si intermatallic phase. In: *EMAG 87, Analytical Electron Microscopy,* ed. G. W. Lorimer, The Institute of Metals, Manchester, pp. 63–65.

Al    Chandrasekaran, M., Lin, Y. P., Vincent, R., and Staniek, G. (1988b). The structure and stability of some intermetallic phases in rapidly solidified Al–Fe, *Scr. Metall.* **22**, 797–802.

Al   Kaufman, M. J. and Fraser, H. L. (1983b). Metastable phase formation in rapidly solidified submicron powders of Al–30.3 at% Ge eutectic alloy, *Mater. Sci. Eng.* **57**, L17.

Al   Lynch, J. P., Brown, L. M., and Jacobs, M. H. (1982a). Microanalysis of age-hardening precipitates in Al-alloys, *Acta Metall.* **30**, 1389.

Al   Muddle, B. C. and Hugo, G. R. (1990). Microdiffraction analysis of precipitate crystallography and morphology in Al alloys, In *Proc. 12th Int. Congr. on Electron Microscopy*, Vol. 4, San Francisco Press, San Francisco, pp. 954–955.

Al   Skjerpe, P., Gjønnes, J., and Langsrud, Y. (1987c). Solidification structure and primary Al–Fe–Si particles in direct-chilled cast Al alloys, *Ultramicroscopy* **22**, 239–250.

Al   Yamashita, T. (1981). Zero order lines in electron channelling patterns from crystals of Al and iron at various temperatures, *J. Electron. Microsc.* **30**(4), 298.

$Al_2CuLi$   Vecchio, K. S. and Williams, D. B. (1988). CBED analysis of the $T_1(Al_2CuLi)$ phase in Al–Li–Cu alloys, *Metall. Trans.* **19A**, 2885–2891.

$Al_2LiCu$   Boom, G., Last, S., Bronsveld, P. M., and De Hossen, J. T. M. (1988). Determination of the space group symmetry of $Al_2LiCu$ by CBED. In *Proc. 9th European Congr. on Electron Microscopy*, eds. P. J. Goodhew and H. G. Dickinson, Inst. Phys. Conf. Ser., Vol. 93, Institute of Physics, London, pp. 503–504.

$Al_2O_3–BaAl_2O_4$   Yamamoto, N. and O'Keefe, M. (1984). Electron microscopy and diffraction of phases in the $Al_2O_3–BaAl_2O_4$, *Acta Crystallogr.* **B40**, 21–26.

$Al_2O_3$   Liu, P. and Skogsmo, J. (1991). Space-group determination and structure model for $\kappa$-$Al_2O_3$ by CBED, *Acta Crystallogr.* **B47**, 425–433.

$Al_2O_3$   Sklad, P. S. (1989). In-situ observation of the amorphous-to-gamma transformation in ion-implanted $Al_2O_3$. In *Proceedings of the 47th Annual Meeting of EMSA*, ed. G. W. Bailey, San Francisco Press, San Francisco, pp. 654–655.

$Al_2O_3$   Sklad, P. S. and Bentley, J. (1990). Analysis of a new type of extended defect in $\alpha$-$Al_2O_3$. In *Proc. 12th Int. Congr. on Electron. Microscopy*, Vol. 4, San Francisco Press, San Francisco, pp. 464–465.

$Al_2O_3$   Skogsmo, J., Liu, P., and Norden, H. (1991). The crystal structure and point group of $\kappa$-$Al_2O_3$, *Micron and Microscopica Acta* **22**(1/2), 171–172.

$Al_2O_3$   Bentley, J. and Sklad, P. S. (1991). Domain boundaries formed during epitaxial growth of $\alpha$-$Al_2O_3$. In *Proc. MRS Symposium*, Vol. 202, Material Research Society, Pittsburgh, pp. 451–456.

$Al_2Ti_7O_{15}$   Kang, Z. C., Boulesteix, C., Nihoul, G., Monnereau, O., and Remy, F. (1988). A study by CBD and HREM of a new mixed valence compound $Al_2Ti_7O_{15}$. In *Proc. 9th European Congr. on Electron Microscopy*, eds. P. J. Goodhew and H. G. Dickinson, Inst. Phys. Conf. Ser., Vol. 93, Institute of Physics, London, pp. 273–274.

$Al_3Fe$   Skjerpe, P. (1987a). An electron microscopy study of the phase $Al_3Fe$, *J. Micros.* **148**, 33.

$Al_3Zr$   Ma, Y., Gjønnes, J., and Taftø, J. (1991). Structure refinement of $Al_3Zr$ using LACBED technique, *Micron and Microscopica Acta* **22**(1/2), 163–164.

$Al_3Zr$   Vecchio, K. S. and Williams, D. B. (1987d). CBED study of $Al_3Zr$ in Al–Zr and Al–Li–Z alloys, *Acta Metall.* **35**, 2959–2970.

$Al_6Li_3Cu$   Vecchio, K. S. and Williams, D. B. (1987a). Microtwinning evidence for the apparent 5-fold symmetry in $T_2(Al_6Li_3Cu)$. In *Proc. of the 45th Annual Meeting of EMSA*, ed. G. W. Bailey, San Francisco Press, San Francisco, pp. 24–25.

$Al_6Li_3Cu$   Vecchio, K. S. and Williams, D. B. (1987b). CBED and microanalysis evidence for the nonicosahedral $T_2(Al_6Li_3Cu)$ phase. In *Analytical Electron Microscopy '87*, San Francisco Press, San Francisco, pp. 129–131.

Al$_6$Li$_3$Cu  Vecchio, K. S. and Williams, D. B. (1988a). The apparent "five-fold" nature of large T$_2$(Al$_6$Li$_3$Cu) crystals, *Metall. Trans.* **19A**, 2875-2884.

Al$_6$Mn  Bourdillon, A. J. (1987). Fine line structure in CBED of icosahedral Al$_6$Mn. *Philos. Mag.* **55**(1), 21-26.

Al$_6$Mn  Schechtman, D. and Blech, I. A. (1985). The microstructure of rapidly solidified Al$_6$Mn, *Metall. Trans.* **16A**, 1005-1011.

Al$_{11}$Ti$_4$Zn  Kolby, P. and Taftø, J. (1991). Determination of atom coordinates in the Al$_{11}$Ti$_4$Zn intermatallic phase by LACBED, *Micron and Microscopica Acta* **22**(1/2), 151-152.

Al$_{13}$Fe$_4$  Fung, K. K. and Zou, X. D. (1986). Tenfold Al$_{13}$Fe$_4$ twins in rapidly cooled Al-Fe alloys, In *Proc. 11th Int. Congr. on Electron Microscopy*, (Kyoto), Maruzen, Tokyo, pp. 1539-1540.

Al$_{60}$Si$_{20}$Cr$_{20}$  Chen, H. S. and Inoue, A. (1987). Formation and structure of new quasicrystals of Ga$_{16}$Mg$_{32}$Zn$_{52}$ and Al$_{60}$Si$_{20}$Cr$_{20}$, *Scr. Metall.* **21**, 527-530.

Al$_{65}$Cu$_{20}$Co$_{15}$  He, L. X., Zhang, Z., Wu, Y. K., and Kuo, K. H. (1988b). Stable decagonal quasicrystals with different periodicities along the tenfold axis in Al$_{65}$Cu$_{20}$Co$_{15}$. In *Proc. 9th European Congr. on Electron Microscopy*, eds. P. J. Goodhew and H. G. Dickinson, Inst. Phys. Conf. Ser., Vol. 93, Institute of Physics, London, pp. 501-502.

Al$_{65}$In$_{20}$M$_{15}$  He, L. X., Wu, Y. K., and Kuo, K. H. (1988a). Decagonal quasicrystals with different periodicities along the tensold axis in rapidly solidified Al$_{65}$In$_{20}$M$_{15}$. *J. Mater. Sci. Lett.* **7**, 1284-1286.

AlAg  Prabhu, N. and Howe, J. M. (1991). Comparison between calculated and experimental displacement fringe intensities on the focus of $\gamma'$ precipitate plates in an Al-4.2 at% Ag alloy, *Philos. Mag.* **63A**(4), 645-655.

AlCeFe  Ayer, R., Angers, L. M., Mueller, R. R., Scanlon, J. C., and Klein, C. F. (1988). Microstructural characterization of the dispersed phases in Al-Ce-Fe system, *Metall. Trans.* **19A**, 1645-1656.

AlCoNi  Tsai, A., Inoue, A., and Masumoto, T. (1989). Stable decagonal Al-Co-Ni and Al-Co-Cu quasicrystals, *Mater. Trans. JIM* **30**(7), 463-467.

AlCr  Bendersky, L. A., Roth, R. S., Ramon, J. T., and Schechtman, D. (1991). Crystallographic characterization of some intermetallic compounds in the Al-Cr system, *Metall. Trans.* **22A**, 5-10.

AlCu  Bouazra, Y. and Reynaud, F. (1984). Determination of the heights of ledges of $\theta'$ plates Al-4 wt% Cu by CBED and lattice plane imaging, *Acta Metall.* **32**(4), 529.

AlCu  Zhu, J. and Cowley, J. M. (1985). Study of early-stage precipitation in Al-4% Cu by microdiffraction and STEM, *Ultramicroscopy* **18**, 419-426.

AlCuCo  He, L. X., Wu, Y. K., Meng, X. M., and Kuo, K. H. (1990). Stable Al-Cu-Co decagonal quasicrystals with decaprismatic solidification morphology, *Philos. Mag. Lett.* **61**(1), 15-19.

AlCuCr  Ebalard, S. and Spaepen, F. (1991). Approximants to the icosahedral and decagonal phases in the Al-Cu-Cr system, *J. Mater. Res.* **6**(8), 1641-1649.

AlCuF  Hiraga, K. (1988). Highly ordered icosahedral quasicrystal of Al-Cu-Fe alloy studied by electron diffraction and HREM, *Jpn. J. Appl. Phys.* **27**, L951-L953.

AlCuFe  Cheng, Y. F., Hui, M. J., and Li, F. H. (1991). A new commensurate phase and its related incommensurate phase in Al-Cu-Fe alloy, *Philos. Mag. Lett.* **63**(1), 49-55.

AlCuFe  Ebalard, S. and Spaepen, F. (1989). The bcc type icosahedral reciprocal lattice of the Al-Cu-Fe quasi-periodic crystal. *J. Mater. Res.* **4**(1), 39-43.

AlCuFe        Ebalard, S. and Spaepen, F. (1990). Long-range chemical ordering in Al–Cu–Fe, Al–Cu–Mn, and Al–Cu–Cr quasicrystals, *J. Mater. Res.* **5**(1), 62–73.
AlCuFe        Liu, P., Tsai, A. P., Inoue, A., and Arnberg, L. (1991). Microstructure of a rapidly solidified 65Al–20Cu–15Fe (at%) alloy, *J. Mater. Sci.* **26**, 963–969.
AlCuMgAg    Garg, A. and Howe, J. M. (1991). CBED analysis of the $\Omega$ phase in an Al–4.0 Cu–0.5 Mg–0.5 Ag alloy, *Acta Metall. Mater.* **39**(8), 1939–1946.
AlCuMgAg    Muddle, B. C. and Polmear, I. J. (1989). The precipitate $\Omega$ phase in Al–Cu–Mg–Ag alloys, *Acta Metall.* **37**(3), 777–789.
AlFe        Dunlop, R. A., Lloyd, D. J., Christie, I. A., Strink, G., and Stadnik, Z. M. (1988). Physical properties of rapidly quenched Al–Fe alloys, *J. Phys. F,* **18**, 1329–1341.
AlFe        Skjerpe, P. (1987b). Intermetallic phases formed during DC-casting of an Al–0.25 wt%, Fe–0.13 wt% Si alloy, *Metall. Trans.* **18A**, 189–200.
AlFe        Vasudevan, V. K. and Fraser, H. (1987). Identification of precipitates in rapidly solidified and heat-treated Al–8 Fe–2 Mo–Si alloys, *Sci. Metall.*, **21**, 1105–1110.
AlFeCe        Angers, L. A., Marks, L. D., Weertman, J. R., and Fine, M. E. (1985). A quasicrystalline decagonal phase in Al–Fe–Ce. In *Proceedings of MRS*, Materials Research Society, Pittsburgh, **62**, 255–262.
AlFeMnSi    Hansen, V., Skjerpe, P., and Gjonnes, J. (1990a). Crystal structures, defects, and transformations in the Al–(Fe,Mn)–Si alloy system. In *Proc. 12th Int. Congr. on Electron Microscopy*, vol. 4, San Francisco Press, San Francisco, pp. 956–957.
AlFeSi        Bendersky, L. A., McAlister, A. J., and Bianacaniello, F. S. (1988). Phase transformation during annealing of rapidly solidified Al-rich Al–Fe–Si alloys, *Metall. Trans.* **19A**, 2893–2900.
AlFeSi        Liu, P. and Dunlop, G. L. (1987). Determination of the crystal symmetry of two Al–Fe–Si phases by CBED, *J. Appl. Crystallogr.* **20**, 425–427.
AlFeSi        Liu, P. and Dunlop, G. L. (1988). Long-range ordering of vacancies in BCC $\alpha$-AlFeSi, *J. Mater. Sci.* **12**, 419–1424.
AlFeV        Skinner, D. J., Ramanan, V. R. V., Zedalis, M. S., and Kim, N. J. (1988). Stability of quasicrystalline phases in Al–Fe–V alloys, *J. Mater. Sci. Eng.* **99**, 407–411.
AlGaAs        Eaglesham, D. J. and Humphreys, C. J. (1986). A new technique for microanalysis using CBED: Model study of (Al/Ga)As. In *Proc. 11th Int. Congr. on Electron Microscopy* (Kyoto), Maruzen, Tokyo, pp. 209–210.
AlGaAs        Eaglesham, D. J., Kiely, C. J., Cherns, D., and Missous, M. (1989e). Electron diffraction from epitaxial crystals—a CBED of the interface structure for $NiSi_2$/Si and Al/GaAs, *Philos. Mag.* **60**(2), 161–175.
AlGaAs        Ou, H.-J., Higgs, A. A., Perkes, P. R., and Cowley, J. M. (1989b). High spatial resolution microanalysis on the (200) nanodiffraction intensity to determine the Al concentration of AlGaAs–GaAs MQWS. In *Proceedings of the 47th Annual Meeting of EMSA*, ed. G. W. Bailey, San Francisco Press, San Francisco, pp. 232–233.
AlGaAs        Ou, H.-J., Cowley, J. M., Chyi, J. I., Salvador, A., and Morkoc, H. (1990). Microanalysis on the (200) diffraction intensity to determine the Al concentrations for AlGaAs–GaAs multiple-quantum-well structures, *J. Appl. Phys.* **67**(2), 698–704.
AlGaAs        Ou, H.-J., Glaisher, R. W., Cowley, J. M., and Morkoc, H. (1989a), Using the (200) thickness contour to measure the Al concentration $Al_xGa_{1-x}As$–GaAs MQWS structures. In *Proceedings of Microbeam Analysis* ed. P. E. Russell, San Francisco Press, San Francisco, pp. 480–482.
AlGaAs        Vincent, R., Cherns, D., Bailey, S. J., and Morkoc, H. (1987b). Structure of AlGaAs/GaAs multilayers images in superlattice reflections, *Philos. Mag.* **56**(1), 1–6.

AlGaAs/GaAs    Cherns, D., Jordan, I. K., and Vincent, R. (1988a), CBED from AlGaAs/GaAs single quantum wells, *Philos. Mag. Lett.* **58**(1) 45–51.
AlGaAs/GaAs    Cherns, D., Jordan, I. K., and Vincent, R. (1989d). Composition profiles in AlGaAs/GaAs and InGaAs/InP structures examined by CBED. In *Proc. MRS Symposium, Characterization of the Structure and Chemistry of Defects in Materials*, eds. B. C. Larson, M. Ruhle, and D. N. Seidman, Vol. 138, Materials Research Society, Pittsburgh, pp. 431–436.
AlGe    Exelby, D. R. and Vincent, R. (1990). Structural analysis of a metastable Al–Ge phase using CBED. In *Proc. 12th Int. Cong. on Electron Microscopy*, Vol. 2, San Francisco Press, San Francisco, pp. 510–511.
AlGe    Kaufman, M. J., Cunningham, Jr., J. E., and Fraser, H. L. (1987b). Metastable phase production and transformation in Al–Ge alloy films by rapid crystallization and annealing treatments, *Acta Metall.* **35**, 1181–1192.
AlGe    Vincent, R. and Exelby, D. R. (1990). Analysis by electron diffraction of a rhombohedral Al–Ge phase. In *Proc. 12th Int. Cong. on Electron Microscopy*, Vol 2, San Francisco Press, San Francisco, pp. 524–525.
Alge    Vincent, R. and Exelby, D. R. (1991). Structure of metastable Al–Ge phases determined from HOLZ patterson transforms, *Philos. Mag. Lett.* **63**(1), 31–38.
AlLi    Kenik, E. A. (1984). HVEM in-situ studies of strain localization in Al–Li alloys. In *Fracture: Measurement of Localized Deformation by Novel Technique*, eds. W. W. Gerberich and D. L. Davidson, TMS-AIME, p. 125.
AlLi    Radmilovic, V., Fox, A. G., Fisher, R. M., and Thomas, G. (1989). Li depletion in PFZ' in Al–Li base alloys, *Scr. Metall.* **23**, 75–79.
AlLi    Sung, C. M., Chan, H. M., and Williams, D. B. (1986). Quantitative microanalysis of Li in binary Al–Li alloys. In: *Al–Li III Proceedings*, Institute of Metals, London, pp. 337–348.
AlLi    Tosten, M. H., Ramani, A., Bartges, C. W., Michel, D. J., Ryba, E., and Howell, P. R. (1989). On the origin and nature of microcrystalline regions in an Al–Li–Cu–Zr alloy. *Scr. Metall.* **23**, 829–834.
AlLiCu    Howell, P. R., Michel, D. J., and Ryba, E. (1989). The nature of microcrystalline regions produced by an in-situ transformation of $T_2$ particles in a ternary Al–2.5 Li–2.5 Cu alloy. *Scr. Metall.* **23**, 825–828.
AlLiCuMg    Ahmad, M. (1991). Effects of environment and microstructure on stress corrosion crack propagation in an Al–Li–Cu–Mg alloy, *Mater. Sci. Eng.* **A125**, 1–14.
AlLiCuMgZr    Xia, X. and Martin, J. (1990). TEM convergent beam measurement of S-phase volume fraction in Al–Li–Cu–Mg–Zr alloy 8090, *Material Characterization* **25**(4), 325–337.
AlMg    Hales, S. J. and McNelley, T. R. (1988). Microstructural evolution by continuous recrystallization in a superplastic Al–Mg alloy, *Acta. Metall.* **36**(5), 1229–1239.
AlMn    Bendersky, L. (1987a). Al–Mn microphase and its relation to quasicrystals, *J. Microsc.* **146**, 303–312.
Almn    Fan, C., Wu, Z., Wang, Y., and Tsien, L. (1987). The substructure of quasicrystalline Al–Mn alloy observed by TEM, *J. Microsc.* **146**, 261–265.
AlMn    Gerald, J. D. F., Withers, R. L., and Stewart, A. M. (1987). Relationships between decagonal and crystalline phases in the Al–Mn systems. In: *Analytical Electron Microscopy '87*, San Francisco Press, San Francisco, pp. 91–93.
AlMn    Gerald, J. D. F., Withers, R. L., Stewart, A. M., and Calka, A. (1988). The Al–Mn decagonal phase, 1. A re-evaluation of some diffraction effects, 2. Relationship to crystalline phases, *Philos. Mag.* **58B**, 15–33.
AlMn    Kaufman, M. J., Biancaniello, F. S., and Kreider, K. G. (1988). The annealing

behavior of sputter-deposited Al–Mn and Al–Mn–Si films, *J. Mater. Res* **3**(6), 1342–1348.

AlMn Ohashi, T., Fukatsu, N., and Asai, K. (1989). Crystallization and precipitation structures of quasicrystalline phase in rapidly solidified Al–Mn–X ternary alloys, *J. Mater. Sci.* **24**, 3717–3724.

AlMnNi Lalla, N. P., Tiwari, R. S., and Srivastava, O. N. (1991). TEM investigation of rapidly solidified Al–Mn–Ni quasicrystalline alloys, *Philos. Mag.* **63B**, 629–640.

AlMnSi Bendersky, L. A. and Kaufman, M. J. (1986). Determination of the point group of the icosahedral phase in an Al–Mn–Si alloy using CBED, *Philos. Mag.* **53B**(3), L75.

AlMnSi Kreider, K. G., Biancaniello, F. S., and Kaufman, M. J. (1987a). Sputter deposition of icosahedral Al–Mn and Al–Mn–Si, *Scr. Metall.* **21**, 657–662.

AlN Bando, Y., Kitami, Y., Sakai, T., and Izumi, F. (1987). Combined use of lattice imaging and microanalysis in structure determination of AlN-related polytype by a 400 kV HRAEM. In: *Analytical Electron Microscopy '87*, San Francisco Press, San Francisco, pp. 109–112.

AlN Berger, A. (1991). Inversion domains in aluminum nitride, *J. Am. Ceram. Soc.* **74**(5), 1148–1151.

AlN Callahan, D. and Thomas, G. (1990). CBED analysis of impurity distribution on AlN. In *Proc. 12th Int. Congr. on Electron Microscopy*, Vol. 4, San Francisco Press, San Francisco, pp. 214–215.

AlN Vallahan, D. L. and Thomas, G. (1990). Impurity distribution in polycrystalline aluminum nitride ceramics, *J. Am. Ceram. Soc.* **73**(7), 2167–2170.

AlN Westwood, A. D. and Notis, M. R. (1991). Inversion domain boundaries in AlN, *J. Am. Ceram. Soc.* **74**(6), 1226–1239.

AlN–$ZrO_2$ Stoto, T., Doukham, J., and Mocellin, A. (1989). AEM investigation of the AlN–$ZrO_2$ system: identification of a quaternary Zr–Al–O–N phase, *J. Am. Ceram. Soc.* **72**(8), 1453–1457.

AlO Jayaram, V. and Levi, C. G. (1989). The structure of $\delta$-alumina evolved from the melt and the $\gamma \rightarrow \delta$ transformation, *Acta Metall.* **37**(2), 569–578.

AlON Dravid, V. P., Sutliff, J. A., Westwood, A. D., Notis, M. R., and Lyman, C. E. (1989b). Centrosymmetric and nonsymmorphic Al–O–N spinel. In *Proceedings of the 47th Annual Meeting of EMSA*, ed. G. W. Bailey, San Francisco Press, San Francisco, pp. 512–513.

AlSi Bardel, A. and Hoier, R. (1991). CBED measurements of residual strains in Al–Si carbide composites, *Micron and Microscopica Acta* **22**(1/2), 111–112.

AlSiC Rozeveld, S. J. (1990). Determination of residual strains in an Al–SiC composite using CBED. In *Proc. 12th Int. Congr. on Electron Microscopy*, Vol. 2, San Francisco Press, San Francisco, pp. 502–503.

AlSiMn Dai, M. and Wang, R. (1990). Comparative investigation of microarea quasilattice parameters of Al–Si–Mn and Al–Cu–Fe icosahedral phases using HOLZ line patterns, *Acta Crystallogr.* **B46**, 455–458.

AlZnMgCu Ayer, R., Koo, J. Y., Steeds, J. W., and Park, B. K. (1985). Microanalytical study of the heterogeneous phases in commercial Al–Zn–Mg–Cu alloys, *Metall. Trans.* **16A**, 1925.

Au Richards, C. J. and Steeds, J. W. (1974). Accurate structure factor dermination for gold using new techniques in HEED. In *Diffraction Studies of Real Atoms and Real Crystals*, Canberra, Australian Academy of Science, pp. 112–113.

Au–GaAs Kim, T. and Chung, D. D. L. (1990). TEM study of the interfacial reactions in gold based contacts to GaAs, *Philos. Mag.* **62A**(3), 283–317.

AuGe  Rackham, G. M. and Steeds, J. W. (1978b). Electron diffraction from a metastable phase of gold and germanium. In *Electron Diffraction 1927-1977, Proc. Conf. on Electron Diffraction*, eds. P. J. Dobson, J. B. Pendry, and C. J. Humprheys, Conf. Ser. Vol. 41, Institute of Physics, Bristol and London, pp. 178-181.

AuGeAs  Vincent, R., Bird, D. M., and Steeds, J. W. (1984a). Structure of AuGeAs determined by CBED, I. Derivation of basic structure, *Philos. Mag.* **50A**(6), 745.

AuGeAs  Vincent, R., Bird, D. M., and Steeds, J. W. (1984b). Structure of AuGeAs determined by CBED, *Philos. Mag.* **50A**(6), 765.

Ba-Y-Cu-O  Sishen, X. and Hanjie, F. (1987). A high Tc superconducting phase in the Ba-Y-Cu-O system, *J. Phys. D* **20**, 809-811.

$Ba_2YCu_3O_{7-x}$  Nakahara, S., Boone, T., Yan, M. F., Fisanick, G. J., and Johnson, Jr., D. W. (1988). Defect structure in $Ba_2YCu_3O_{7-x}$. *J. Appl. Phys.* **63**(2), 451-455.

$Ba_2YCu_3O_{7-x}$  Moodie, A. F. and Whitfield, H. J. (1988). Structure and microstructure in $Ba_2YCu_3O_{7-x}$, *Ultramicroscopy* **24**, 329-338.

BaTiO  Amstrong, T. R. and Buchanan, R. C. (1990). Influence of core-shell grains on the internal stress state and permittivity response of zirconia-modified barium titanate, *J. Am. Ceram. Soc.* **73**(5), 1268-1273.

BaTiO  Tanaka, M. and Lehmpfuhl, G. (1972). Study of barium titantate by CBED, *Jpn. J. Appl. Phys.* **11**, 1755-1756.

BeTi  Liu, D. R., and Williams, D. B. (1989c). Accurate composition determination of Be-Ti alloys by EELS, *J. Microsc.* **156**(2), 201-210.

$Bi_2Sr_2CaCu_2O_x$  Sung, C. M., Harmer, M. P., Smyth, D. M., and Williams, D. B. (1988). Microstructure of the superconducting phase (85K) in the $Bi_2Sr_2CaCu_2O_x$ system. In *Proceedings of the 46th Annual Meeting of EMSA*, ed. G. W. Bailey, San Francisco Press, San Francisco, pp. 876-877.

$Bi_2Sr_2CaCu_2O_x$  Kan, X. B., Kulik, J., Chow, P. C., Moss, S. C., Yan, Y. F., Wang, J. H., and Zhao, Z. X. (1990). X-ray and electron diffraction study of single crystal $Bi_2Sr_2CaCu_2O_x$, *J. Mater. Res.* **5**(4), 731-736.

$Bi_2Sr_2CuO_6$  Wen, J. G., Liu, Y., Ren, Z. F., Tan, Y. F., Zhou, Y. Q., and Fung, K. K. (1989). Structural study of undoped and doped $Bi_2Sr_2CuO_6$ phases by TEM, *Appl. Phys. Lett.* **55**(26), 2775-2777.

$Bi_3TiNbO_9$  Thompson, J. G., Rae, A. D., Withers, R. L., and Craig, D. C. (1991). Revised structure of $Bi_3TiNbO_9$, *Acta Crystallogr.* **B47**, 174-180.

BiCa-Sr-Cu-O  Ramesh, R. and Thomas, G. (1988a). Structure and composition of the 115K superconducting phase in the Bi-Ca-Sr-Cu-O system, *Appl. Phys. Lett.* **53**(6), 520-522.

Bica-Sr-Cu-O  Ramesh, R., Hetherington, C. J. D., and Thomas, G. (1988b). Further evidence for the presence of c = 38.2 A phase in a Bi-Ca-Sr-Cu-O superconductor, *Appl. Phys. Lett.* **53**(7), 615-617.

BiSr-Ca-Cu-O  Gai, P. L., Subramanian, M. A., and Sleight, A. W. (1990). Origin of superstructure modulations in Bi-Sr-Ca-O superconductors. In *Proc. 12th Int. Congr. on Electron Microscopy*, San Francisco Press, San Francisco, pp. 36-37.

BiSr-Ca-Cu-O  Lin, Y. P., Barbier, J., Greedan, J. E., Wang, Z., Lee, M. J. G., and Statt, B. W. (1989a). Electron diffraction investigation of Pb-stabilized Bi-Sr-Ca-Cu-O high Tc superconductor, *Physica C* **158**, 241-246.

BiSr-Ca-Cu-O  Tan, N. X., Bourdillon, A. J., Dou, S. X., Liu, H. K., and Sorrell, C. C. (1989). Superlattices in Pb-doped bi-Sr-Ca-Cu-O and in a non-superconducting Sr-Ca-Cu-O precursor, *Philos. Mag.* **59**(5) 213-217.

BiSrCaCuO  Zhao, Z. X. (1988). Some problems of superconducting Bi-Sr-Ca-Cu-O system, *Physica C* 153-155, 1144-1147.

BiTiO$_3$    Weiss J. and Rosenstein, G. (1988). Addition of Ba$_2$TiSi$_2$O$_8$ to Mn-doped BiTiO$_3$. Effect on oxygen diffusion and grain boundary composition. *J. Mater. Sci.* **23**, 3263-3271.
C    Goodman, P. (1976a). Examination of the graphite structure by CBED, *Acta Crystallogr.* **A32**, 793.
C    Silva, E. and Riquelme, J. (1980). ZAPs maps for graphite, *J. Appl. Crystallogr.* **13**, 364-367.
C    Walmsley, J. C. and Lang, A. R. (1990). The value of electron microscope studies of imperfect natural diamonds. In: *Proc. 12th Int. Congr. on Electron Microscopy*, San Francisco Press, San Francisco, pp. 194-195.
Ca-Si-O-N    Pugar, E. A., Kennedy, J. H. Morgan, P. E. D., and Porter, J. R. (1988). New cubic structure in the Ca-Si-O-N system. *J. Am. Ceram. Soc.* **71**(6), C288-C291.
Ca$_3$Ti$_2$O$_7$    Elcombe, M. M., Kisi, E. H., Hawkins, K. D., White, T. J., Goodman, P., and Matheson, S. (1991). Structure determinations for Ca$_3$Ti$_2$O$_7$, Ca$_4$Ti$_3$O$_{10}$, Ca$_{3.6}$Sr$_{0.4}$Ti$_3$O$_{10}$ and a refinement os Sr$_3$Ti$_2$O$_7$, *Acta Crystallogr.* **B47**, 305-314.
CaF$_2$    Ichimiya, A. and Lehmpfuhl, G. (1988). Imaging potential for electrons of CaF$_2$ from a Bloch-wave analysis, *Acta Crystallogr.* **A44**, 806-809.
CaSrBiCu    Zaluzec, N. J. (1988). Incommensurate ordering in Ca-Sr-Bi-Cu oxide superconductors. In *Proceedings of the 46th Annual Meeting of EMSA*, ed. G. W. Bailey, San Francisco Press, San Francisco, pp. 872-873.
CaTiO$_3$    Rossouw, C. J., Turner, P. S., and White, T. J. (1988a). Axial electron-channelling analysis of perovskite, I. Theory and experiment for CaTiO$_3$, *Philos. Mag.* **57B**(2), 209-225.
CaTiO$_3$    Rossouw, C. J., Turner, P. S., and White, T. J. (1988b). Axial electron-channelling analysis of perovskite, II. Site Identification of Sr, Zr, and Cl impurities, *Philos. Mag.* **57B**(2), 227-241.
CaZrO    Dravid, V. P., Sung, C. M., Notis, M. R., and Lyman, C. E. (1989a). Crystal symmetry and coherent twin structure of calcium zirconate, *Acta Crystallogr.* **B45**, 218-227.
CdCu    Bendersky, L. A. and Biancaniello, F. S. (1987b). TEM observation of icosahedral new crystalline and glassy phases in rapidly quenched Cd-Cu alloys, *Scr. Metall.* **21**, 531-536.
CdS    Goodman, P. and Lehmpfuhl, G. (1964). Verbotene elektronenbergungreflexe von CdS, *Z. Naturforsch*, **19a**, 818-820.
CdTe    Spellward, P. and Preston, A. R. (1988a). Determination of the polarity of twins in epitaxial CdTe using CBED. In *Proc. 9th European Congr. on Electron Microscopy*, eds. P. J. Goodhew and H. G. Dickinson, Conf. Ser., Vol. 93, Institute of Physics, London, pp. 29-30.
CdTe    Aindow, M., Eaglesham, D. J., and Pond, R. C. (1988). Misorientation effects in MOCVD CdTe on sapphire. In *Proc. 9th European Congr. on Electron Microscopy*, eds. P. J. Goodhew and H. G. Dickinson, Inst. Phys. Conf. Ser., Vol. 93, Institute of Physics, London, pp. 405-406.
CdTe$_2$O$_5$    Silva, E. and Scozia, R. (1987). CdTe$_2$O$_5$ space group determination by Kossel patterns, *J. Mater. Sci. Lett.* **6**, 451-452
CeO$_2$-Y$_2$O$_3$    Wallenberg, R., Withers, R., Bevan, D. J. M., Thompson, J. G., Barlow, P., and Hyde, B. E. (1990). The fluoride-related solid solutions of CeO$_2$-Y$_2$O$_3$I: A re-examination by electron microscopy and diffraction, *J. Less-Common Met.* **156**, 1-16.
Cr    Matsuhata, H. and Steeds, J. W. (1987b), Observation of Cr$\langle 111 \rangle$ zone-axis critical-voltage effect, *Philos. Mag.* **55B**, 17-38.

Cr₇O₃   Morniroli, J. P., Ayatti, H., Gantois, M., Kuypers, S., Mahy, J., and Van Landuyt, J. (1988). Characterization of twins and APBs in $Cr_7O_3$ carbide by electron diffraction and HREM. In *Proc. 9th European Congr. on Electron Microscopy*, eds. P. J. Goodhew and H. G. Dickinson, Inst. Phys. Conf. Ser. Vol. 93, Institute of Physics, London, pp. 419–420.

CrNiNb   Bennett, M. J., Desport, J. A., Houlton, M. R., Labun, P. A., and Titchmarsh, J. M. (1988). Inhibition of scale growth on 20Cr–25Ni–Nb stabilized stainless steel by yttrium ion implanation revealed by AEM, *Mater. Sci. Technol.* **4**, 1107.

Cu   Matsuhata, H., Tomokiyo, Y., Watanabe, H., and Euchi, T. (1984). Determination of the structure factors of Cu and $Cu_3Au$ by the intersecting Kikuchi line method, *Acta Crystallogr.* **B40**, 544–549.

Cu   Schwarzer, R. (1979). The determination of structure potentials from Kossel Möllenstedt diagrams of epitaxially evaporated, self-supporting single-crystal layers of copper, *Optik* **54**(3), 193–199.

Cu   Tomokiyo, Y. and Matsmura, S. (1986). Application of HOLZ patterns to lattice parameter determination in Cu-based alloys, *J. Electron Microsc.* **35**(4), 359–364.

$Cu_2ZnGeS_4$   Moodie, A. F. and Whitfield, H. J. (1986b). Determination of the structure of $Cu_2ZnGeS_4$ polymorphs by lattice imaging and CBED, *Acta Crystallogr.* **B42**, 236–247.

$Cu_3As_2S_3I$   Goodman, P. and Whitfield, H. J. (1980b). The space-group determination of GaS and $Cu_3As_2S_3I$ by CBED, *Acta Crystallogr.* **A36**, 219.

$Cu_3Au$   Chon, T. C. and Tu, K. N. (1988). Secondary grain growth and formation of antiphase domains in ordered $Cu_3Au$ thin films, *J. Appl. Phys.* **64**(5), 2375–2379.

CuAgS   Baker, C. L., Lincoln, F. J., and Johnson, A. W. S. (1990). A CBED study of CuAgS. In *Proc. 12th Int. Congr. on Electron Microscopy*, Vol. 2, San Francisco Press, San Francisco, pp. 484–485.

CuAl   Okuyama, T., Matsumura, S., Kuwano, N., Oki, K., and Tomokiyo, Y. (1989b), HOLZ patterns in CBED and determinations of local lattice parameters in α- and $\alpha_2$-phase of Cu–20 at% Al alloy, *ISIJ International* **29**(3), 191–197.

CuAl   Okuyama, T. Tomokiyo, Y., Mutsumura, S., Kuwano, N., and Oki, K. (1988b). Estimation of local lattice strain in a Cu–Al alloy by HOLZ patterns. In *Annual Reports*, HVEM Lab. Kyushu Univ., pp. 35–36.

CuAl   Sellar, J. R. and Imeson, D. (1988). Observation of size effect diffuse electron scattering in disordered α-CuAl near the critical voltage, *Acta Crystallogr.* **A44**, 768–772.

$CuAsSe_{1-x}S_x$   Goodman, P., Olson, A., and Whitfield, H. J. (1985a). Mirror-sysmmetric and non-mirror symmetric glide planes in $CuAsSe_{1-x}S_x$, I. CBED study, *Ultramicroscopy* **16**, 227.

CuAu   Matsuhata, H., Watanabe, H., Tomokiyo, Y., and Eguchi, T. (1982). Application of intersecting Kikuchi-line method to pure Cu and $Cu_3Au$ alloy. In *Proc. Electron Microscopy 1932–82*, Deutsche Gesellschaft für Electronemikroskopie, Hamburg, p. 635.

CuAuPd   Matsumuar, S., Morimura, T., and Oki, K. (1990). A combined IKL–ALCHEMI technique for atom configurations in multinary ordering alloys and its application to Cu–Au–Pd. In *Proc. 12th Int. Congr. on Electron Microscopy*, Vol. 2, San Francisco Press, San Francisco, pp. 488–489.

CuInSe   Kiely, C. J., Pond, R. C., Kenshole, G., and Rockett, A. (1991). A TEM study of the crystallography and defect structures of single crystal and polycrysalline copper indium diselemide, *Philos. Mag.* **63A**(6), 1249–1273.

CuNi    Jepson, J. R. and Dingley, D. J. (1988). Structure and strain measurement in compositionally modulated CuNi and CuFe foils. In *EMAG 1987, Analytical Electron Microscopy*, ed. G. W. Lorimer, The Institute of Metals, Manchester, pp. 71–75.

CuZn    Guan, Z. M., Liu, G. X., Notis, M. R., and Williams, D. B. (1988). Detection of slight crystal symmetry change accompanying DIGM in the Cu–Zn system by CBED, *Scr. Metall.* **22**, 985–990.

CuZn    Sun, Y. S., Lorimer, G. W., and Ridley, N. (1989). Microstructure of high tensile strength brasses containing Si and Mn, *Metall. Trans.* **20A**, 1199–1206.

CuZr    Marshall, A. F., Steeds, J. W., Bouchet, D., Shinde, S. L., and Walmsley, R. G. (1981), Crystal structure analysis of Cu–Zr alloys by CBED. In *Proceedings of the 39th Annual Meeting ot the EMSA*, ed. G. W. Bailey, San Francisco Press, San Francisco, pp. 354–355.

Fe      James, A. W. and Shepherd, C. M. (1989). Some effects of heat treatment on grain boundary chemistry and precipitation in type 316 steel, *Mater. Sci. Technol.* **5**, 333–345.

Fe      Reaney, I. M. and Lorimer, G. W. (1988). Observations of a Mo, Fe-rich phase in an oxidized 9 wt% Cr, 1 wt% Mo steel which exhibits 5-fold symmetry, *Proc. Institute of Physics Short Meeting*, Institute of Physics, London, pp. 113–122.

Fe      Redjaimia, A., Morniroli, J. P., Metauer, G., and Gantois, M. (1990). Identification by electron microdiffraction of intermetallic phases in a duplex stainless steel. In *Proc. 12th Int. Congr. on Electron Microscopy*, Vol. 2, San Francisco Press, San Francisco, pp. 494–495.

Fe      Ricks, R. A. (1984). Duplex grain boundary precipitation in austenitic stainless steels containing Al and Ti, *Acta Metall.* **32**, 1105–1115.

Fe      Ricks, R. A. (1985). Pre-precipitation phenomena associated with $\gamma'$ formation in stainless steels. In: *Electron Microscopy and Analysis*, ed. G. J. Tatlock, Inst. Phys. Conf. Ser., Vol., 78, Institute of Physics, London and Bristol, pp. 253–256.

Fe      Stoter, L. P. (1981a). Thermal aging effects in AISI type 316 stainless steel, *J. Mater. Sci.* **16**, 1039–1051.

Fe      Yamashita, T. (1981). Zero order lines in electron channelling patterns from crystals of Al and Iron at various temperatures, *J. Electron Microsc.* **30**(4), 298.

$Fe_2O_3$   Ho, H., Goo, E., and Thomas, G. (1985). Crystal structure of acicular $\gamma$-$Fe_2O_3$ particles used in recording media, *J. Appl. Phys.* **59**(5), 1606–1610.

$Fe_3O_4$   Lin, I., Mishra, R. K., and Thomas, G. (1984). Electron microscopy of the structure and composition of grain boundary in $(Mn, Zn)Fe_2O_4$. In: *Advances in Materials Characterization*, Proc. Mater. Sci. Res., Vol. 15, pp. 351–358.

$Fe_3Al$    Nourbaksh, S., Margolin, H., and Liang, F. L. (1990). Microstructure of a pressure-cast $Fe_3Al$ intermetallic alloy composite reinforced with $ZrO_2$-toughened alumina fibers, *Metall. Trans.* **21A**, 2881–2889.

FeAlZr  Cheng, C. C., King, W. E., and McNallan, M. J. (1989). Characterization and crystallization studies of melt-spun glassy Fe-22.5Al-10Zr metal by AEM. In EMAG-MICRO 89, Inst. Phys. Conf. Ser. Vol. 98, Institute of Physics Publ. Ltd., London, pp. 3399–3407.

FeBe    Burke, M. G. and Miller, M. K. (1989). A combined TEM/APFIM approach to the study of phase transformations: phase idenification in the Fe–Be system, *Ultramicroscopy* **30**, 199–209.

FeCrAlY Ayer, R. (1986b). Convergent beam diffraction study of intermetallic phase in a Fe–20Cr–4Al–0.5Y alloy, In: *Proc. XIth Int. Congr. on Electron Microscopy* (Kyoto) Maruzen, Tokyo, pp. 715–716.

FeCrAlY      Ayer, R. and Scanlon, J. C. (1987). Crystal structure of Intermatallic phase in Fe–20Cr–4Al–0.5Y alloy by CBED, *J. Mater. Res.* **2**(1), 16–27.
FeMgO      Tanaka, N., Mihama, K., Ou, H., and Cowley, J. M. (1988b). Nanometer-area diffraction of small iron crystallites in single crystalline Fe–MgO composite films. In *Proceedings of the 46th Annual Meeting of EMSA*, ed. G. W. Bailey, San Francisco Press, San Francisco, pp. 706–707.
FeNi      Kim, Y., Lin, H., and Kelly, T. F. (1988). Solidification structures in submicron spheres of Fe–Ni: Experimental observation, *Acta. Metall.* **36**(9), 2525–2536.
FeO      Zuo, J. M., Spence, J. C. H., and Petuskey, W. (1990). The low-temperature structure of magnetite studied by CBED. In *Proc. 12th Int. Congr. on Electron Microscopy*, Vol. 2, San Francisco Press, San Francisco, pp. 508–509.
$Ga_{16}Mg_{31}Zn_{52}$      Chen, H. S. and Inoue, A. (1987). Formation and structure of new quasicrystals of $Ga_{16}Mg_{31}Zn_{52}$ and $Al_{60}Si_{20}Cr_{20}$, *Scr. Metall.* **21**, 527–530.
GaAs      Bleloch, A. L., Howie, A., and Lanzerotti, M. Y. (1990). Convergent-beam RHEED of GaAs (110) in a STEM: theory and experiment. In *Proc. 12th Int. Congr. on Electron Microscopy*, Vol. 2, San Francisco Press, San Francisco, pp. 394–395.
GaAs      Britton, E. G., and Stobbs, W. M. (1985). The applicability of CBED methods for the determination of local alloy compositions in GaAs heterostructure. In *Electron Microscopy and Analysis*, ed. G. J. Tatlock, Inst. Phys. Conf. Ser. Vol. 78, Institute of Physics, London and Bristol, p. 45.
GaAs      Chu, S. N. G., Nakahara, S., Pearton, S. J., Boone, T., and Vernon, S. M. (1988). Antiphase domains in GaAs grown by metalorganic CVD on Si-on-insulator, *J. Appl. Phys.* **64**(6), 2981–2989.
GaAs      Eaglesham, D. J. (1987a). Defects in MBE and MOCVD grown GaAs on Si. In *Electron Microscopy of Semiconducting Materials*, Inst. Phys. Conf. Ser. Vol. 87, Institute of Physics, London, pp. 105–110.
GaAs      Gjønnes, K. and Gjønnes, J. (1988b). Quasi two-beam description of the systematic row Kossel pattern—Applications to determination of low order structure factors in GaAs and $Ga_{1-x}Al_xAs$. In *Proc. 9th European Congr. on Electron Microscopy*, eds. P. J. Goodhew and H. G. Dickinson, Inst. Phys. Conf. Ser., Vol. 93, Institute of Physics, London, pp. 41–42.
GaAs      Gjønnes, K., Gjønnes, J., Zuo, J., and Spence, J. C. H. (1988a). Two-beam features on electron diffraction pattern—application to refinement of low order structure factors in GaAs, *Acta Crystallogr.* **A44**, 810–820.
GaAs      Liliental-Weber, Z. (1990). Application of electron microscopy for the detection of point defects in MBE GaAs layers grown at low temperature. In *Proc. 12th Int. Congr. on Electron Microscopy*, Vol. 4, San Francisco Press, San Francisco, pp. 588–589.
GaAs      Liliental-Weber, Z., Ishikawa, A., Teriauchi, M., and Tanaka, M. (1991). Microscopic determination of stress distribution in GaAs grown at low temperatures on GaAs (100). In *Proc. of MRS Symposium*, Vol. 202, Materials Research Society, Pittsburgh, pp. 305–310.
GaAs      Liliental-Weber, Z., Weber, E. R., Parechanian-Allen, L., and Washburn, J. (1988). On the use of CBED for identification of antiphase boundary in GaAs grown on Si, *Ultramicroscopy* **26**, 59–64.
GaAs      Matsumura, S., Morimura, T., and Oki, K. (1988). CBED and EDX method for the polarity of GaAs crystal. In *Annual Report*, HVEM Lab., Japan, pp. 37–38.
GaAs      Penncock, G. M. and Schapink, F. W. (1987). CBED from GaAs/AlAs multilayers. In *Electron Microscopy of Semiconducting Materials*, Inst. Phys. Conf. Ser. Vol. 87, Institute of Physics, Bristol and London, pp. 219–224.

GaAs    Preston, A. R. and Spellward, P. (1988). Polarity determination in GaAs by matching of [110] CBED patterns and simulations. In *Proc. 9th European Congr. on Electron Microscopy*, eds. P. J. Goodhew and H. G. Dickinson, Inst. Phys. Conf. Ser., Vol. 93, Institute of Physics, London, pp. 27–28.

GaAs    Steeds, J. W., Rackham, G. M., and Merton-Lyn, D. (1981f), Microanalysis of metal contacts to GaAs and InP. In *Microscopy of Semiconducting Materials 1981*, Proc. of the Royal Microscopical Soc. Conf. Oxford, eds. A. G. Cullis and D. C. Joy, Inst. Phys. Conf. Ser. Vol. 60, Institute of Physics, Bristol and London, pp. 387–396.

GaAs    Xie, Q. H., Fung, K. K., Ding, A. J., Cai, L. H., Huang, Y., and Zhou, J. M. (1990c). Asymmetric distribution of microtwins in a GaAs/Si heterostructure grown by MBE, *Appl. Phys. Lett.* **57**(26), 2803–2805.

GaAs    Zuo, J. M., O'Keefe, M. O., and Spence, J. C. H. (1988c). On the accuracy of structure factor measurement in GaAs by CBED. In *Proceedings of the 46th Annual Meeting of EMSA*, ed. G. W. Bailey, San Francisco Press, San Francisco, pp. 824–825.

GaAs    Zuo, J. M., Spence, J. C. H., and O'Keefe, M. (1988a). Bonding in GaAs, *Phys. Rev. Lett.* **61**(3), 353–356.

GaAsAlAs    Gat, R. and Schapink, F. W. (1987). CBED from GaAs/AlAs superlattices, *Ultramicroscopy* **21**, 289–392.

GaAsAlAs    Gong, H. and Schapink, F. W. (1989b). Comparison of features in the HOLZ reflections from GaAs/AlAs superlattices and GaAlAs single crystals. In *EMAG-MICRO 1989*, Inst. Phys. Conf. Ser., Vol. 98, Institute of Physics, London, pp. 263–266.

GaAsAlAs    Gong, H. and Schapink, F. W. (1989c). Structure of HOLZ reflections from a GaAs/AlAs superlattices. In *EMAG-MICRO 1989*, Inst. Phys. Conf. Ser., Vol. 98, Institute of Physics, London, pp. 91–94.

GaAsInGaAs    Fung, K. K., York, P. K., Fernandez, G. E., Eades, J. A., and Coleman, J. J. (1988a). CBED study of strain modulation in GaAs/InGaAs superlattices grown by metal-organic chemical vapor deposition, *Philos. Mag. Lett.* **57**(4), 221–227.

GaInAsP    Bons, A. J., Gong, H., and Schapink, F. W. (1991). The interpretation of the FOLZ reflection fine structure in CBED patterns of GaInAsP, *Philos. Mag.* **63B**(6), 1395–1408.

GaInAsP    Preston, A. R. (1987). CBED investigation of LPE grown layers of GaInAsP. In *Electron Microscopy of Semiconducting Materials*, Inst. Phys. Conf. Ser. Vol. 87, Institute of Physics, Bristol and London, pp. 159–164.

GaP    Chung, D. W. (1985). Lattice misfit between GaP and Si, *Scr. Metall.* **19**, 817.

Gap    Chung, D. W. (1987a). Antiphase domain in GaP grown on Si by MBE, *Mater. Res. Bull.* **22**, 61–68.

GaP    Chung, D. W. (1987b). Analysis of defect displacements in GaP film using CBED, *Mater. Res. Bull.* **22**, 211–218.

GaS    Blasi, C. D., Manno, D., and Rizzo, A. (1990b). CBEd characterization of dislocation in GaS single crystals, *Ultramicroscopy* **33**, 143–149.

GaS    Goodman, P. and Whitfield, H. J. (1980b). The space-group determination of GaS and $Cu_3As_2S_3I$ by CBED, *Acta Crystallogr.* **A36**, 219.

GaS    Goodman, P., Olson, A., and Whitfield, H. J. (1986). An investigation of a metastable form of GaS by CBED and HREM, *Acta Crystallogr.* **B41**, 292.

GaSe    Blasi, C. D. and Manno, D. (1990a). CBED analysis of extended defects in melt-grown GaSe single crystals. In *Proc. 12th Int. Congr. on Electron Microscopy*, Vol. 4, San Francisco Press, San Francisco, pp. 494–495.

GaSe    Blasi, De C., Mancini, A. M., Manno, D., and Rizzo, A. (1989). CBED study of structural modification in vapor grown GaSe crystals. In *Microscopy of Semiconducting*

*Materials Conference,* Inst. Phys. Conf. Ser. vol. 100, Institute of Physics, London, pp. 451–456.

GaSe  De Blasi, C. and Manno, D. (1991). Analysis of extended defects in melt-grown GaSe single crystals by CBED techniques. *Ultramicroscopy* **35**, 71–76.

GaSe  De Blasi, C., Mancini, A. M., Manno, D., and Rizzo, A. (1988a). Structural modifications induced by iodine in melt grown GaSe. In *Proc. 9th European Congr. on Electron Microscopy,* eds. P. J. Goodhew and H. G. Dickinson, Inst. Phys. Conf. Ser., Vol. 93, Institute of Physics, London, pp. 123–124.

$GaS_xSe_{1-x}$  Blasi, C., Manno, D., Micocci, G., and Rizzo, A. (1986). Selected area and CBED in vapor grown $GaS_xSe_{1-x}$ single crystals. In *Proc. 11th Int. Congr. on Electron Microscopy* (Kyoto), Maruzen, Tokyo, pp. 725–726.

GdBaCu-O  Ramesh, R., Thomas, G., Meng, R. L., Hor, P. H., and Chu, C. W. (1989). Electron microscopy of a Gd–Ba–Cu–O superconductor, *J. Mater. Res.* **4**(3), 515–520.

$Ge_{0.5}Si_{0.5}/Si$  Duan, X. F., Fung, K. K., Chu, Y. M., Sheng, C., and Zhou, G. L. (1991). CBED study of $Ge_{0.5}Si_{0.5}/Si$ strained layer superlattices grown by molecular beam epitaxy, *Philos. Mag. Lett.* **63**(2), 79–85.

GeGaAs  Kiely, C. J., Tavitian, V., and Eden, J. G. (1989). Microstructural studies of epitaxial Ge films grown on [100] GaAs by laser photochemical vapor deposition, *J. Appl. Phys.* **65**(10), 3883–3895.

GeSi  Kvam, E. P., Eaglesham, D. J., Humphreys, C. J., Maher, D. M., Bean, J. C., and Fraser, H. L. (1987). Heteroepitaxial strains and interface structure of Ge–Si alloy layers on Si(100). In *Electron Microscopy of Semiconducting Materials,* Inst. Phys. Conf. Ser. Vol. 87, Institute of Physics, Bristol and London, pp. 165–168.

$Ge_xSi_{1-x}$  Eaglesham, D. J., Maher, D. M., Fraser, H. L., Humphreys, C. J., and Bean, J. C. (1989d). Tetragonal and monoclinic forms of $Ge_xSi_{1-x}$ epitaxial layers, *Appl. Phys. Lett.* **54**(3), 222–224.

$Ge_xSi_{1-x}$  Xiaofeng, D., Fung, K. K., and Chu, Y. M. (1990). A study of lattice distortion in $Ge_xSi_{1-x}$ superlattice by CBED. In *Proc. 12th Int. Congr. on Electron Microscopy,* Vol. 4, San Francisco Press, San Francisco, pp. 614–615.

$HaTaSe_2$  Fung, K. K., McKernan, S., Steeds, J. W., and Wilson, J. A. (1981). Broken hexagonal symmetry in the locked-in state of $2Ha-TaSe_2$ and the discommensurate microstructure of its incommensurate CDW states, *J. Phys. C.* **14**, 5417–5432.

HgCdTe  Spellward, P. and Cherns, D. (1987). Determination of composition in $Cd_xHg_{1-x}Te$ by measurement of the [111] zone axis critical voltage. In *Electron Microscopy of Semiconducting Materials,* Inst. Phys. Conf. Ser., Vol. 87, Institute of Physics, Bristol and London, pp. 153–158.

InGaAs  Sung, C. M., Ostreicher, K. J., Abdalla, M., Kenneson, D. G., Powazinik, W., and Hefter, J. (1990b). TEM investigation of lattice-matched InGaAs layers grown on InP by LPMVP epitaxy. In *Proc. 12th Int. Congr. on Electron Microscopy,* Vol. 4, San Francisco Press, San Francisco, pp. 674–675.

InGaAs  Xie, Q. H. and Fung, K. K. (1990b). CBED study of strain in InGaAs/GaAs strained layer superlattices. In *Proc. 12th Int. Congr. on Electron Microscopy,* Vol. 4, San Francisco Press, San Francisco, pp. 670–671.

InGaAs  Xie, Q. H., Fung, K. K., York, P. K., Fernandez, G. E., Eades, J. A., and Coleman, J. J. (1990a). CBED study of lattice distortion in InGaAs/GaAs strained layer superlattices grown by MOCVD, *Appl. Phys. Lett.* **57**(19), 1978–1980.

InGaAsInP  Cherns, D., Jordan, I. K., and Vincent, R. (1989d). Composition profiles in AlGaAs/GaAs and InGaAs/InP structures examined by CBED. In *Proc. MRS Symposium, Characterization of the Structure and Chemistry of Defects in Materials,* eds. B. C. Larson, M. Ruhle, and D. N. Seidman, Vol. 138, Materials Research Society, Pittsburgh, pp. 431–436.

InGaAsP     Twigg, M. E., Chu, S. N. G., Joy, D. C., Maher, D. M., Macrander, A. T., and Chin, A. K. (1987b). Relative lattice parameter measurement of submicron quaternary (InGaAsP) device structures grown on InP substrates, *J. Appl. Phys.* **62**(8), 3156.

InGaAsP     Twigg, M. E., Chu, S. N. G., Joy, D. C., Maher, D. M., Macrander, A. T., Nakahara, S., and Chin, A. K. (1986). Relative lattice parameters measurement in quaternary (InGaAsP) layers on InP substrates using CBED. In *MRS Symposia Proceedings*, Vol 62, Materials Characterization, Materials Research Society, Pittsburgh, p. 147.

InP     Jordan, I. K., Cherns, D., Hockly, M., and Spurdens, P. C. (1990). Investigation of the uniformity of well thickness in InP/InGaAs single QWs by LACBED and dark-field imaging. In *Proc. 12th Int. Congr. on Electron Microscopy*, Vol. 4, San Francisco Press, San Francisco, pp. 680–681.

InP     Jordan, I. K., Cherns, D., Hockly, M., and Spurdens, P. C. (1989a). CBED from InP/InGaAs single quantum wells. In *Microscopy of Semiconducting Materials Conference*, Inst. Phys. Conf. Ser. Vol. 100, Institute of Physics, London, pp. 293–298.

InP     Jordan, I. K., Preston, A. R., Qin, L. C., and Steeds, J. W. (1989). Analysis of LACBED patterns from InP/InGaAs multiple quantum well samples by dynamical theory. In *EMAG-MICRO 1989*, Inst. Phys. Conf. Ser., Vol. 98, Institute of Physics, London, pp. 131–134.

InP     Steeds, J. W., Rackham, G. M., and Merton-Lyn, D. (1981f). Microanalysis of metal contacts to GaAs and InP. In *Microscopy of Semiconducting Materials 1981*, Proc. of the Royal Microscopical Soc. Conf. Oxford, eds. A. G. Cullis and D. C. Joy, Inst. Phys. Conf. Ser., Vol. 60, Institute of Physics, Bristol and London, pp. 387–396.

InP     Twigg, M. E. and Chu, S. N. G. (1988). CBED measurement of lattice mismatch for sub-micron quaternary structures grown on InP substrates, *Ultramicroscopy* **26**, 51–58.

IrNb     Fleischer, R. L., Field, R. D., Denike, K. K., and Zabala, R. J. (1990). Mechanical properties of alloys of IrNb and other high-temperature intermetallic compounds, *Metall. Trans.* **21A**, 3063–3074.

$La_{1.8}Sr_{0.2}CuO_y$     Eaglesham, D. J. (1988b). A TEM zone-axis critical voltage in $La_{1.8}Sr_{0.2}CuO_y$ superconductors: A method for local composition microanalysis, *Philos. Mag.* **57**(1), 1–9.

$La_2CuO_{4-\delta}$     Yamada, K. and Kudo, E. (1988). Determination of space group and refinement of structure parameters for $La_2CuO_{4-\delta}$ crystals, *Jpn. J. Appl. Phys.* **27**(7), 1132–1137.

$La_2Ti_2O_7$     Tanaka, M., Sekii, H., and Ohi, K. (1985c). Structural study of $La_2Ti_2O_7$ by CBED and electron microscopy, *Jpn. J. Appl. Phys.* **24**, 814–816.

LaAl     Yang, C. Y., Huang, Z. R., Zhou, Y. Q., Li, C. Z., Yang, W. H., and Fung, K. K. (1990). TEM study of lathanum aluminate. In *Proc. 12th Int. Congr. on Electron Microscopy*, Vol. 4, San Francisco Press, San Francisco, pp. 1060–1061.

$LaBa_2Cu_3O_{5+x}$     Sishen, X. and Cuiying, Y. (1988). Crystal structure and superconductivity of $LaBa_2Cu_3O_{5+x}$, *Phys. Rev.* **36**(4), 2311–2312.

LiTa     Loveluck, J. E. and Steeds, J. W. (1977a). Crystallography of lithium tantalate and quartz. In *Developments in Electron Microscopy and Analysis*, ed. D. L. Misell, Inst. Phys. Conf. Ser., Vol. 36, Institute of Physics, Bristol and London, pp. 293–296.

$LiZnTa_3O_9$     Yang, C. Y., Zhou, Y. Q., and Fung, K. K. (1983). Determination of space groups of $LiZnTa_3O_9$ by CBED, *Acta Crystallogr.* **A39**, 531–533.

LnNbO$_4$  Tanaka, M., Saito, R., and Watanabe, D. (1980c). Symmetry determination of the room-temperature form of LnNbO$_4$ (Ln = La, Nd) by CBED, *Acta Crystallogr.* **A36**, 350–352.

MgAl$_2$O$_4$  Ballesteros, C., Lain, L. S., Pennycook, S. J., Gonzalez, R., and Chen, Y. (1989). Optical and analytical TEM characterization of thermochemically reduced MgAl$_2$O$_4$ spinel, *Philos. Mag.* **59A**(4), 907–916.

MgAl$_2$O$_4$  Hwang, L., Heuer, A. H., and Mitchell, T. E. (1973). On the space group of MgAl$_2$O$_4$ spinel, *Philos. Mag.*, **14**, 241–243.

MgAl$_2$O$_4$  Liu, D. R. (1990). Observation of dynamic extinction lines in the {420} diffraction disks of the MgAl$_2$O$_4$ spinel ⟨100⟩ CBED pattern. In *Proc. 12th Int. Congr. on Electron Microscopy*, Vol. 2, San Francisco Press, San Francisco, pp. 506–507.

MgAl$_2$O$_4$  Takonami, M. and Horiuchi, H. (1980). On the space group of spinel, MgAl$_2$O$_4$, *Acta Crystallogr.* **A36**, 122–126.

MgO  Gjønnes, J. and Watanabe, D. (1966). Dynamical diffuse scattering from MgO single crystals, *Acta Crystallogr.* **21**, 297–302.

MgO  Goodman, P. (1971). The influence of temperature on the MgO zone axis pattern, *Acta Crystallogr.* **27A**, 140.

MgO  Goodman, P. and Lehmpfuhl, G. (1968b). Electron diffraction study of MgO $h$00-systematic interactions, *Acta Crystallogr.* **22**, 14.

MgZn  Liuying, W. and Dunlop, G. L. (1988). The structure of intergranular phases in a cast Mg-8Zn-1.5RE alloy. In *Proc. 9th European Congr. on Electron Microscopy*, eds. P. J. Goodhew and H. G. Dickinson, Inst. Phys. Conf. Ser., Vol. 93, Institute of Physics, London, pp. 511–512.

Mn-Si-Al  Wang, N., Fung, K. K., and Kuo, K. H. (1988). Symmetry study of the Mn-Si-Al octagonal quasicrystal by CBED, *Appl. Phys. Lett* **52**(25), 2120–2121.

MnAlC  Gau, J. S. (1983). A new ordered phase in Mn-Al-C alloys. In *Proceedings of the 41st Annual Meeting of EMSA*, ed. G. W. Bailey, San Francisco Press, San Francisco, pp. 252–253.

MnSi  Tanaka, M., Takayoshi, H., Ishida, M., and Endoh, Y. (1985f). Crystal chirality and helicity of the helical spin density wave in MnSi, I. CBED, *J. Phys. Soc. Jpn.* **54**(8), 2970–2974.

MoO  Tatlock, G. J. (1975). The simulation and interpretation of ZAPs in molybdenite, *Philos. Mag.* **31**, 1159.

MoO$_3$  Bursill, L. A., Dowell, W. C. T., Goodman, P., and Tate, N. (1974). Investigation of thin MoO$_3$ crystals by CBED, *Acta Crystallogr.* **A34**, 296–308.

MoS  Vazquezz, A., Dominguez, J. M., Pina, C., Jaidar, A., and Fuentes, S. (1990). Surface defects in MoS$_2$ crystals synthesized by vapor phase transport methods, *J. Mater. Sci. Lett.* **9**, 712–714.

MoS$_2$  Jesson, D. E. and Steeds, J. W. (1990b). An investigation of three-dimensional diffraction from 2H$_6$-MoS$_2$, *Philos. Mag.* **61A**(3), 363–384.

MoSe$_2$  Quandt, E., Weickenmeier, A., Kohl, H., and Niedrig, H. (1990b). Investigation of energy spectroscopic intensity distributions in large angle CBED patterns of MoSe$_2$. In *Proc. 12th Int. Congr. on Electron Spectroscopy*, Vol. 2, San Francisco Press, San Francisco, pp. 44–45.

Na$_2$Ti$_9$O$_{19}$  Bando, Y. (1982). Combination of CBED and 1 MeV structure imaging in a structure determination of Na$_2$Ti$_9$O$_{19}$, *Acta Crystallogr.* **A38**, 211.

NaLiSi  Withers, R. L., Thompson, J. G., and Hyde, B. G. (1989a). A TEM study of modulated sodium lithium metasilicates, *Acta Crystallogr.* **B45**, 136–141.

NbFe  Ecob, E. C., Lobb, R. C., and Kohler, V. L. (1987). The formation of G-phase in 20/25 Nb stainless steel AGR fuel cladding alloy and its effect on creep properties, *J. Mater. Sci.* **22**, 2867–2880.

$Nb_2Zr_{x-2}O_{2x+1}$ Thompson, J. G., Withers, R. L., Sella, J., Barlow, P. J., and Hyde, B. G. (1990). Incommensurate composite modulated $Nb_2Zr_{x-2}O_{2x+1}$: $x = 7.1$–$10.3$, *J. Solid State Chem.* **88**, 465–475.

$Nb_2Zr_{x-2}O_{2x+1}$ Withers, R. L., Thompson, J. G., and Hyde, B. G. (1991). A modulation-wave approach to the structural description of the $Nb_2Zr_{x-2}O_{2x+1}$ ($x = 7.1$–$10.3$) solid-solution field, *Acta Crystallogr.* **B47**, 166–174.

NbGe Evans, N. D. (1990). Morphologies of hypereutectic, rapidly solidified Nb–Ge alloys. In *Proc. 12th Int. Congr. on Electron Spectroscopy*, Vol. 4, San Francisco Press, San Francisco, pp. 1016–1017.

NbSi Miller, D. J., Sears, J. W., and Fraser, H. L. (1989). Characterization of metastable phases in a rapidly solidified Nb–22Si alloy, *Acta Metall.* **37**(4), 999–1007.

$Nd_{0.66}Sr_{0.205}Ce_{0.135}$–$Cu_{4-\delta}$ Tsuda, K., Tanaka, M., Sankanoue, J., Sawa, H., Suzuki, S., and Akimitsu, J. (1989). Space-group determination of a new superconductor $(Nd_{0.66}Sr_{0.205}Ce_{0.135})_2Cu_{4-\delta}$ using CBED, *Jpn. J. Appl. Phys.* **28**(3), L389–391.

NdFeCoB Koestler, C., Ramesh, R., Echer, C. J., Thomas, G., and Wecker, J. (1989). Microstructure of melt spun Nd–Fe–Co–B magnets, *Acta Metall.* **37**(7), 1945–1955.

Ni Crompton, J. S. and Hertzberg, R. .W (1986). Analysis of second phase particles in a powder metallurgy HIP nickel-base superalloy, *J. Mater. Sci.* **21**, 3445–3454.

Ni Ecob, R. C., Ricks, R. A., and Porter, A. J. (1982). The measurement of precipitate/matrix lattice mismatch in Ni-base superalloys, *Scr. Metall.* **16**, 1085–1090.

Ni Fraser, H. L. (1983). Applications of CBED to Ni-base superalloys, *J. Microsc. Spectrosc. Electron.* **8**, 421–441.

Ni Gjønnes, K. and Anderson, A. (1987). Scale structures during oxidation and sulphidation of Ni and a Ni–Cr alloy, *Ultramicroscopy* **22**, 289–296.

Ni Kaufman, M. J., Pearson, D. D., and Fraser, H. L. (1986c). The use of CBED to determine local lattice distortion in Ni-base superalloys, *Philos. Mag.* **54A**, 79–92.

Ni Lahrman, D. F., Field, R. D., Darolia, R., and Fraser, H. L. (1988). Investigation of techniques for measuring lattice mismatch in a rhenium containing Ni-base superalloy, *Acta Metall.* **36**(5), 1309–1320.

Ni Porter, A. J. and Ricks, R. A. (1983a). Temperature dependence of precipitate/matrix lattice mismatch in Ni-base superalloy. In *Electron Microscopy and Analysis*, ed. P. Doig, Inst. Phys. Conf. Ser., Vol. 68, Institute of Physics, London, p. 51.

Ni Randle, V. and Ralph, B. (1987a). Microstructural characterization of a Ni-base superalloy (Nimonic PE16), *Electron Optics Bull.* **124**, 28–38.

$Ni_3Al$ Baker, I. and Schulson, E. M. (1989). On the grain boundaries in nickel-rich $Ni_3Al$, *Scr. Metall.* **23**, 1883–1886.

$Ni_3Al$ Baker, I., Schulson, E. M., Michael, J. R., and Pennycook, S. J. (1990). The effects of both deviation from stoichiometry and boron on grain boundary in $Ni_3Al$, *Philos. Mag.* **62B**(6), 659–676.

$Ni_3Mo$ Kaufman, M. J., and Fraser, H. L. (1982). Structure determination of $Ni_3Mo$ using CBED, In *Proceedings of the 40th Annual Meeting of EMSA*, ed. G. W. Bailey, San Francisco Press, San Francisco, pp. 686–687.

NiAl Webber, J. G., and Van Aken, D. C. (1989). Studies of a quasi-binary $\beta$-NiAl and $\alpha$-Re eutectic, *Scr. Metall.* **23**, 193–196.

NiAlCrHf Sircar, S., Ribaudo, C., and Mazumder, J. (1989b). Laser-clad $Ni_{70}Al_{20}Cr_7Hf_3$ alloys with extended solid solution of Hf: Part I. Microstructure evolution, *Metall. Trans.* **20A**, 2267–2277.

NiAlMo Kaufman, M. J., Eades, J. A., Loretto, M. H., and Fraser, H. L. (1983a). A

BIBLIOGRAPHY OF CBED APPLICATIONS 345

study of a cellular phase transformation in the ternary Ni-Al-Mo alloy system, *Metall. Trans.* **14A**, 1561.
NiAs Vincent, R. and Withers, R. L. (1987c). Analysis of a dispersive superlattice in Ni arsenide, *Philos. Mag.* **56**(2), 57-62.
NiAu Butler, E. P. (1972). Application of the critical voltage effect to the study of compositional change in Ni-Au alloys, *Philos. Mag.* **16**, 33.
NiCrFeCB Kruger, R. M., Was, G. S., Mansfield, J. F., and Martin, J. R. (1988). A quantitative model for the intergranular precipitation of $M_7X_3$ and $M_{23}X_6$ in Ni-16Cr-9Fe-C-B, *Acta Metall.* **36**(12), 3163-3176.
NiFe Matsuhata, H., Kuroda, K., Tomokiyo, Y., and Eguchi, T. (1983). Critical voltage effect in $Ni_3Fe$ and FeCo ordering alloys, *Jpn. J. Appl. Phys.* **22**(3), 404-407.
NiGeAu Rackham, G. M. and Steeds, J. W. (1980). Nickel germanium gold contacts to GaAs. In *Electron Microscopy and Analysis,* ed. T. Mulvey, Inst. Phys. Conf. Ser., Vol. 52, Institute of Physics, Bristol and London, pp. 157-60.
NiGeP Vincent, R. and Pretty, S. F. (1986c). Phase analysis in the Ni-Ge-P system by electron diffraction, *Philos. Mag.* **53A**, 843-862.
NiGeP Withers, R. L. and Bird, D. M. (1986b). Electron diffraction from modulated structures: II. Application to the ternary modulated phase NiGeP, *J. Phys. C* **19**, 3507-3516.
NiHf Sircar, S., Singh, J., and Mazumder, J. (1989a). Microstructure evolution and nonequilibrium phase diagram for Ni-Hf binary alloy produced by laser cladding, *Acta Metall.* **37**(4), 1167-1176.
NiMoFeB Kim, Y. W., Rabenberg, L., and Bourell, D. L. (1988). Identification of a boride phase as a crystallization product of a NiMoFeB amorphous alloy, *J. Mater. Res.* **3**(6), 1336-1342.
NiNb Bertero, G. A., Hofmeister, W. H., Evans, N. D., Wittig, J. E., and Bayuzick, R. J. (1990). Devitrification products of rapidly solidified Ni-Nb amorphous alloys. In *Proc. 12th Int. Congr. on Electron Microscopy,* Vol. 4, San Francisco Press, San Francisco, pp. 950-951.
NiO Eaglesham, D. J., Kvan, E. P., and Humphreys, C. J. (1988c). CBED study of the antiferromagnetic distortion of NiO. *In Proc. 9th European Congr. on Electron Microscopy,* eds. P. J. Goodhew and H. G. Dickinson, Inst. Phys. Conf. Ser., Vol. 93, Institute of Physics, London, pp. 33-34.
$NiOZrO_2$(CaO) Dravid, V. P., Lyman, C. E., Notis, M. R., and Revcolevschi, A. (1990). Low-energy interfaces in $NiO-ZrO_2$(CaO) eutectic, *Metall. Trans.* **21A**, 2309-2315.
$NiOZrO_2$(CaO) Dravid, V. P., Notis, M. R., Lyman, C. E., and Revcolevschi, A. (1990a). Plan-view CBED studies of $NiO-ZrO_2$(CaO) interfaces. In *Atomic Scale Structure of Interfaces,* MRS Symp. Proc., Vol. 159, ed. R. D. Brigans, R. M. Feenstra, and J. M. Gibson, Materials Research Society, Pittsburgh, pp. 95-100.
$NiPS_3$ Baba-Kishi, K. Z. and Dingley, D. J. (1988). A contribution to the point group symmetry of $NiPS_3$. In *EMAG 87, Analytical Electron Microscopy,* ed. G. W. Lorimer, The Institute of Metals, Manchester, Sept. 1987, pp. 89-92.
$NiSi_2Si$ Eaglesham, D. J., Kiely, C. J., Cherns, D., and Missous, M. (1989e). Electron diffraction from epitaxial crystals—a CBED of the interface structure for $NiSi_2$/Si and Al/GaAs, *Philos. Mag.* **60**(2), 161-175.
$NiSi_2Si$ Kiely, C. J. and Cherns, D. (1985). Studies of misfit strains in $NiSi_2$/Si bicrystals. In *Microscopy of Semiconducting Materials,* Inst. Phys. Conf. Ser. Vol. 76, Institute of Physics, London, pp. 183-188.
NiZr Allem, R., L'Esperance, G., Altounian, Z., and Strom-Olsen, J. O. (1991). A

TEM study of the microstructures formed during the crystallization of Ni–Zr metallic glasses, *J. Mater. Res.* **6**(4), 755–759.

$Pb_{1-x}Ca_x$–$TiO_3$  King, G., Goo, E., Yamamoto, T., and Okazaki, K. (1988). Crystal structure and defects of ordered $(Pb_{1-x}Ca_x)TiO_3$, *J. Am. Ceram. Soc.* **71**(6), 454–460.

$Pb_2Sr_2Y/CaCu_3O_{9-\delta}$  Lin, Y. P., Xue, J. S., and Greedan, J. E. (1990). Microstructure and crystal symmetry of $Pb_2Sr_2(Y/Ca)Cu_3O_{9-\delta}$. In *Proc. 12th Int. Congr. on Electron Microscopy*, Vol. 4, San Francisco Press, San Francisco, pp. 64–65.

PbBiSrCaCuO  Rong, T. S. and Shi, F. (1989). Crystal structure of the high Tc (107 K) superconducting phase in Pb–Bi–Sr–Ca–Cu–O, *Appl. Phys. Lett.* **54**(21), 2157–2158.

PbBSrCaCuO  Aurisicchio, C., Calestani, G. L., De Blasi, C., Fiorani, D., Manno, D., and Palmisano, M. (1990). Structural and superconducting study of the (Pb, Bi)–Sr–Ca–Cu–O. In *Proc. 12th Int. Congr. on Electron Microscopy*, Vol. 4, San Francisco Press, San Francisco, pp. 98–99.

$PbSc_{1/2}Ta_{1/2}O_3$  Baba-Kishi, K. Z. and Barber, D. J. (1990). TEM studies of phase transitions in single crystals and ceramics of ferroelectric $Pb(Sc_{1/2}Ta_{1/2})O_3$, *J. Appl. Crystallogr.* **23**, 43–54.

$PbTiO_3$  Fox, G. R., Breval, E., and Newnham, A. E. (1991). Crystallization of nanometer-size coprecipitated $PbTiO_3$ powders, *J. Mater. Sci.* **26**, 2566–2572.

$PbZn_{1/3}Nb_{2/3}O_3$  Sung, C. M., Gorton, A. J., Chen, J., Harmer, M. P., and Smyth, D. M. (1989). AEM study of perovskite $Pb(Zn_{1/3}Nb_{2/3})O_3$ ferroelectric relaxor. In *Proceedings of the 47th Annual Meeting of EMSA*, ed. G. W. Bailey, San Francisco Press, San Francisco, pp. 514–515.

$PbZrO_3$  Tanaka, M., Saito, R., and Tsuzuki, K. (1982c). Determination of space group and oxygen coordinates in the antiferroelectric phase of $PbZrO_3$ by conventional and CBED, *J. Phys. Soc. Jpn*, **51**(8), 2635–2640.

$PbZrO_3$  Tanaka, M. and Saito, R. (1980e). Crystal symmetry and domain structure of $PbZrO_3$, In *Proc. 7th European Congr. on Electron Microscopy*, Vol. I, Hague, eds. P. Brederoo and G. Boom, Deutsche Gesellschaft fur Elektronmikroscopie, Frankfurt, pp. 402–403.

Pd–MgO  Giorgio, S., Henry, C. R., Chapon, C., and Penisson, J. M. (1990). Structure and morphology of small Pd particles (2–6 nm) supported on MgO micro-cubes, *J. Cryst. Growth* **100**, 254–260.

$PrCO_5$  Shen, Y. and Laughlin, D. E. (1990). Magnetic effects on the symmetry of CBED patterns of ferromagnetic $PrCO_5$, *Philos. Mag. Lett.* **62**, 187–193.

PtSiO  Lamber, R. and Jaeger, N. I. (1991). On the reaction of Pt with $SiO_2$ substrates: Observation of the $Pt_3Si$ phase with the $Cu_3Au$ superstructure, *J. Appl. Phys.* **70**(1), 457–461.

Pt catalyst particles on C, alumina, zeolite supports  Liu, J. and Cowley, J. M. (1991), H.A.A.D. and S.E. imaging of supported catalyst clusters, *Ultramicroscopy* **34**, 119.

PZT  Dass, M. L. A. (1986a). Convergent beam diffraction studies on PZT ceramics. In *Proceedings of the 44th Annual Meeting of EMSA*, ed. G. W. Bailey, San Francisco Press, San Francisco, pp. 482–483.

PZT  Dass, M. L. A. (1987a). CBED studies of domains in tetragonal phase of PZT ceramics. In *Proceedings of the 45th Annual Meeting of EMSA*, ed. G. W. Bailey, San Francisco Press, San Francisco, pp. 26–27.

PZT  Dass, M. L. A. and Thomas, G. (1987b). CBED studies of domain in rhombohedral phase of PZT ceramics. In *Analytical Electron Microscopy '87*, San Francisco Press, San Francisco, pp. 132–134.

PZT  Dass, M. L. A., Dahmen, U., Thomas, G., Yamamoto, T., and Okazaki, K.

(1986b). Electron microscopy characterization of PZT ceramics at the morphotropic phase boundary composition, *IEEE* **14**, 146–149.

PZT  Demczyk, B. G. (1987b). In-situ TEM study of the paraelectric to ferroelectric phase transformation in La-modified PT ceramics. In *Proceedings of the 45th Annual Meeting of EMSA*, ed. G. W. Bailey, San Francisco Press, San Francisco, pp. 172–173.

PZT  Demczyk, B. G., Rai, R. S., and Thomas, G. (1990). Ferroelectric domain structure of La-modified lead titanate ceramics, *J. Am. Ceram. Soc.* **73**(3), 615–620.

PZT  Yamamoto, T., Okazaki, K., Dass, M. L., and Thomas, G. (1988). Microstructures and dielectric properties at the MPB of PT–PZ system, *Ferroelectrics* **81**, 331–334.

RuTa  Fleischer, R. L., Field, R. D., and Briant, C. L. (1991). Mechanical, elastic, and structural properties of alloys of Ru–Ta high temperature intermetallic compounds, *Metall. Trans.* **22A**, 129–137.

Sc  Vujic, D. R., Lohemeier, D. A., and Whang, S. H. (1990). Occurrence of glassy phases in Sc–Co and Sc–Ni system, *Int. J. Rapid Solidification* **5**, 277–296.

SiAl  Bando, Y., Mitomo, M., Kitami, Y., and Izumi, F. (1986c). Structure and composition analysis of SiAl oxynitride polytypes by combined use of structure imaging and microanalysis, *J. Microsc.* **142**, 235–246.

Si  L'Ecuyer, J. D., Loretto, M. H., and L'Esperrance, G. (1987). Measurement of strain at Si-porous interfaces by use of CBIM and CBD. In *Analytical Electron Microscopy '87*, San Francisco Press, San Francisco, pp. 155–158.

Si  Lu, G., Wen, J. G., Zhang, W., and Wang, R. (1990). Simulation and application of the distorted ZOLZ patterns from dislocations in Si, *Acta Crystallogr.* **46A**, 103–112.

Si  Pribat, D., Chazelas, J., and DuPuy, M. (1990). Structural characterizations of conformally grown (100) Si films, *J. Jpn. Appl. Phys.* **29**(11), 1943–1946.

Si  Terasaki, O. and Watanabe, D. (1979). Determination of crystal structure factors of Si by the intersecting Kikuchi line method, *Acta Crystallogr.* **A35**, 895.

Si  Voss, R., Lehmpfuhl, G., and Smith, P. J. (1980). Influence of doping on the crystal potential of Si investigated by the CBED technique, *Z. Naturforsch.* **35a**, 973–984.

Si  Wen, J., Wang, R., and Lu, G. (1989). Distortion of the zeroth-order Laue-zone pattern caused by dislocations in a Si crystal, *Acta Crystallogr.* **A45**, 422–427.

$Si_3N_4$  Sung, C. M., Ostreicher, K. J., Buljan, S. T., Baldoni, J. G., and Hefter, J. (1990a). Microstructural analysis of $Si_3N_4$ ceramics containing various sintering aids and additional $Si_3N_4$. In *Proc. 12th Int. Congr. on Electron Microscopy*, Vol. 4, San Francisco Press, San Francisco, pp. 1074–1075.

SIALON  Drennan, J. and Trigg, M. B. (1987). AEM of the glass phases observed in yttrium O-SIALON. In *Analytical Electron Microscopy '87*, San Francisco Press, San Francisco, pp. 41–44.

SiC  Edmond, J. A., Davis, R. F., Withrow, S. P., and More, K. L. (1988). Ion implantation in $\beta$-SiC: effect of channeling direction and critical energy for amorphization, *J. Mater. Res.* **3**(2), 321–328.

SiC  Hangas, J., Shinozaki, S., and Donlon, W. T. (1986). Lattice parameter measurements of SiC materials using CBED HOLZ Lines. In *Proceedings of the 44th Annual Meeting of EMSA*, ed. G. W. Bailey, San Francisco Press, San Francisco, pp. 464–465.

SiC  Peteves, S. D., Tambuyser, P., Helbach, P., Audier, M., Laurent, V., and Chatain, D. (1990). Microstructure and microchemistry of the Al/SiC interface, *J. Mater. Sci.* **25**, 3765–3772.

SiC  Yaney, D. L. and Joshi, A. (1990). Reaction between Nb and silicon carbide at 1373 K, *J. Mater. Res.* **5**(10), 2197–2208.

SiGe  Eguchi, T., Tomokiyo, Y., and Matsuhata, H. (1987). Measurement of atomic

mean square displacement in Si–Ge solid solutions by critical voltage method, *J. Microsc. Spectrosc. Electron* **12**, 559–568.
SiGe  Mansfield, J. F., Lee, D. M., and Rozgonyi, G. A. (1987a). Examination of anomalous fringes in a Si–3% Ge alloy layer—evidence of a superlattice. In *Electron Microscopy of Semiconducting Materials*, Inst. Phys. Conf. Ser., Vol. 87, Institute of Physics, London, pp. 169–174.
SiGe  Okuyama, T., Matsumura, S., Yasunaga, T., Tomokiyo, Y., Kuwano, N., and Oki, K. (1990). Dynamical diffraction effect on HOLZ pattern geometry for semiconductor alloys of $Si_{1-x}Ge_x$ and $Ga_{1-x}In_xAs$. In *Proc. 12th Int. Congr. on Electron Microscopy*, Vol. 2, San Francisco Press, San Francisco, pp. 486–487.
SiGe  Pike, W. T. (1990). Microdiffraction from cleaved $Si–Si_{1-x}Ge_x$ multilayers. In *Layered structures—Heteroepitaxy, superlattices, strain, and metastability*, MRS Proc., eds. B. W. Dodson, L. J. Schowalter, J. E. Cunningham, and F. H. Pollak, Materials Research Society, Pittsburgh, pp. 111–116.
SiGe  Pike, W. T. and Brown, L. M. (1989). Microanalysis of $Si–Ge_xSi_{1-x}$ superlattices in a dedicated STEM. In *Microscopy of Semiconding Materials Conference*, Inst. Phys. Conf. Ser., Vol. 100, Institute of Physics, London, pp. 69–74.
SiMn  Zhang, L. and Ivey, D. G. (1991). Low temperature reactions of thin layers of Mn with Si, *J. Mater. Res.* **6**(7), 1518–1531.
SiN  Bando, Y., Mitomo, M., and Kitami, Y. (1986a). Grain boundary analysis of silicon nitrides by 400 kV AEM, *J. Electron Microsc.* **35**(4), 371–377.
SiN  Bando, Y. (1983). Weak asymmetry in $\beta$-$Si_3N_4$ as revealed by CBED, *Acta Crystallogr.* **B39**, 185–189.
SiN  Iturriza, I., Echeberria, J., Gutierrez, I., and Castro, F. (1990). Densification of silicon nitride ceramics under sinter-HIP conditions, *J. Mater. Sci.* **25**, 2539–2548.
SiO  Kim, Y., Shindo, S., LOng, N., and Spence, J. C. H. (1987). Three-dimensional coherent electron microdiffraction applied to oxygen precipitates in Si. In *Proceedings of the 45th Annual Meeting of EMSA*, ed. G. W. Bailey, San Francisco Press, San Francisco, pp. 28–29.
SiO  Loveluck, J. E. and Steeds, J. W. (1977a). Crystallography of lithium tantalate and quartz. In *Developments in Electron Microscopy and Analysis*, Inst. Phys. Conf. Ser., vol. 36, ed. D. L. Misell, Institute of Physics, Bristol and London, pp. 293–296.
$SnO_2$  Fujimoto, M., Urano, T., Murai, S., and Nishi, Y. (1989). Microstructure and x-ray study of preferentially oriented $SnO_2$ thin film prepared by pyrohydrolytic decomposition, *Jpn. J. Appl. Phys.* **28**(12), 2587–2593.
$Sr_2Bi_2CuO_6$  Roth, R. S., Rawn, C. J., and Bendersky, L. A. (1990). Crystal chemistry of the compound $Sr_2Bi_2CuO_6$, *J. Mater. Res.* **5**(1), 46–52.
$Sr_2Nb_2O_7$  Yamamoto, N. and Ishizuka, K. (1983). Analysis of the incommensurate structure in $Sr_2Nb_2O_7$ by electron microscopy and CBED, *Acta Crystallogr.* **B39**, 210–216.
$Sr_3Ti_2O_7$  Elcombe, M. M., Kisi, E. H., Hawkins, K. D., White, T. J., Goodman, P., and Matheson, S. (1991). Structure determinations for $Ca_3Ti_2O_7$, $Ca_4Ti_3O_{10}$, $Ca_{3.6}Sr_{0.4}ti_3O_{10}$ and a refinement of $Sr_3Ti_2O_7$, *Acta Crystallogr.* **B47**, 305–314.
$TaS_3$  Tanaka, M. and Kaneyama, T. (1986d). Space group determination of pseudo one-dimensional compound $TaS_3$ by CBED. In *Proc. 11th Int. Congr. on Electron Microscopy* (Kyoto), Maruzen, Tokyo, pp. 723–724.
$TaSe_2$  Bird, D. M. (1985a). HOLZ diffraction from zone axes containing zigzagged strings: Theory and application to the commensurate superlattice state of $2H–TaSe_2$, *J. Phys. C.* **18**, 481–498.
Ti  Court, S. A., Stanley, J. T., Konitzer, D. G., Loretto, M. H., and Fraser, H. L.

(1988). The microstructure of rapidly solidified and heat-treated Ti alloys containing La, *Acta. Metall.* **36**, 1585–1594.

Ti    Emiliani, M., Richman, M., and Brown, R. (1990b). Diffusion of sputtered INCONEL 61 coatings in Ti, *Metall. Trans.* **21A**, 1613–1625.

TiN    Hibbs, M. K., Sundgren, J. E., Jacobson, B. E., and Johansson, B. O. (1983). The microstructure of reactively sputtered Ti–N films. *Thin Solid Films* **107**, 149–157.

Ti    Shelton, C. G., Porter, A. J., and Ralph, B. (1984). Examination of the crystal symmetry in thin foils of Ti by CBED. In *Electron Microscopy and Analysis*, Inst. Phys. Conf. Ser., Vol. 68, Institute of Physics, Bristol and London, p. 47.

Ti    Snow, D. B. (1987). Precipitation and coarsening of rare-earth oxides in rapidly solidified alpha-Ti alloys. In *Analytical Electron Microscopy '87*, San Francisco Press, San Francisco, pp. 38–40.

Ti    Woodfield, A. P. and Loretto, M. H. (1987). Structural determination of silicides in Ti 5331S, *Scr. Metall.* **21**, 229–232.

Ti–Al    Konitzer, D. G., Muddle, B. C., and Fraser, H. L. (1983). A comparison of the microstructure of As-cast and laser surface melted Ti–8Al–4Y, *Metall. Trans.* **14A**, 1979–1988.

$Ti_{1-x}V_xNi$    Zhang, Z. and Kuo, K. H. (1986). Orientation relationship between the icosahedral crystalline phases in $(Ti_{1-x}V_x)_2$ Ni alloys, *Philos. Mag.* **54B**(3), L83–87.

$Ti_2Al$    Kaufman, M. J., Konitzer, D. G., Shull, R. D., and Fraser, H. L. (1986d). An AEM study of the recently reported $Ti_2Al$ phase in γ-TiAl alloys, *Scr. Metall.* **20**, 103–108.

$Ti_3Al$    Chen, D., Es-Souni, M., Beaven, P. A., and Wager, R. (1991). Microstructure of $Ti_3Al$ based alloys containing Nb and Si, *Scr. Metall.* **25**(6), 1363–1368.

$Ti_3Al$–Nb    Banerjee, D., Gogia, A. K., Nandi, T. K., and Joshi, V. A. (1988). A new ordered orthorhombic phase in a $Ti_3Al$–Nb alloy, *Acta Metall.* **36**(4), 871–882.

$Ti_3Al$–Nb    Muraleedharan, K., Naidu, S. V. N., and Banerjee, D. (1990). Orthorhombic distortion of the $\alpha_2$ phase in $Ti_3Al$–Nb alloys: Artifacts and facts, *Scr. Metall.* **24**, 27–32.

$Ti_4Al_3Nb$    Bendersky, L. A., Boettinger, W. J., Burton, B. P., and Biancaniello, S. F. (1990). The formation of ordered ω-related phases in alloys of composition $Ti_4Al_3Nb$, *Acta Metall.* **38**(6), 931–943.

TiAl    Graves, J. A., Bendersky, L. A., Biancaniello, F. S., Perepezko, J. H., and Boettinger, W. J. (1988). Pathways for microstructural development in TiAl, *Mater. Sci. Eng.* **98**, 265–268.

TiAl    Jones, C., Kiely, C. J., and Wang, S. S. (1989). The characterization of an SCS6/Ti–6Al–4V MMC interphase, *J. Mater. Res.* **4**(3) 327–335.

TiAl    Jones, C., Kiely, C. J., and Wang, S. S. (1990). The effect of temperature on the chemistry and morphology of the interphase in an SCS6/Ti–6Al–4V metal composte, *Mater. Res. Soc. Symp. Proc.* **170**, 179–184.

TiAl    Mahon, G. J. and Howe, J. M. (1990). TEM investigation of interface on two phase TiAl alloy, *Metall. Trans.* **21A**, 1655–1662.

TiAl    Morris, M. A. and Morris, D. G. (1991). Strain localization, slip-band formation and twinning associated with deformation of a Ti–24 at% Al–11 at% Nb alloy, *Philos. Mag.* **64A**(6), 1175–1194.

TiAl    Muraleedharan, K. and Banerjee, D. (1989a). Alloy partitioning in Ti–24 Al–11 Nb by AEM, *Metall. Trans.* **20A**, 1139–1142.

TiAlNb    Bendersky, L. A. and Boettinger, W. J. (1989). Omega-related phases in a Ti–Al–Nb alloy. In *Proceedings of the 47th Annual Meeting of EMSA*, ed. G. W. Bailey, San Francisco Press, San Francisco, pp. 324–325.

TiAlNb  Muraleedharan, K. and Banerjee, D. (1989c). Alloy partitioning in Ti–24Al–11Nb by AEM, *Metall. Trans.* **20A**, 1139–1142.

$TiBe_2$  Liu, D. R. and Williams, D. B. (1989b). Problems in determining the structure of $TiBe_2$ by CBED, *Philos. Mag.* **59A**(2), 199–216.

$TiBe_2$  Liu, D. R., Wall, M., and Williams, D. B. (1988a). Possible ambiguity in the structural analysis of $TiBe_2$ by CBED. In *Microbeam Analysis, 1988* ed. D. E. Newbury, San Francisco Press, San Francisco, pp. 87–90.

TiC  Burbery, A. J., Konitzer, D. G., and Loretto, M. H. (1989). Assessment of the local C-content of TiC particulate in heat treated TiC–Ti6Al4V. In *EMAG-MICRO 89*, Inst. Phys. Conf. Ser., Vol. 98, Institute of Physics, Bristol and London, pp. 83–86.

TiC  Hansen, P. L. and Madsen, K. (1990b). Determination of the lattice parameter for $TiC_{1-x}$ and $TiN_{1-x}$ by use of HOLZ lines in CBED [114] zone axis patterns. In *Proc. 12th Int. Congr. on Electron Microscopy*, Vol. 2, San Francisco Press, San Francisco, pp. 482–483.

TiC  Konitzer, D. G. and Loretto, M. H. (1989). Microstructural assessment of interaction zone in titanium aluminide/TiC metal matrix composite, *Mater. Sci. Technol.* **5**, 627–631.

TiCuAlN  Carim, A. H. (1989). Identification and characterization of $(Ti, Cu, Al)_6N$, a new nitride phase, *J. Mater. Res.* **4**(6), 1456–1461.

TiN  Pelton, A. R., Dabrowski, B. W., Lehman, L. P., Ernsberger, C., Miller, A. E., and Mansfield, J. F. (1989). Microstructural studies of sputter-deposited TiN ceramic films, *Ultramicroscopy* **29**, 50–59.

TiO  Berry, F. J., Gogarty, P. M., and Loretto, M. H. (1987). Twin boundaries in antimony-doped rutile Ti-oxide, *Solid State Commun.* **64**(2), 273–275.

$TiO_3$  Fujimoto, M., Tanaka, J., and Shirasaki, S. (1988). Planar faults and grain boundary precipitation in non-stoichiometric $(Sr, Ca) TiO_3$ ceramics, *Jpn. J. Appl. Phys.* **27**(7), 1162–1166.

$TiO_xAl_2O_3$  Alexander, K. B., Walker, F. J., McKee, R. A., and List III, F. A. (1990). Multilayer ceramic thin films, *J. Am. Ceram. Soc.* **73**, 1737–1743.

TiSi  Banerjee, D. (1987). On the structural determination of silicides in Ti alloys, *Scr. Metall.* **21**, 1615–1617.

TiBaCaCuO  Fung, K. K., Zhang, Y. L., Xie, S. S., and Zhou, Y. Q. (1988b). TEM study of high-temperature superconducting phases in the Tl–Ba–Ca–Cu–O system, *Phys. Rev. B* **38**(7), 5028–5030.

TiBaCaCuO  Iijima, S., Ichihashi, T., and Kubo, Y. (1988). Crystal growth faults in superconductor Tl–Ba–Ca–Cu–O oxide crystals, *Jpn. J. Appl. Phys.* **27**(7), L1168–1171.

TlCaBaCu  Liang, J. M., Liu, R. S., and Chang, L. (1988). Structural characterization of TlCaBaCu oxide in $T_c$ onset = 155 K and $T_c$ zero = 123 K superconducting specimens. *Appl. Phys. Lett* **53**(15), 1434–1436.

TlCaBaCuO  Wu, P. T., and Liu, R. S. (1988). Synthesis, transport, magnetization and structural characterizations of Tl–Ca–Ba–Cu–O specimens with $T_0 = 123$ K and $T_{onset} = 155$ K, *Physica C* **156**, 109–112.

V  Tafto, J. and Metzger, T. H. (1985b). LACBED; a simple technique for the study of modulated structures with application to $V_2D$, *J. Appl. Crystallogr.* **18**, 110.

$VSe_2$  Bird, D. M., Eaglesham, D. J., and Withers, R. L. (1985b). Convergent beam diffraction of the distorted structure of $1T–VSe_2$. In *Electron Microscopy and Analysis*, ed. G. J. Tatlock, Inst. Phys. Conf. Ser. Vol. 78, Institute of Physics, London and Bristol, pp. 39–40.

WNiFe  Muddle, B. C. (1984). Interphase boundary precipitation in liquid phase sintered W–Ni–Fe and W–Ni–Cu alloys, *Metall. Trans.* **15A**, 1089–1098.

WNiFe  Posthill, J. B. and Edmonds, D. V. (1986). Matrix and interfacial precipitation in the W–Ni–Fe system, *Metall. Trans.* **17A**, 1921–1934.

WO  Krause, H. B., Vincent, R., and Steeds, J. W. (1986). Diffraction symmetry of an incommensurate tungsten bronze structure. In *Proc. 11th Int. Congr. on Electron Microscopy* (Kyoto), Maruzen, Tokyo, pp. 719–720.

$Y_2BaCuO_{5-x}$  Mansfield, J. F., Chevacharoenkal, S., and Kingon, A. I. (1987c). Space groups and chemical analysis of $Y_2BaCuO_{5-x}$ by CBED, *Appl. Phys. Lett.* **51**(13), 1035–1037.

$Y_2O_3$–$SiO_2$–AlN  Dinger, T. R., Rai, R. S., and Thomas, G. (1988). Crystallization behavior of a glass in the $Y_2O_3$–$SiO_2$–AlN system. *J. Am. Ceram. Soc.* **71**(4), 236–244.

$YAlO_3$  Schaffer, G. B., Loretto, M. H., and Smallman, R. E. (1988). A new $YAlO_3$ polymorph. In *Proc. 9th European Congr. on Electron Microscopy*, eds. P. J. Goodhew and H. G. Dickinson, Inst. Phys. Conf. Ser., Vol. 93, Institute of Physics, London, pp. 275–276.

$YBa_2Cu_{3-x}Al_xO_y$  Kuroda, K., Matsuhata, H., and Saka, H. (1990). CBED analysis of $YBa_2Cu_{3-x}Al_xO_y$ single crystals. In *Proc. 12th Int. Congr. on Electron Microscopy*, Vol. 2, San Francisco Press, San Francisco, pp. 512–513.

$YBa_2Cu_3O_7$  Midgley, P. A., Vincent, R., and Cherns, D. (1989). Structural transformation in $YBa_2Cu_3O_7$. In *EMAG-MICRO 1989*, Inst. Phys. Conf. Ser., Vol. 98, Institute of Physics, Bristol and London, pp. 499–502.

$YBa_2Cu_3O_7$  Wang, Z. Z., Ong, N. P., and McGinn, J. T. (1989). Observation of large, untwinned orthorhombic domains in $YBa_2Cu_3O_7$ single crystals, *J. Appl. Phys.* **65**(7), 2794–2798.

$YBa_2Cu_3O_{7-x}$  Boe, N. and Gjønnes, K. (1991). Determination of electron distribution in $YBa_2Cu_3O_{7-x}$ superconductor by CBED intensities, *Micron and Microscopica Acta* **22**(1/2), 113–114.

$YBa_2Cu_3O_{7-x}$  Brokman, A., Weger, M., Schapink, F. W., and De Haan, C. D. (1988). Electron diffraction comparative study of the tetragonal phases in $YBa_2Cu_3O_{7-x}$ compounds, *Solid State Commun.* **65**(6), 473–465.

$YBa_2Cu_3O_{7-x}$  Bursill, L. A., Goodman, P., Patterson, J. D., and Tonner, S. (1987). An electron microscopy study of defect free $YBa_2Cu_3O_{7-x}$ ($x = 0.15$) with superconducting at 93 K, *Aust. J. Phys.* **40**, 635–641.

$YBa_2Cu_3O_{7-x}$  Sarikaya, M. and Thiel, B. L. (1988). Identification of intergranular $Cu_2O$ in polycrystalline $YBa_2Cu_3O_{7-x}$ superconductors, *J. Am. Ceram. Soc.* **71**(6), C305–C309.

$YBa_2Cu_3O_{7-x}$  Zou, J., Cockayne, D. J. H., and Auchterlonie, H. (1988). Twin structures, transformation and symmetry of superconducting $YBa_2Cu_3O_{7-x}$ observed by TEM, *Philos. Mag. Lett.* **57**(3), 157–163.

$YBa_2Cu_3O_{7-y}$  Tanaka, M., Terauchi, M., Tsuda, K., and Ono, A. (1987c). Electron diffraction study of $YBa_2Cu_3O_{7-y}$, *Jpn. J. Appl. Phys.* **26**(7), L1237–1239.

$YBa_2Cu_3O_{9-x}$  Beyers, R. and Lim, G. (1987). Crystallography and microstructure of $YBa_2Cu_3O_{9-x}$, a perovskite-based superconducting oxide, *Appl. Phys. Lett.* **50**(26), 1918–1920.

$YBa_2Cu_3O_{9-x}$  Sung, C. M., Peng, P., Gorton, A., and Harmer, M. P. (1987b). Microstructure, crystal symmetry and possible new compounds in the system $YBa_2Cu_3O_{9-x}$. In *Advanced Ceramic Materials*, American Ceramic Society, p. 678.

$YBa_2Cu_3O_x$  Ou, H.-J., and Cowley, J. M. (1988). High resolution STEM imaging study on high $T_c$ superconductors $YBa_2Cu_3O_x$. In *Proceedings of the 46th Annual Meeting of EMSA*, ed. G. W. Bailey, San Francisco Press, San Francisco, pp. 882–883.

$YBa_2Cu_3O_y$    Eaglesham, D. J., Humphreys, C. J., McN Alford, N., Clegg, W. J., Harmer, M. A., and Birchall, J. D. (1987e). The orthorhombic to tetragonal phase transformation in $YBa_2Cu_3O_y$ superconducing ceramics. In *Electron Microscopy and Analysis* ed. L. M. Brown, Inst. Phys. Conf. Ser. Vol. 90, Institute of Physics, Bristol and London, pp. 295–298.

$YBaCu_2O_3$    Midgley, P. A., Vincent, R., and Cherns, D. (1990). A study of twinning in $YBaCu_2O_3$ by large angle CBED. In *Proc. 12th Int. Congr. on Electron Microscopy*, Vol. 4, San Francisco Press, San Francisco, pp. 20–21.

YBaCuO    Eaglesham, D. J. and Humprheys, C. J. (1987d). New phases in the superconducting Y–Ba–Cu–O system, *Appl. Phys. Lett.* **51**(6), 81–94.

YYbBaCuO    Yuechao, Z. and Shuyuan, Z. (1987). A TEM study of phases in the superconducting Y(Yb)–Ba–Cu–O system, *Solid State Commun.* **64**(4), 493–496.

ZnNbFe    Woo, O. T. and Carpenter, G. J. C. (1990). Microanalysis identification of a new Zr–Nb–Fe phase. In *Proc. 12th Int. Congr. on Electron Microscopy*, Vol. 2, San Francisco Press, San Francisco, pp. 132–133.

ZnO    Kim, J. C. and Goo, E. (1989). Morphology and formation mechanism of the pyrochlore phase in ZnO varistor materials, *J. Mater. Sci.* **24**, 76–82.

ZuO    Suyama, Y. (1988). Shape and structure of zinc oxide particles prepared by vapor-phase oxidation of zinc vapor, *J. Am. Ceram. Soc.* **71**(5), 391–395.

$ZnOBi_2O_3$–CoO    Suyama, Y., Tomokiyo, Y., and Terasaka, K. (1990). Grain boundary structure of a $ZnO$–$Bi_2O_3$–CoO varistor. In *Proc. 12th Int. Congr. on Electron Microscopy*, Vol. 4, San Francisco Press, San Francisco, pp. 374–375.

ZnTe    Niu, F., Wang, R., and Lu, G. (1991). Determination of the sign of a dislocation in a ZnTe crystal by CBED, *Acta Crystallogr.* **A47**, 36–39.

ZrO    Hannink, R. H. J. (1988). Significance of microstructure in transformation toughening zirconia ceramics, *Mater. Forum.* **11**, 43–60.

ZrO    Pandolfelli, V. C., Rainforth, M., and Stevens, R. (1990). Sintering and microstructural studies in the system $ZrO_2$–$TiO_2$–$CeO_2$, *J. Mater. Sci.* **25**, 2233–2244.

ZrO    Silva, E. and Scozia, R. (1990a). CBED ZAP map of zirconium. In *Proc. 12th Int. Congr. on Electron Microscopy*, Vol. 2, San Francisco Press, San Francisco, pp. 496–497.

ZrO    Silva, E., Riquelme, J., and Robinson, J. W. (1984a). ZAP maps from zirconium, *Metallography* **17**, 103–107.

ZrO    Silva, E. G. and Scozia, R. A. (1990b). CBED zone axis pattern map of zirconium. In *Microbeam Analysis—1990*, eds. J. R. Michael and P. Ingram, San Francisco Press, San Francisco, pp. 351–353.

ZrO    Tendeloo, G. V., Anders, I., and Thomas, G. (1983). Electron microscopy investigation of the $ZrO_2$–ZrN system—II. Tetragonal and monoclinic $ZrO_2$ precipitation, *Acta Metall.* **31**, 1619–1625.

$ZrO_2$    Koo, J. K. and Anderson, M. P. (1985). Stress-induced ordering in $Y_2O_3$-stabilized tetragonal $ZrO_2$. In *Proceedings of the 43rd Annual Meeting of EMSA*, ed. G. W. Bailey, San Francisco Press, San Francisco, pp. 222–223.

$ZrS_3$    Gjønnes, K. (1985a). CBED from faulted crystals of $ZrS_3$, *Ultramicroscopy* **17**, 133–140.

$ZrTe_3$    Eaglesham, D. J., Steeds, J. W., and Wilson, J. A. (1984b). Electron microscope study of superlattices in $ZrTe_3$, *J. Phys. C.* **17**, L697–698.

# References for the Appendixes

Bird, D. (1989). *J. Electron Microsc.* **13**, 77.
Boisen, M. B. and G. V. Gibbs (1985). "Mathematical Crystallography," in *Reviews in Mineralogy,* Vol. 15, Mineralogical Society of America.
Buseck, P., J. M. Cowley, and L. Eyring (1989). *High Resolution Transmission Electron Microscopy and Related Techniques,* Oxford University Press, New York.
Buxton, B. F., J. A. Eades, J. W. Steeds, and G. M. Rackham (1976). *Phil. Trans. R. Soc.* **281**, 181.
Cheng, Y. and R. Wang (1989). *Phys. Status Solidi B* **152**, 33.
Cherns, D. (1974). *Philos. Mag.* **30**, 549.
Cowley, J. M. (1981). *Diffraction Physics,* North-Holland, New York.
Eades, J. A. (1992). *Ultramicroscopy,* in press.
Goodman, P. and A. F. Moodie (1974). *Acta Crystallogr.* **30**, 280.
Head, A., P. Humble, L. Clareborough, A. Morton, and C. Forwood (1973). *Computed Electron Micrographs and Defect Identification,* North-Holland, Amsterdam.
Hirsch, P. S., A Howie, R. B. Nicholson, D. W. Pashley, and M. J. Whelan (1977). *Electron Microscopy of Thin Crystals,* Robert E. Krieger Publ. Co. Inc., Florida.
Howie, A. (1971). "The Theory of Electron Diffraction Image Constrast," in *Electron Microscopy in Materials Science,* U. Valdre, ed. Academic Press, New York.
Humphreys, C. J. (1979). *Rep. Prog. Phys.* **42**, 1825.
Ishizuka, K. and N. Uyeda (1977). *Acta Crystallogr.* **A33**, 740.
Kilaas, R., M. A. O'Keefe, and K. M. Krishnan (1987). *Ultramicroscopy* **21**, 47.
Kirkland, E. J., R. Loane, P. Xu, and J. Silcox (1989). "Multislice simulation of ADF-STEM and CBED images," in *Computer Simulation of Electron Microscope Diffraction and Images,* The Minerals Metals and Materials Society, New York.
Krakow, W. and M. O'Keefe (1989). *Computer Simulation of Electron Microscope Diffraction and Images,* The Minerals, Metals and Materials Society, New York.
Lewis, A. L., R. E. Villagrana, and A. J. F. Metherall (1978). *Acta Crystallogr.* **A34**, 138.
Loane, R., E. J. Kirkland, and J. Silcox (1988). *Acta Crystallogr.* **A44**, 912.
Loane, R., P. Xu, and J. Silcox (1991). *Acta Crystallogr.* **A47**, 267.
Lynch, D. F. (1971). *Acta Crystallogr.* **A27**, 399.
Lynch, J. P., E. Lesage, H. Dexpert, and E. Freund (1981). *EMAG 81,* Institute of Physics, London, p. 67.
McKie, D. and C. McKie (1986). *Essentials of Crystallography,* Blackwell Scientific, Oxford.
Metherall, A. J. F. (1975). *Electron Microscopy and Materials Science, Third Course of the International School of Electron Microscope,* Commission of Euro Communities, Directorate General, "Scientific and Technical Information," Luxembourg.

Pearson, W. B. (1967). *A Handbook of Lattice Spacings for Metals and Alloys*, Pergamon Press, Oxford.
Reimer, L. (1984). *Transmission Electron Microscopy*, Springer-Verlag, New York.
Self, P. G., M. A. O'Keefe, P. R. Buseck, and A. E. C. Spargo (1983). *Ultramicroscopy* **11**, 25.
Smith, B. T. (1976). *Lecture Notes in Computer Science*, Springer-Verlag, New York.
Spence, J. C. H. (1978). *Acta Crystallogr.* **A34**, 112.
Spence, J. C. H. (1988). *Experimental High Resolution Electron Microscopy*, Oxford University Press, New York.
Stadelmann, P. A. (1987). *Ultramicroscopy* **21**, 131.
Steeds, J. W. (1979). In *Introduction to Analytic Electron Microscopy*, eds. J. J. Hren, J. I. Goldstein, and D. Joy, Plenum Press, New York, p. 387.
Van Dyck, D. (1983). *J. Microsc.* **132**, 31.
Wang, Z. L. and J. M. Cowley (1989). *Ultramicroscopy* **31**, 437.
Zu. J. M., K. Gjønnes, and J. C. H. Spence (1989). *J. Electron Microsc. Technique* **12**, 29.
Zuo, J. M. and J. C. H. Spence (1991). *Ultramicroscopy* **35**, 185.

# Index

Absorption, 39, 75–82
Acentric crystals, 89
  polarity, 131
Alloys, 123, 329–352
Amorphous materials, CBED, 209
Anomolous dispersion, 81, 148
Anomolous transmission effect, 77
Antiphase domains in $Cu_3Au$, by CBED, 189
Astigmatism, 219
Atomic positions, finding, 108

Background scattering in microdiffraction, 227
BeO, 115
Bethe, 1
  potentials, 40–44
Bird, 77, 81, 112
Bloch waves, 1, 36–40
Borrman effect, 78
Bound states, 82, 109
Boundary conditions, 38–41
Bragg law, 8
Bravais lattice
  definition, 146
  determination, 158
Brightness, of electron sources compared, 236–238
Buerger cell, 159
Buxton, 67, 150

Calibration of CBED pattern, 15
Camera length, 9
CBIM, 208
CCD, 222
CdS, 112
Cell constant, measurement of, 125
Centering, 159
Channeling, 82, 109

Charge density, 33
Charge density, role in total energy, 103
Cherns, 207
Chi-squared analysis, 95
Chirality, 165
Coherence, 171–172
Coherent CBED
  compared with "incoherent," 173
  computing, 183–185, 227–322
  resolution limit, 187
  showing dynamical extinctions, 188
  summary of applications, 189
Coherent current, 176
Cold stages, 216
Conditional potential, 109
Contamination, 215
Conventional cell, 146
Correlation energy, crystal electrons, 103
Cowley, 195
Critical voltages, 120
Crystal systems, 147
Crystallography, books on, 146
Crystallography, useful results. Appendix 3.
$Cu_3Au$, 123
Current in electron probe, 241

Darwin representation, 39
Debye-Waller factor, definitions, 32–34
Debye–Waller factor, measurement, 100
Defects
  coherent CBED, 181–191
  "incoherent" CBED, 200
Degeneracy, in three-beam, 110
Detectors for microdiffraction, 219
Deviation parameter, 11
Diamagnetic susceptibility, 105
Diffraction groups, 147

*355*

Dislocations, in CBED, 201
Dispersion equation, 36
Dispersion surface, 44–48
  direct observation of, 49
Double-rocking method, 138
DQE, 223–224
Drift, 217
Dynamically forbidden reflections, 161–166

Eades, 145
Eigenvalues, 58
Eigenvectors, 58
Einstein model, 76
Elastic relaxation, 128
Electron sources, 233
Electron spectroscopic diffraction, 229
Enantiamorphs, 120, 165
Energy consevation, 18
Energy-filtered diffraction patterns, 225–233
Error propagation in dynamical refinement, 98
Errors in lattice spacings from HOLZ lines, 127
Ewald, 1
Exchange energy, crystal electrons, 103
Excitation error, 13
Experimental technique, 217
Extinction distance, 33

Field of view, angular, 216
Field-emission guns, 237–240
Field-emission sources, 235–244
Filtering, elastic, 219–233
Focus, best for CBED, 175
Forbidden reflections
  definitions, 146
  effect of multiple scattering, 161
FORTRAN source code, 277–322
Fourier coefficients, 30
Fourier images in CBED patterns, 192, 197
Fowler–Nordheim plot, 241
Friedel's law, 147

G–M lines, 160–161
GaAs, 111
GaAs, polarity, 131
Ge, 132
Geometry of CBED, 9
Gjonnes, 54
Gjonnes–Moodie extinction, 160–166
Glasses, 209

Glide plane, 147
Goodman, 145

Handedness of a crystal, 165
Hartree energy, 103
Helmholtz–Lagrange theorem, 236
Hermann–Mauguin symbol, 148
History of CBED, 1
Hoier, 54
Hologram, Gabor, in line. Relation to coherent CBED, 187
Holography, 196–200
HOLZ lines
  applications for strain measurement, 125
  displacement due to multiple scattering, 63–75
  geometry, 14, 20–28
  indexing, 24, 267–322
  for structure determination, 119
Howie, 83
Hydrogen, in structure analysis, 86

IKL method, 122–123
Ichimiya, 112
Image plate, 225
Inelastic scattering, 227
Inner potential
  mean, 35, 104–106
  sensitivity to bonding, 106
Instrumentation for microdiffraction, 214
Intermetallic alloys, 142. Appendix 7.
Intersection Kikuchi (or HOLZ) line method, 123
Inversion problem, 134
  uniqueness, 135
Kambe, 53
  approximation, 62
Kikuchi lines, 13, 20
Kossel, 1

$LaB_6$, 235–244
LACBED, 139
  energy filtering effect of aperure, 141
  shadow imaging, 207
Lattice spacings, measurement of, 125
Laue
  circle, 18
  classes, 148
Lehmpfuhl, 145

MacGillavry, 1

Many-beam theory, 35
Mean inner potential, 104
Minerals, structure analysis, 85
Minimization, search, 96
Mirror plane, 147
Miyake, 106
Mollenstedt, 1
Moodie, 110
Mott formula, 30
Multilayers, normal to beam, effect on CBED, 207
Multiple scattering via HOLZ, 133

Nanoprobes, 169
Nanotips, abberrations of, 196
Niggli cell, 160
Nonsystematics, three-beam, 57–62
Noncentrosymmetric crystals, 37, 131

Omega filter, 225
Optical potential, 77
Overlapping CBED orders, consequences, 179

PEELS, 221
Perturbation theory, 43, 48–52
  for bonding effect, 97
  for error analysis, 99
  weak beams, 000
Phase identification, 166
Phase problem, 56–63
  solved by three-beam diffraction, 110–120
Phonon scattering, 77
Phosphors, 233
Photomultipliers, dynamic range, 221
Planar faults, in CBED, 205
Point group determination, 148–158
Point projection images, 191
Poisson's equation, 31
Polarity determination, 131
Positron diffraction, 83
Primitive cell, 146
Probe current, 176, 233–244
Probe size, factors affecting, 174–177

Quasi crystals, 158

RDF methods, 210
Reciprocity, 154
REFINE/CB, 92
Refinement, automated, least squares, 88

Rez, P., 37
RHEED, 38
Rocking curve, 11
Ronchigrams, 191
Rutherford scattering, 35

Scherzer focus, 179
Schottky source, 243
Schrödinger equation, 36
Screw axis, 147
Shadow imaging
  relationship to CBED, 182, 191
  relationship to HREM, 195
Si, 129, 329–352
SiC, 123
Sidebands from semiconductor multilayers, 208
Sommerfeld, 1
Sources of electrons compared, 236–238
Space-group determination from forbidden reflections, 160
Space groups for common crystals, 148
Specimen preparation, 213
Speckle, 211
Spherical aberration coefficient, measuring, 192, 219
Steeds, 145, 150
STEM instruments for CBED, 219
STEM lattice imaging, relationship to CBED, 177
STEM, bakeable electronic detector systems, 230–233
Strain, measurement of using HOLZ lines, 125
Structure analysis, 108
Structure factors
  definitions, 30
  measurement, 85–106, 116–124
Structure invariant, for phases, 54, 110
Structure matrix, 39
Symmetry determination, 145, 150
Symmetry operations, 147
Symmorphic, 147

Tafto, 119, 131
Tanaka patterns, 142
Tanaka, 119, 148, 158
Temperature factors, 32
Thickness, measurement, 87
Three-beam theory centrosymmetric, 52
Three-beam theory, noncentrosymmetric, 56–63, 110–120
Tilt range, 216

Total crystal energy (in terms of structure factors), 101
Twin image problem, 199
Two-beam theory, 40–43

Uncertainty principle, 18
Unit cell dimensions, measurement of, 125

Vegard's law, 128
Vincent, 143
Virtual inelastic scattering, 37, 104

Virtual source, 235

Wavelength of electron, 7
Wavevector of electron, 8
Wide-angle methods, 137–142
Wyckoff symbol, 147

YAG, 222
$YBa_2Cu_3O_7$, 128, 143, 329–352

ZOLZ, 16